Lecture Notes in Physics

Springer
Berlin
Heidelberg
New York
Barcelona
Hong Kong
London
Milan
Paris
Tokyo

Editorial Policy

The series *Lecture Notes in Physics* (LNP), founded in 1969, reports new developments in physics research and teaching -- quickly, informally but with a high quality. Manuscripts to be considered for publication are topical volumes consisting of a limited number of contributions, carefully edited and closely related to each other. Each contribution should contain at least partly original and previously unpublished material, be written in a clear, pedagogical style and aimed at a broader readership, especially graduate students and nonspecialist researchers wishing to familiarize themselves with the topic concerned. For this reason, traditional proceedings cannot be considered for this series though volumes to appear in this series are often based on material presented at conferences, workshops and schools (in exceptional cases the original papers and/or those not included in the printed book may be added on an accompanying CD ROM, together with the abstracts of posters and other material suitable for publication, e.g. large tables, colour pictures, program codes, etc.).

Acceptance

A project can only be accepted tentatively for publication, by both the editorial board and the publisher, following thorough examination of the material submitted. The book proposal sent to the publisher should consist at least of a preliminary table of contents outlining the structure of the book together with abstracts of all contributions to be included.
Final acceptance is issued by the series editor in charge, in consultation with the publisher, only after receiving the complete manuscript. Final acceptance, possibly requiring minor corrections, usually follows the tentative acceptance unless the final manuscript differs significantly from expectations (project outline). In particular, the series editors are entitled to reject individual contributions if they do not meet the high quality standards of this series. The final manuscript must be camera-ready, and should include both an informative introduction and a sufficiently detailed subject index.

Contractual Aspects

Publication in LNP is free of charge. There is no formal contract, no royalties are paid, and no bulk orders are required, although special discounts are offered in this case. The volume editors receive jointly 30 free copies for their personal use and are entitled, as are the contributing authors, to purchase Springer books at a reduced rate. The publisher secures the copyright for each volume. As a rule, no reprints of individual contributions can be supplied.

Manuscript Submission

The manuscript in its final and approved version must be submitted in camera-ready form. The corresponding electronic source files are also required for the production process, in particular the online version. Technical assistance in compiling the final manuscript can be provided by the publisher's production editor(s), especially with regard to the publisher's own Latex macro package which has been specially designed for this series.

Online Version/ LNP Homepage

LNP homepage (list of available titles, aims and scope, editorial contacts etc.):
http://www.springer.de/phys/books/lnpp/

LNP online (abstracts, full-texts, subscriptions etc.):
http://link.springer.de/series/lnpp/

M. Lässig A. Valleriani (Eds.)

Biological Evolution and Statistical Physics

Springer

Editors

Michael Lässig
University of Cologne
Institute for Theoretical Physics
Zülpicher Straße 77
50937 Köln, Germany

Angelo Valleriani
MPI for Colloids and Interfaces
Theory Division
14424 Potsdam, Germany

Cover Picture: The illustration on the front cover shows a typical form of an evolutionary tree. New phenotypes are created continuously through mutations and are suppressed through competition. For viral evolution, such tress can be reconstructed from sequence data collected over a few years only; see, for example, R.M. Bush et al., Proc. Natl. Acad. Sci. 97, 6974 (2000).

Library of Congress Cataloging-in-Publication Data.
Die Deutsche Bibliothek - CIP-Einheitsaufnahme

Biological evolution and statistical physics / M. Lässig ; A. Valleriani (eds.). - Berlin ; Heidelberg ; New York ; Barcelona ; Hong Kong ; London ; Milan ; Paris ; Tokyo : Springer, 2002
(Lecture notes in physics ; Vol. 585)
(Physics and astronomy online library)

ISSN 0075-8450
ISBN 978-3-642-07743-2 e-ISBN 978-3-540-45692-6

Springer-Verlag Berlin Heidelberg New York
a member of BertelsmannSpringer Science+Business Media GmbH

http://www.springer.de

Printed in Germany

Camera-data conversion by Steingraeber Satztechnik GmbH Heidelberg
Cover design: *design & production*, Heidelberg

Printed on acid-free paper
54/3111 - 5 4 3 2 1

Preface

People have always asked what distinguishes the living from the inanimate world and what unifies the two. The fields of biology and physics have a long history of exchange. Milestones at the molecular level were the discoveries of the structure of DNA, RNA, and proteins.

It is not by coincidence that this exchange has intensified in recent years. Laboratory experiments reach down to the level of single molecules. Moreover, there is now a vast amount of genomic information, which is still growing exponentially due to the various sequencing projects. Biologists increasingly feel the need for theoretical models to interpret these data in a quantitative way. At the same time, theoretical physics has made significant progress in areas likely to be relevant for the understanding of biological systems. Some important examples are cooperative phenomena, statistics far from thermodynamic equilibrium, systems with quenched disorder, and soft matter.

Some forms of biological matter have indeed become established areas of research within physics, such as biomembranes, heteropolymers, molecular motors, microtubules, neural systems etc. This volume is focused on a different aspect of the living world that can be called *biological information*, its coding, reproduction, and evolution. Biological information is translated into structures and patterns over an enormous range of scales, from single biomolecules to species networks coupled over entire continents.

The *statistical theory* of biological information lives not only in three-dimensional space. It involves various abstract spaces in which this information is encoded and evolves, such as nucleotide sequences, gene networks, or topologies of the 'tree of life'. The articles collected highlight a few directions of research that may become important parts of this emerging field.

The first part of the book, *Molecular Information and Evolution*, starts with two articles on sequence similarity analysis, a central theme in bioinformatics which has surprisingly deep connections to statistical physics. The genetic code, RNA, and proteins are three examples of the intricate interplay of sequence, structure, and function in evolution.

Phylogeny is the inference of evolutionary relationships from genomic or phenotypic data. The articles in the second part of the book move beyond the traditional goal of reconstructing the unique historical tree of life and emphasize statistical aspects of a tree ensemble. Examples are biological properties of an ancestor inferred from phylogeny, the correlation between trees in host–parasite systems, the evolution of traits along trees, and the statistics of tree topologies.

The *Evolution of Populations and Species* results from an interplay of randomness and Darwinian selection. The topics covered in part III include quasispecies and fitness landscapes, evolutionary optimization and the age structure in a population. Fitness values and populations may have spatial structure as well, as shown by the articles on pattern formation and morphogenesis and on spatio-temporal modes of species formation. These examples have an interesting connection to field theories of nonequilibrium systems.

Large–Scale Evolution, the final part of the book, is governed by the ecological interactions between many species organized in a food web or species network. This dynamics, whose temporal scales are measured in millions of years, is described by 'effective models' that neglect many details of the lower levels of molecules, individuals, and populations. Which are the important evolutionary forces at the largest scales remains a matter of active debate. In different systems and at different scales of space and time, various articles ask very similar questions on evolution, although there is no common language or agenda of research yet. It is the goal of theory to develop unifying concepts that relate different pictures of evolution in a quantitative way and bridge the gaps between scales. In some of the research areas mentioned, concepts and methods from statistical physics have already led to significant contributions, in others they are likely to do so in the future. The offspring of this new encounter between the two disciplines may not be physics or biology in the traditional sense. What matters, however, is if it is interesting science.

Most of the contributions in this book are based on invited talks given at the workshop *Biological Evolution and Statistical Physics* last year in Dresden. We are grateful to the Max Planck Institute for the Physics of Complex Systems for making this workshop possible and creating a unique environment of scientific exchange. Particular thanks are due to Peter Fulde, Sergej Flach, and Katrin Lantsch. We are also indebted to Reinhard Lipowsky and the Max Planck Institute for Colloids and Interfaces for encouragement and support to produce this volume. Martin Brinkmann is gratefully acknowledged for his help in the editorial process, and Susann Valleriani for designing the conference poster.

Köln, Potsdam, *Michael Lässig, Angelo Valleriani*
November 2001

Contents

List of Contributors

Ugo Bastolla
Centro de Astrobiología, INTA-CSIC,
Ctra. de Ajalvir Km. 4,
28850 Torrejón de Ardoz,
Madrid, Spain
bastollau@inta.es

Ralf Bundschuh
The Ohio State University,
Department of Physics,
174 West 18th Avenue,
Columbus, OH 43210-1106, USA
bundschuh@mps.ohio-state.edu

Michael A. Charleston
Department of Zoology,
University of Oxford,
South Parks Road,
Oxford OX1 3PS, UK
michael.charleston@zoo.ox.ac.uk

Dannie Durand
Department of Biological
Sciences,
Carnegie Mellon University,
Pittsburgh, PA 15213, USA
durand@cmu.edu

Werner Ebeling
Physics Institute,
Humboldt University of Berlin,
Invalidenstr. 110,
10115 Berlin, Germany
Ebeling@physik.hu-berlin.de

Martin Farach-Colton
Department of Computer Science,
Rutgers University,
Piscataway, NJ 08855, USA
farach@cs.rutgers.edu

Nicolas Galtier
UMR 5000, Génome, Populations,
Interactions,
Université Montpellier 2,
Place E. Bataillon CC 63 34095,
Montpellier, France
galtier@crit1.univ-montp2.fr

Richard A. Goldstein
Biophysics Research Division
and Dept. of Chemistry,
University of Michigan Ann Arbor,
MI 48109-1055, USA
richardg@umich.edu

Paul Higgs
School of Biological Sciences,
University of Manchester,
Stopford Building, Oxford Road,
Manchester M13 9PT, UK
paul.higgs@man.ac.uk

Paulien Hogeweg
Theoretical Biology
and Bioinformatics Group,
Utrecht University Padualaan 8,
3584CH Utrecht, The Netherlands
P.Hogeweg@bio.uu.nl

Terence Hwa
Department of Physics,
University of California
at San Diego, La Jolla,
CA 92093-0319, USA
hwa@matisse.ucsd.edu

Vincent Jansen
School of Biological Sciences,
Royal Holloway, University of London,
Egham, Surrey, TW20 0EX, UK
vincent.jansen@sun.rhul.ac.uk

Alan J. McKane
Department of Physics
and Astronomy,
The University of Manchester,
Oxford Road, M13 9PL
Manchester, England
alan.mckane@man.ac.uk

David C. Krakauer
Institute for Advanced Study,
Princeton NJ 08540, USA
krakauer@ias.edu

Joachim Krug
Fachbereich Physik,
Universität Essen,
45117 Essen, Germany
jkrug@theo-phys.uni-essen.de

Michael Lässig
Institut fuer Theoretische Physik
Universität zu Köln,
Zülpicher Str. 77,
50937 Köln, Germany
lassig@thp.uni-koeln.de

François Lutzoni
Department of Biology,
Duke University, Box 90338,
Durham, NC 27708 USA
flutzoni@fmnh.org

Susanna C. Manrubia
Centro de Astrobiología, INTA-CSIC,
Ctra. de Ajalvir Km. 4,
28850 Torrejón de Ardoz,
Madrid, Spain
susanna@complex.ups.es

Lutz Molgedey
Physics Institute,
Humboldt University of Berlin,
Invalidenstr. 110, 10115 Berlin
molgedey@physik.hu-berlin.de

Martin Nowak
Institute for Advanced Study,
Princeton NJ 08540, USA
nowak@ias.edu

Mark Pagel
School of Animal and Microbial
Sciences,
University of Reading,
Reading RG6 6AJ, UK
m.pagel@reading.ac.uk

Christopher Quince
Department of Physics
and Astronomy,
The University of Manchester,
Oxford Road, M13 9PL
Manchester, England
chris@theory.ph.man.ac.uk

Axel Reimann
Physics Institute,
Humboldt University of Berlin,
Invalidenstr. 110, 10115 Berlin
reimann@physik.hu-berlin.de

Martin Rost
Institut für Theoretische Physik,
Universität zu Köln,
50937, Köln, Germany
mar@thp.uni-koeln.de

Peter Schuster
Institute für Theoretische Chemie
und Molekulare Strukturbiologie,
Währingerstraße 17,
A-1090 Wien, Austria
pks@tbi.univie.ac.at

Peter F. Stadler
Institute für Theoretische Chemie
und Molekulare Strukturbiologie,
Universität Wien,
Währingerstrasse 17,
A-1090 Wien, Austria
studla@tbi.univie.ac.at

Dietrich Stauffer
Institute for Theoretical Physics,
Cologne University,
50923 Köln, Germany
stauffer@thp.uni-koeln.de

Mike Steel
Biomathematics Research Centre,
University of Canterbury,
Christchurch, New Zealand
M.Steel@math.canterbury.ac.nz

Ricard V. Solé
Complex Systems Research Group,
Department of Physics,
FEN-UPC, Campus Nord B4,
08034 Barcelona, Spain
ricard@complex.upc.es

Angelo Valleriani
Max Planck Institute
of Colloids and Interfaces,
Theory Division,
14424 Potsdam, Germany
valleriani@mpikg-golm.mpg.de

Yi-Kuo Yu
Department of Physics,
Florida Atlantic University,
Boca Raton, FL 33431-0991, USA
yyu@fau.edu

Part I

Molecular information and evolution

Statistical significance and extremal ensemble of gapped local hybrid alignment

Yi-Kuo Yu[1], Ralf Bundschuh[2], and Terence Hwa[2]

[1] Department of Physics, Florida Atlantic University, Boca Raton, FL 33431-0991
[2] Department of Physics, University of California at San Diego, La Jolla, CA 92093-0319

Abstract. A "semi-probabilistic" alignment algorithm which combines ideas from Smith-Waterman and probabilistic alignment is proposed and studied in detail. It is predicted that the score statistics of this "hybrid" algorithm is of the universal Gumbel form, with the key Gumbel parameter λ taking on a *fixed* asymptotic value for a wide variety of scoring parameters. We have also characterized the "extremal ensemble", i.e., the collection of sequence pairs exhibiting similarities that a given scoring system is most sensitive to. Based on this extremal ensemble, a simple recipe for the computation of the "relative entropy", and from it the correction to λ due to finite sequence length is also given. This allows us to assign p-values to the alignment results for arbitrary scoring parameters and gap costs. The predictions compare well with direct numerical simulations for a broad range of sequence lengths with various choices of the substitution scores and affine gap parameters.

Key words: sequence alignment; statistical significance; maximum likelihood; hidden Markov model

1 Introduction

Computer-assisted sequence comparison tools such as BLAST (Altschul *et al.*, 1990) and FASTA (Pearson, 1988) have become an integral part of modern molecular biology. They reveal evolutionary relationships between protein sequences and therefore provide a basis for the functional identification of new genes and for the construction of phylogenic trees. Two types of algorithms have been used: those which search for the *optimal* alignment (as exemplified by the algorithm of Smith and Waterman (1981)), and those which identify *likely* alignments (as exemplified by the hidden-Markov model (HMM) based "Sequence Alignment Modules" (Hughey & Krogh, 1996)). In each case, the quality of the alignment is summarized by an alignment score S; the latter is typically taken to be the logarithm of the total likelihood in the probabilistic approaches. However, such an alignment score is assigned to *any* pair of sequences, also to biologically completely unrelated ones (e.g., to pairs of random sequences.) In order to be able to distinguish *true evolutionary relationships* from *random similarities* it is an important goal common to all algorithms to understand the probability distribution function pdf(S) of the score S for the appropriate null models. The knowledge of this distribution gives the possibility of assigning p-values, i.e., the probabilities that a high score could have arisen by chance, to alignment results. These *p*-values quantify the amount of surprise behind a given alignment score. Only if a score is sufficiently surprising, the underlying alignment is considered to be of evolutionary origin.

Rigorous results on such background statistics are known only for the gapless alignment, whose score distribution follows the so-called Gumbel form (Gumbel, 1958),

$$\text{pdf}(\mathsf{S}) = KMN\lambda \exp\left[-\lambda\mathsf{S} - KMNe^{-\lambda\mathsf{S}}\right], \tag{1}$$

for long sequence lengths M and N (Arratia *et al.*, 1988; Karlin & Altschul, 1990, 1993; Karlin & Dembo, 1992). Explicit formulae relating the hundreds of alignment parameters to the two Gumbel parameters λ and K are available (Karlin & Altschul, 1990). For gapped sequence alignment with large enough gap cost, the score distribution is also empirically known to obey Gumbel statistics (Smith *et al.*, 1985; Collins *et al.*, 1988; Mott, 1992; Waterman & Vingron, 1994a, 1994b; Altschul & Gish, 1996, Olsen *et al.*, 1999). However, the dependence of the two Gumbel parameters on the hundreds of scoring parameters is generally so complicated that it is very difficult to determine the Gumbel parameters in an efficient enough manner to render them useful.

This problem is partially overcome in (gapped) BLAST by pre-computing the null statistics for a fixed set of scoring parameters. However, this is a severe restriction on the flexibility of the method and leads, e.g., to wrong predictions of *p*-values for query sequences with unusual amino acid compositions. Even more importantly, the restriction to a small set of scoring parameters for which the null statistics is pre-computed becomes prohibitive for the use of *position-specific* scoring functions (Henikoff & Henikoff, 1994) as they are needed for detailed modeling of protein families, folds, etc. Because of this problem the iterative similarity search algorithm PSI-BLAST (Altschul *et al.*, 1997) is currently limited to *uniform* gap costs which is an unfortunate drawback.

Position-specific scoring systems are naturally incorporated in probabilistic (e.g., the HMM-based) alignment algorithms. However, only very little is known about the statistics of the log-likelihood score, even at the empirical level.

In this paper, we describe a "semi-probabilistic" alignment algorithm which is a *hybrid* of the Smith-Waterman and the probabilistic alignment algorithms. Our hybrid algorithm has the same computational complexity as the Smith-Waterman and the probabilistic algorithms, with computation time scaling as $O(M \cdot N)$; also, its sensitivity in detecting sequence homology is comparable to or better than the existing algorithms (Yu & Hwa, 1999). The key advantage of the hybrid algorithm is that its score statistics can be characterized theoretically. Moreover, the ensemble of rare sequence pairs responsible for the high-scoring events can be characterized. This "extremal ensemble" consists of sequence pairs exhibiting similarities that a given scoring system is most sensitive to. The knowledge of the connection between scoring systems and their extremal ensembles is very useful in *constructing* the optimal scoring parameters for a given model of sequence evolution. This is analogous to how the Karlin-Altschul theory of gapless alignment can be used to guide the selection of the appropriate amino-acid substitution score (Karlin & Altschul, 1990). In the following sections, we will first describe the algorithm followed by characterization of score distribution and extremal ensemble. A numerical test of the prediction of the theory will be presented at the end. More technical issues are relegated to the appendices.

2 Algorithms

Consider two sequences $\mathbf{a} = [a_1, a_2, ..., a_M]$ and $\mathbf{b} = [b_1, b_2, ..., b_N]$ of lengths M and N, with elements a_i and b_j taken from a finite character set χ. We will employ a frequently used null model with independently identically distributed letters where the probability of having two sequences is given by the distribution function

$$P_0[\mathbf{a}, \mathbf{b}] = \prod_{\substack{1 \le m \le M \\ 1 \le n \le N}} p(a_m) \cdot p(b_n), \tag{2}$$

with $p(a)$ being the background frequency of the element a, and $\sum_{a \in \chi} p(a) = 1$.

2.1 Probabilistic global alignment

We first review probabilistic global alignment of the sequences $\mathbf{a}$ and $\mathbf{b}$. We adopt the approach of Bishop and Thompson (1986), evaluating probabilistic global alignment as the likelihood of observing the sequences $\mathbf{a}$ and $\mathbf{b}$ given a fictitious evolution model producing pairs of *related* sequences $\mathbf{a}$ and $\mathbf{b}$. For the purpose of illustration, let us consider the following simple version of this evolution model: Start with empty sequences $\mathbf{a}$ and $\mathbf{b}$ and go through the hidden Markov model illustrated in Fig. 1.

- Until one of the sequences reaches the desired length N, there is a probability ν for a "deletion step" and the same probability ν for an "insertion step".
  - if the "insertion mode" is selected, generate a new element a according to the background frequencies $p(a)$ and append it to sequence $\mathbf{a}$.
  - if the "deletion mode" is selected, generate a new element b according to the background frequencies $p(b)$ and append it to sequence $\mathbf{b}$.
  - if neither deletion nor insertion is selected, generate a *pair of elements* (a, b) according to some *joint probability distribution* $\mathcal{P}(a, b)$[1] and append a to sequence $\mathbf{a}$ and b to sequence $\mathbf{b}$.
- If one of the sequences reaches the desired length N generate random elements according to the background frequencies $p(a)$ and append them to the shorter of the two sequences until they both have the length N.

[1] The joint probability distribution $\mathcal{P}(a, b)$ is often chosen as $\mathcal{T}(b|a)p(a)$ where $\mathcal{T}(b|a)$ is the *transition probability* for a mutation from element a into element b.

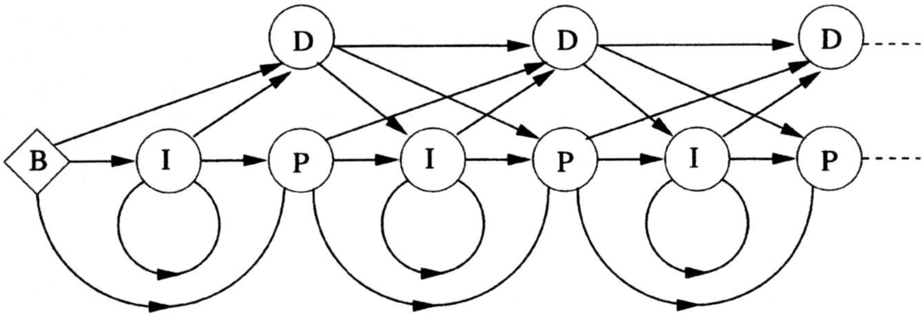

Fig. 1. Schematics of the hidden Markov model $\mathcal{M}$ for sequence evolution. The different states are B for the "begin" state, I for the "insertion" states, D for the "deletion" states, and P for the "pair emission" state. The arrows indicate the allowed transitions between the states, with transition probabilities as given in the text. Sequence elements are "emitted" according to the following rules: An element a is emitted into sequence **a** with probability $p(a)$ every time the state I is visited, and an element b is emitted into sequence **b** with probability $p(b)$ every time the state D is visited. In state P a pair (a, b) is emitted according to some joint distribution $\mathcal{P}(a, b)$ and elements a and b are appended to sequences **a** and **b** respectively.

The "weight" $W[\mathbf{a}, \mathbf{b}]$ for a specific random sequence **a** mutating into a sequence **b** can be computed iteratively (Bishop & Thompson, 1986) by introducing an auxiliary variable $\mathcal{W}_{i,j}$:

$$\begin{aligned} \mathcal{W}_{i,j} &= w(a_i, b_j) \cdot \mathcal{W}_{i-1,j-1} \\ &\quad + \nu \cdot [\mathcal{W}_{i-1,j} + \mathcal{W}_{i,j-1}], \end{aligned} \tag{3}$$

with

$$w(a, b) = (1 - 2\nu)\frac{\mathcal{P}(a, b)}{p(a)p(b)} \tag{4}$$

being the net substitution probability. W is then obtained as

$$\begin{aligned} W[\mathbf{a}, \mathbf{b}] &= [2\nu + w(a_M, b_N)] \cdot \mathcal{W}_{M-1,N-1} \\ &\quad + \sum_{i=1}^{M-1} [\nu + w(a_i, b_N)] \cdot \mathcal{W}_{i-1,N-1} \\ &\quad + \sum_{j=1}^{N-1} [\nu + w(a_M, b_j)] \cdot \mathcal{W}_{M-1,j-1}. \end{aligned} \tag{5}$$

where $\mathcal{W}_{1\le m\le M, 1\le n\le N}$ is obtained by iterating the recursion relation (3) from the initial conditions

$$\mathcal{W}_{i\ge 0, j=0} = \nu^i \qquad 1 \le i \le M \text{ and} \tag{6}$$

$$\mathcal{W}_{i=0, j\ge 0} = \nu^j \qquad 1 \le j \le N. \tag{7}$$

The recursion relation (3) is nothing but the probabilistic version of the Needleman-Wunsch global alignment algorithm (Needleman & Wunsch, 1970) with linear gap cost. Alternatively, the Needleman-Wunsch algorithm is just the Viterbi version of Eq. (3). In the context of alignment, ν controls the gap penalty and $w(a, b)$ the substitution cost. Note that due to the condition (4), we have

$$\sum_{[\mathbf{a},\mathbf{b}]} W[\mathbf{a}, \mathbf{b}] \cdot P_0[\mathbf{a}, \mathbf{b}] = 1, \tag{8}$$

which is nothing but the statement of probability conservation for the different ways sequences can be mutated into each other. Note that $W[\mathbf{a}, \mathbf{b}] \cdot P_0[\mathbf{a}, \mathbf{b}]$ is the likelihood of generating the sequence pair by the mutation model (Thorne *et al.*, 1991, 1992).

2.2 Probabilistic local alignment

Local alignment identifies subsequences, e.g., $\hat{\mathbf{a}} = [a_{m'}, a_{m'+1}, ..., a_m]$ and $\hat{\mathbf{b}} = [b_{n'}, b_{n'+1}, ..., b_n]$ with $1 \leq m' \leq m \leq M$ and $1 \leq n' \leq n \leq N$, whose mutual global alignment score $S(m', n'; m, n)$ is the highest, especially in cases where the global alignment of the complete sequences yields negative total scores, i.e., $S(1, 1; M, N) < 0$. Instead of directly optimizing over the four variables m', n', m, and n, the Smith-Waterman algorithm (Smith & Waterman, 1981) proceeds by first computing an auxiliary score $H(m, n) = \max_{m',n'} S(m', n'; m, n)$ by slightly modifying the Needleman-Wunsch algorithm. Then it maximizes over the H's. This procedure maintains the computational complexity at $O(M \cdot N)$ as in global alignment.

The same strategy can be adopted in probabilistic local alignment if in order to be really looking for local similarities the scoring parameters are chosen such that the total weight of global alignment is small, i.e., such that $\ln \mathcal{W}_{M,N} < 0$.

Let $\mathcal{W}_{m',n';m,n}$ be the likelihood of the global probabilistic alignment of the subsequences $\hat{\mathbf{a}}$ and $\hat{\mathbf{b}}$. It is computed by applying the recursion Eq. (5) with the intial conditions Eqs. (6) and (7) to the sequences $[a_{m'}, a_{m'+1}, ..., a_m]$ and $[b_{n'}, b_{n'+1}, ..., b_n]$. The likelihood is then given by $\mathcal{W}_{m',n';m,n} \equiv \mathcal{W}_{m-m'+1,n-n'+1}$.

Then, we introduce an auxiliary variable

$$Z_{m,n} \equiv 1 + \sum_{m'=1}^{m} \nu^{m'} + \sum_{n'=1}^{n} \nu^{n'} + \sum_{\substack{1 \leq m' \leq m \\ 1 \leq n' \leq n}} \mathcal{W}_{m',n';m,n}, \tag{9}$$

which is the total weight of all subsequence pairs including (a_m, b_n), plus the null alignments. $Z_{m,n}$ can be computed according to the probabilistic version of the Smith-Waterman algorithm,

$$Z_{i,j} = 1 + w(a_i, b_j) \cdot Z_{i-1,j-1} + \nu \cdot [Z_{i-1,j} + Z_{i,j-1}], \tag{10}$$

with the boundary conditions $Z_{0,j} = Z_{i,0} = 1$, for $1 \leq i \leq m$ and $1 \leq j \leq n$. Alternatively, the Smith-Waterman algorithm can be viewed as the Viterbi version of (10).

The total weight $\mathcal{Z}$ characterizing the fully probabilistic alignment is obtained as the sum of the Z's, i.e., as

$$\mathcal{Z} = \sum_{\substack{1 \le m \le M \\ 1 \le n \le N}} Z_{m,n}. \tag{11}$$

However, even the shape of the distribution of the fully probabilistic local alignment score $\ln \mathcal{Z}$ is not very well understood rendering the calculation of p-values for this score very hard. To overcome this difficulty, we introduce a maximum-log-likelihood (MLL) score

$$\mathsf{S} = \max_{\substack{1 \le m \le M \\ 1 \le n \le N}} \ln Z_{m,n} \tag{12}$$

to characterize the quality of the alignment. Eqs. (10) and (12) define the hybrid algorithm which we will focus on from here on.

3 Statistics of hybrid alignment

In this section we will show that the score distribution of hybrid alignment is a Gumbel distribution with $\lambda = 1$ independent of the scoring system. We will moreover characterize the extremal ensemble of the algorithm, i.e., the sequence pairs exhibiting similarities that a given scoring system most sensitive to. This knowledge about the extremal ensemble will help us to characterize the deviation of the Gumbel parameters due to the finite sequence length.

3.1 Score landscape and islands

For the Smith-Waterman algorithm, Olsen *et al.* (1999) utilized the "score landscape" $H(m, n)$ to characterize the tail of the Gumbel distribution. The landscape consists of a collection of *essentially uncorrelated* positive-scoring "islands", separated by a "sea" at $H = 0$. The peak scores of the islands are found to follow Poisson statistics, from which the Gumbel parameters λ and K can be directly derived. Olsen *et al.*'s study indicates clearly that the key to understanding the Gumbel distribution is to characterize the probability tail of obtaining a *single* large island.

Due to our definition (12) of the MLL score as a maximum, the distribution of S is again expected to be of the Gumbel form. (This is not true for the score $\ln \mathcal{Z}$ of the fully probabilistic local alignment, since the score $\mathcal{Z}$ is defined in Eq. (11) via the *sum* rather than the *max* operation.) Similar to what was found by Olsen *et al.* (1999), the pertinent score landscape $\ln Z_{m,n}$ for hybrid alignment consists of islands which are *essentially uncorrelated.* Instead of being separated by the "sea" of zero scores, our MLL islands do have some minor positive score background in between. Since we are interested in the high scoring islands, the minor positive-score background does not affect the identification of the high-scoring islands, see Fig. 2 for example. Since the MLL score is the maximum of

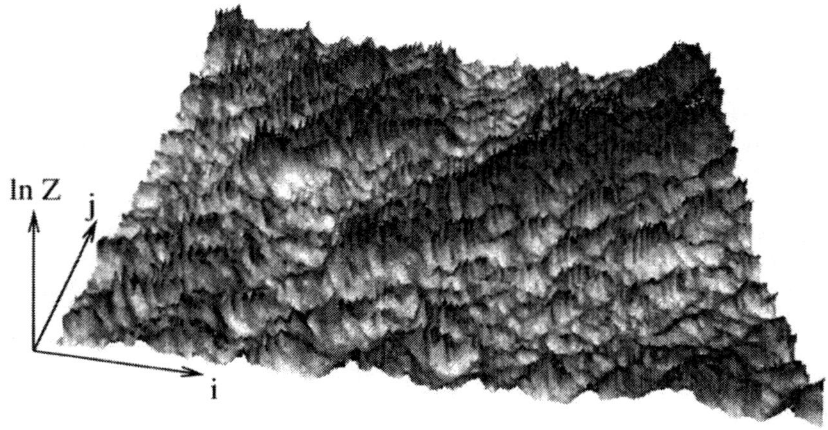

Fig. 2. The $\ln Z$ landscape from aligning two random sequences. This figure is a projection of a three dimensional plot. One sequence is laid along the i direction while the other is laid along the j direction. The MLL score $\ln Z_{i,j}$ is then plotted along the third direction labeled by $\ln Z$. The gray scale is used in such a way that the larger the MLL score, the darker the point $(i, j, \ln Z)$. As shown in this figures, a sea of small ripples separate one medium-sized island from a less significant one.

many of these uncorrelated island peak scores, the statistics of the MLL score (i.e., the Gumbel parameters) can be deduced if the statistics

$$G(h) \equiv \Pr\{\text{peak island score} > h\}$$

of the individual island peak scores is known.

Thus, it is our goal to calculate this peak score distribution. We will do so in several steps. First, we compute the auxiliary quantity

$$D(h|L) \equiv \sum_{\{\mathbf{a},\mathbf{b}\}} \delta(h - \ln W_L[\mathbf{a},\mathbf{b}]) \cdot P_0[\mathbf{a},\mathbf{b}], \tag{13}$$

that a *global* probabilistic alignment of two sequences of length L will have the score about h. Then, we will relate this distribution to the island peak score distribution. Basically, we will establish that high-scoring global alignments which contribute to $D(h|L)$ correspond to high-scoring islands which contribute to $G(h)$. We find that

$$G(h) \sim e^{-\lambda h} \tag{14}$$

with $\lambda = 1$. Since S is the maximum of a large number of independent random island peak scores obeying the distribution (14), it has Gumbel statistics (Gumbel, 1958) with the same λ. This strategy of computing local alignment score statistics using the statistics of *global* alignment has been examined in detail in the context of Smith-Waterman alignments and applied to the special scoring system corresponding to the Longest Common Subsequence problem (Bundschuh, 2000).

3.2 Large-score statistics

The first step in the derivation outlined above is very important since it does not only give us an expreession for the auxiliary quantity $D(h|L)$. It also gives us information on the *extremal ensemble*, i.e., on the *typical* sequence pairs that lead to high-scoring global alignments and thus high-scoring islands. Therefore, we devote this whole section to this computation.

We start by characterizing the distribution $D(h|L)$ for $h \gg 1$ by a simple maximization principle. Instead of computing $D(h|L)$ directly, let us first consider a different (but related) quantity, the probability $\mathcal{D}$ that the sum of the score $\ln W_\ell$ from $\mathcal{N} \gg 1$ independent global alignments of random sequences of length ℓ is $\mathcal{H}$, i.e.,

$$\mathcal{D}_\ell(\mathcal{H}|\mathcal{N}) = \sum_{\{h_j\}} \delta\left(\mathcal{H} - \sum_{j=1}^{\mathcal{N}} h_j\right) \prod_{j=1}^{\mathcal{N}} D(h_j|\ell), \tag{15}$$

where h_j is the score of the jth draw. Using the definition (13) for $D(h|\ell)$ above, we find (with the help of Stirling's formula)

$$\mathcal{D}_\ell(\mathcal{H}|\mathcal{N}) \approx \sum_{\{Q_\ell\}} \exp\left[\mathcal{N} \sum_{\{\mathbf{a},\mathbf{b}\}} Q_\ell[\mathbf{a},\mathbf{b}] \ln\left(\frac{P_0[\mathbf{a},\mathbf{b}]}{Q_\ell[\mathbf{a},\mathbf{b}])}\right)\right]$$
$$\cdot\, \delta\left(\mathcal{H} - \mathcal{N} \sum_{\{\mathbf{a},\mathbf{b}\}} Q_\ell[\mathbf{a},\mathbf{b}] \ln W_\ell[\mathbf{a},\mathbf{b}]\right), \tag{16}$$

where $Q_\ell[\mathbf{a},\mathbf{b}]$ is the fraction among the $\mathcal{N}$ draws that contains a particular sequence pair $[\mathbf{a},\mathbf{b}]$ of lengths ℓ.

For $\mathcal{H} \gg 1$, the right-hand side of Eq. (16) can be evaluated in saddle point approximation with the result

$$\mathcal{D}_\ell(\mathcal{H}|\mathcal{N}) \approx e^{-\lambda\mathcal{H}} \cdot \delta(\mathcal{H} - \mathcal{N}\ell\alpha) \tag{17}$$

where

$$\alpha \equiv \ell^{-1} \sum_{\{\mathbf{a},\mathbf{b}\}} Q_\ell^*[\mathbf{a},\mathbf{b}] \ln W_\ell[\mathbf{a},\mathbf{b}] \tag{18}$$

and the $Q_\ell^*[\mathbf{a},\mathbf{b}]$ are given by the saddle point condition

$$Q_\ell^*[\mathbf{a},\mathbf{b}] = W_\ell^\lambda[\mathbf{a},\mathbf{b}] \cdot P_0[\mathbf{a},\mathbf{b}]. \tag{19}$$

The value of λ is fixed by the normalization condition for Q^*:

$$1 = \sum_{[\mathbf{a},\mathbf{b}]} Q_\ell^*[\mathbf{a},\mathbf{b}] = \sum_{[\mathbf{a},\mathbf{b}]} W_\ell^\lambda[\mathbf{a},\mathbf{b}] P_0[\mathbf{a},\mathbf{b}]. \tag{20}$$

Comparing (20) to the normalization condition (8) for the correlated sequence pairs generated from the evolution model, we see that the solution to Eq. (20) is

$$\lambda = 1. \tag{21}$$

This implies that

$$Q^*_\ell[\mathbf{a},\mathbf{b}] = W_\ell[\mathbf{a},\mathbf{b}] \cdot P_0[\mathbf{a},\mathbf{b}], \tag{22}$$

which describes the sequence configurations contributing significantly to the large-$\mathcal{H}$ events. We shall refer to this ensemble of sequence pairs as the *extremal ensemble*. Inserting $W_\ell = Q^*_\ell/P_0$ into Eq. (18), we see that $\alpha > 0$ is nothing but the relative entropy (per length) between the extremal and the random ensemble. Since $W_\ell[\mathbf{a},\mathbf{b}] \cdot P_0[\mathbf{a},\mathbf{b}]$ is also the likelihood of obtaining the *correlated* sequence pair $[\mathbf{a},\mathbf{b}]$ from the evolution model, Eq. (18) shows that α is also the average score (per length) of the global probabilistic alignment of the correlated sequence pair $[\mathbf{a},\mathbf{b}]$. Eq. (17) states that there exists a *preferred* number of draws, $\mathcal{N}^* = \mathcal{H}/(\ell \cdot \alpha)$, which maximizes the probability of observing high scores $\mathcal{H}$.

On first sight, the distribution Eq. (17) seems not to be normalized. However, Eq. (17) describes only the high-$\mathcal{H}$ component of the full distribution $\mathcal{D}_\ell(\mathcal{H}|\mathcal{N})$. Most of the weight of this distribution is by our assumptions in the region of $\mathcal{H} < 0$. Eq. (17) only applies to the region $\mathcal{H} \gg 1$ and therefore does not have to be normalized by itself.

What does $\mathcal{D}_\ell(\mathcal{H}|\mathcal{N})$ have to do with the quantity of interest $D(h|L)$, which describes the probability of obtaining a score h from the global alignment of a *single* sequence pair of length L ? The statistics of the global alignment of correlated sequences has been studied (see, e.g., Hwa & Nattermann, 1995) in terms of the related problem of directed polymers in a random medium (Hwa & Lässig, 1996), and the results have been elaborated in the context of sequence alignment by Drasdo *et al.* (1998). These studies found that $\ln W_L$ can be decomposed into a sum of essentially independent pieces of some length ξ. Thus, the score $\ln W_L$ of the high-scoring sequence pairs can be broken into a sum of *statistically-independent* pieces, each corresponding to the score $\ln W_l$ of a pair of subsequences of length $l \ll L$ as long as $l > \xi$. Then

$$D(h|L) \approx \mathcal{D}_l(h|L/l) \tag{23}$$

$$= \sum_{\{h_j\}} \delta\left(h - \sum_{j=1}^{L/l} h_j\right) \prod_{j=1}^{L/l} D(h_j|l)$$

$$\approx e^{-h} \cdot \delta(h - L\alpha), \tag{24}$$

where the last line follows from Eq. (17). The approximation (23) can be further justified as explained in Appendix A. Here, we advertise that we indeed get the announced Poisson statistics with $\lambda = 1$ for the probabilistic global alignment scores which generate the islands together with the information that these high scores are created by pairs of correlated subsequences $(\mathbf{a},\mathbf{b})$ as given by the extremal ensemble $Q^*_\ell[\mathbf{a},\mathbf{b}]$.

3.3 Island peak scores

We now turn to derive the island peak score distribution $G(h)$ from the result Eq. (24). This will still require several steps. First, we will calculate the statistics of $Z_{m,n}$ for a fixed choice of (m,n). Afterwards, we will use this statistics to obtain $G(h)$. The connection between the distribution $D(h|L)$ and the statistics of $Z_{m,n}$ for a given (m,n) is made by the observation that $Z_{m,n}$ for a given (m,n) is for large values statistically equivalent to the quantity

$$\overline{Z} \equiv 1 + \sum_{L=1}^{\infty} \overline{W}_L \tag{25}$$

where

$$\overline{W}_L \equiv W[a_1 \dots a_L, b_1 \dots b_L]. \tag{26}$$

The derivation of this equivalence is quite technical and therefore relegated to appendix B. It also shows, how large values of Zm,n are generated by sequence configurations which can be described in terms of the extremal ensemble $Q^*[\mathbf{a},\mathbf{b}]$. Here, we will exploit it by noting that $\overline{W}_L$ is distributed according to the distribution $D(h|L)$. For each L Eq. (24) tells us that roughly $\overline{W}_L$ takes the value $\exp(\alpha L)$ with probability $\exp(-\alpha L)$ and the value 0 with the remaining probability $1 - \exp(-\alpha L)$. Of course, the $\overline{W}_L$ are not statistically independent from each other. However, due to the exponential separation of the possible values of the $\overline{W}_L$ for different L, the probability $\Pr\{\overline{Z} = z\}$ for some large enough z is very well described by the probability that $\overline{W}_{L_0}$ or one of the $\overline{W}_L$ with L very close to $L_0 \equiv \frac{1}{\alpha} \ln z$ takes its non-zero value independently of the other $\overline{W}_L$. The values of $\overline{W}_L$ with $L < L_0$ do not matter since they contribute only little to the sum Eq. (25) and the probability for $\overline{W}_L$ with $L > L_0$ being different from zero is exponentially smaller. Thus, $\Pr\{\overline{Z} = z\} \sim \frac{1}{z}$ or equivalently

$$\Pr\{\ln Z_{m,n} = \overline{h}\} \approx \Pr\{\ln \overline{Z} = \overline{h}\} \sim e^{-\overline{h}}. \tag{27}$$

Finally, we want to relate this distribution of $Z_{m,n}$ at fixed (m,n) to the island peak score distribution $G(h)$. Again, the exponential dependence of the score distribution on the score $\overline{h}$ is essential. The fact, that the probability to find a score $\overline{h}$ at a given point (m,n) on the scoring lattice is an exponential in the score implies that increments in score from one lattice point to the next are essentially independent of the actual score $\overline{h}$ at this lattice point. Specifically, the probability of finding an even higher score at some other neighboring point (m',n') is essentially independent of the score $\overline{h}$ itself either. Thus, for any (m,n) with $Z_{m,n}$ sufficiently large, the probability of (m,n) being an island peak point is some number which depends on the value of $Z_{m,n}$ at most very weakly. Therefore, the probability for an island peak score being h is approximately proportional to the probability of the score $\ln Z_{m,n}$ being h for a fixed (m,n). We calculated the latter with the result given in Eq. (27). Therefore we have the exponential statistics $G(h) \sim e^{-h}$ for the island peak score distribution which implies the Gumbel statistics of S with $\lambda = 1$ as discussed above.

3.4 Sequence length correction

In order to verify the arguments and derivations leading to this result and the extremal ensemble $Q^*_\ell[\mathbf{a},\mathbf{b}]$ we have performed extensive numerical simulations.

However, as presented thus far, our results pertain only to the asymptotic limit of infinitely long sequences. To compare to the numerics performed at *finite* sequence lengths, it is necessary to compute the magnitude of the *corrections* to this result due to the finite sequence length which we will turn to now.

For sequences of finite length we expect a deviation from the asymptotic value $\lambda = 1$ predicted by the above considerations. In order to assess the significance of an alignment of two sequences of finite length we therefore have to characterize this deviation as well.

It was pointed out by Altschul (1991) in the context of gapless local alignment and more recently by Altschul and Gish (1996) for gapped alignment that in using the Gumbel distribution (1) for finite length sequences, one should "correct" the lengths M and N which appear in (1) by a score-dependent amount $L(\mathsf{S})$, and use instead the *effective* sequence lengths $M' = M - L(\mathsf{S})$ and $N' = N - L(\mathsf{S})$. It results from the fact that the available area to launch an island is *reduced* by the size of the island on the alignment lattice, which is the $M \times N$ square obtained by putting one of the two sequences along the $\hat{x}$ direction and the other sequence along the $\hat{y}$ direction in the plane.

As an extreme example, one notes that to have an island of the size of the entire alignment lattice, the island must be launched near the tip of the lattice; in this case, the correction term $L(\mathsf{S})$ is nearly the size of the lattice. Generally, one should take $L(\mathsf{S})$ to be the average island length[2] $\overline{\ell}(\mathsf{S})$ corresponding to the score of the maximum island peak S. Including this correction, the Gumbel statistics becomes

$$\Pr\{\mathsf{S} < x\} = \exp\left[-K \cdot (N - \overline{\ell}(x))^2 e^{-\lambda x}\right], \tag{28}$$

where we have used the more convenient accumulated distribution, and have taken the two sequences to be of equal length N for simplicity. Using the linear island profile $\overline{\ell}(x) = \alpha^{-1} x$ for large islands where α is the "relative entropy", Altschul *et al.* (2001) noted that the terms in Eq. (28) can be rearranged into the classic Gumbel form, i.e.,

$$\Pr\{\mathsf{S} < x\} = \exp\left[-K(N) \cdot N^2 e^{-\lambda(N) \cdot x}\right], \tag{29}$$

with the effective size-dependent parameter $\lambda(N) = \lambda + 2/(\alpha N)$ to leading order in $1/N$. More generally, one has the relation

$$\lambda(N) = \lambda + 2/\overline{\sigma}(N), \tag{30}$$

where $\overline{\sigma}(\ell)$ is the inverse of the function $\overline{\ell}(\sigma)$, and gives the average score for islands of length ℓ. Note that correction formulae such as (30) are applicable as long as the number of islands in the alignment lattice is large. They should however not be applied to very small N's where the sequence lengths are of the same order as the island sizes, and the Gumbel distribution itself breaks down.

---

2 The island width is typically much smaller than its length and hence does not contribute to leading orders.

It is also possible to extend the analysis discussed above for the parameter K. Using the form $\bar{\ell}(x) = \alpha x + c$ in Eq. (28) and rearranging terms into the Gumbel form Eq. (29), we find the result

$$K(N) = K \cdot \left(1 + \frac{c}{\alpha N}\right)^2 \tag{31}$$

which is analogous to Eq. (30) for $\lambda(N)$. Unlike the case for λ, we have not yet developed a theory to compute the asymptotic value of K. It is however still possible to check the form of the correction formula (31) using the numerically obtained values of $K(N)$; see below.

4 Numerics

Although we presented our statistical theory here only based on the simplest linear gap function, it can be easily generalized to incorporate affine gap costs as well. Since affine gap costs are much more frequently used, we present our numerics based on affine gap costs only. In our affine gap function, we have used the symbol μ to denote the weight of gap initiation and ν to denote the weight of a gap extension. It turns out that an equation similar to (4) still exists in which the scaling constant for the transition matrix now depends on both μ and ν. The joint probability distribution $\mathcal{P}(a,b)$ is given in terms of a scoring matrix $s(a,b)$ by $\mathcal{P}(a,b) = e^{\lambda_{\rm ug} s(a,b)} p(a)p(b)$ with $\lambda_{\rm ug}$ defined by $\sum_{a,b} e^{\lambda_{\rm ug} s(a,b)} p(a)p(b) = 1$. For the $s(a,b)$, we use Dayhoff's PAM substitution matrices (Dayhoff *et al.*, 1978).

We use two sets of scoring parameters described by PAM distance $d = 120$, $\mu = 2^{-5.5}$, $\nu = 2^{-0.5}$ and PAM distance $d = 250$, $\mu = 2^{-6}$, $\nu = 2^{-0.5}$ respectively. For brevity, we refer to the first set of parameters as "PAM-120" and the second set as "PAM-250".

We start with the numerical verification that the MLL score obeys Gumbel statistics. We use the two sets of scoring systems PAM-120 and PAM-250 satisfying the conservation condition for affine gaps. Figs. 3(a) and (b) show the pdf's of S obtained from the alignment of 50,000 pairs of random sequences of lengths 300 each, generated according to the null model (2). We see that the pdf's are well-fitted by the Gumbel distribution (1).

In order to measure $\bar{\sigma}(N)$ in a very effective way, we made use of our knowledge of the extremal ensemble. Instead of aligning random sequences and waiting for large islands, we directly generated typical island score landscapes using the correlated ensemble $Q^*_N[\mathbf{a},\mathbf{b}]$ and read off the average score for each length N. The results corresponding to the PAM-120 and PAM-250 scoring systems are shown in Fig. 4. They are well fitted by the form $\bar{\sigma} = \alpha N + c$, with statistical uncertainties in α and c well under 1%. The most striking thing about this result is that the data points in Fig. 4 were averaged over only 15 pairs of alignments and took practically no time to generate, while determining α to such precision using direct simulation or island counting will take weeks on the same computer.

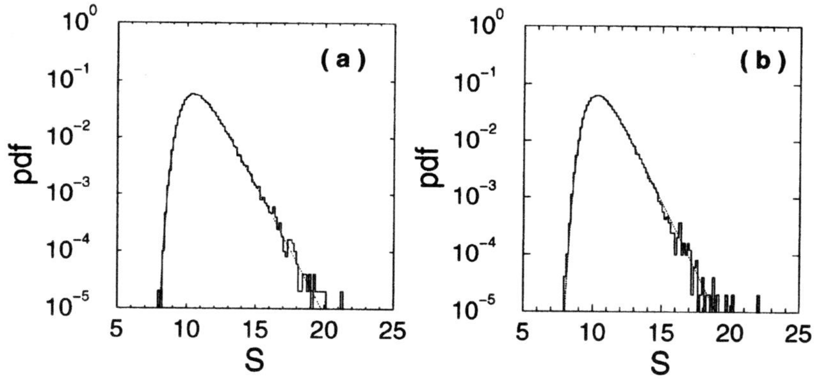

Fig. 3. The pdf's for the semi-probabilistic alignment of random sequences using the two parameter sets (a) PAM-120 and (b) PAM-250. The pdf's are obtained by normalizing histograms of 50, 000 pairs of random sequences of length 300 each.

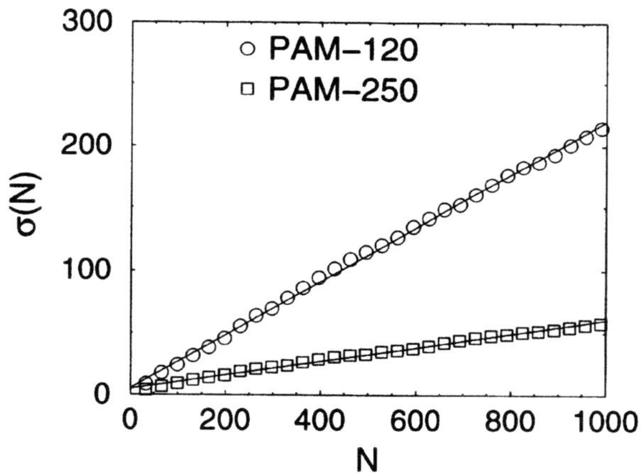

Fig. 4. The circles and squares represent the alignment score of correlated sequences taken from the extremal ensemble, averaged over only 15 pairwise alignments, corresponding to the PAM-120 and PAM-250 scoring systems respectively. The lines represent the respective least-square fits to $\overline{\sigma}(N) = \alpha N + c$. The fits, which are excellent down to $N = 50$, give $\alpha = 0.0554, c = 4.85$ for PAM-250 and $\alpha = 0.2144, c = 5.22$ for PAM-120.

With the accurate determination of $\overline{\sigma}(N)$, we are now in a position to test the prediction of the sequence length dependence (30), and with it, the prediction of the asymptotic result $\lambda = 1$. The predicted expression of $\lambda(N)$ using the numerically obtained $\overline{\sigma}(N)$'s in Eq. (30) is plotted as the line in Figs. 5(a) and (b) for the PAM-120 and PAM-250 scoring systems respectively. Also plotted are the data points obtained from fitting the pdf's for sequences of different lengths to the Gumbel form (1). We find very good agreement between theory and

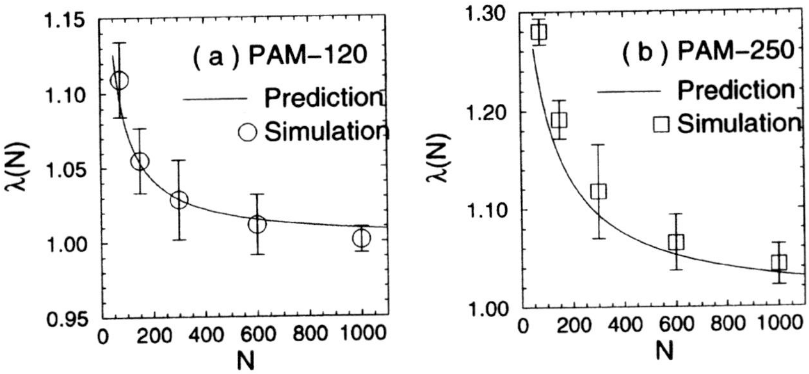

Fig. 5. Direct comparisons of the numerical values of λ obtained from fitting pdf's to Gumbel form (1) and the theoretical prediction for (a) PAM-120 and (b) PAM-250 scoring systems.

measurements down to sequence length of $N = 75$ for PAM-120 and $N = 150$ for PAM-250. (For smaller N's, the pdf's are no longer well-described by the Gumbel distribution for reasons explained earlier.) The striking agreement found lends strong support to the theory presented.

To test the prediction of the sequence length dependence of $K(N)$, we simply plot on the vertical axis $K(N)/(1+c/(\alpha N))^2$, using the values of c and α determined from Fig. 4 for the corresponding scoring system. According to Eq. (31), this simple transformation should render the data points N-independent, and give the value of the asymptotic K. We applied this transformation to $K(N)$ separately for the PAM-120 and PAM-250 scoring systems; see Figs. 6 (a) and

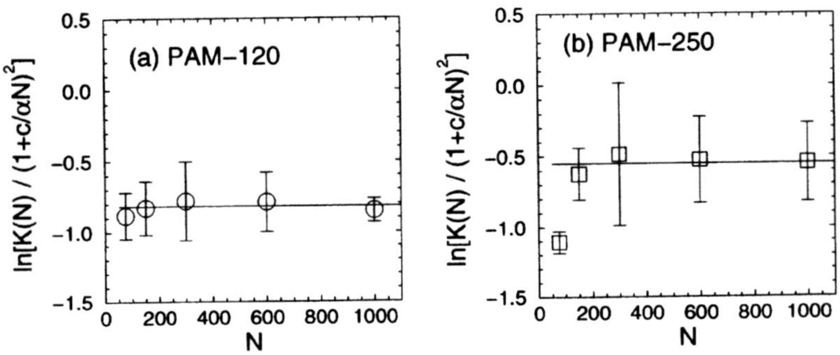

Fig. 6. Dependence of $K(N)$ on the sequence length for (a) PAM-120 and (b) PAM-250 scoring systems. The horizontal lines indicate the values of the asymptotic K's obtained from data at larger N's (not shown).

(b). Other than the smallest size of $N = 75$, the data points are approximately[3] N-independent, hovering around the asymptotic values indicated by the horizontal lines. The results suggest that Eq. (31) does capture the dependence of $K(N)$ on the sequence length correctly. Consequently, it is only necessary to determine $K(N)$ for one sequence length, say, at $N = 300$ by an island counting method similar to Olsen *et al.* (1999), or at $N \to \infty$ if the present theory can be extended to compute K. From this, the value of the effective K for all other N's can be deduced from the sequence length dependence formula (31).

5 Summary

In this paper, we studied the extremal statistics of probabilistic sequence alignment both analytically and numerically. We find that while the statistics of straightforward probabilistic alignment is not understood, the slightly modified semi-probabilistic alignment is well described by Gumbel statistics. For the semi-probabilistic alignment, we can predict the Gumbel parameter λ, including its sequence length dependence, for different scoring functions and parameters. Moreover, for a given scoring scheme, we have characterized the corresponding extremal ensemble of most detectable sequence-pairs. This allows for an optimal choice of scoring parameters for a given search goal. Our results are verified numerically by using various PAM substitution matrices and affine gap functions.

In our study, we have not focused on the behavior of the other Gumbel parameter, K, which is more difficult to compute analytically than λ. It is however straightforward to determine K numerically by extending the island method of Olsen *et al.* (1999) to the semi-probabilistic alignment (Bundschuh *et al.*, to be published.) With the help of a precisely determined λ, and the formula for the sequence length dependence of K, it is possible to fix the value of K for all sequence lengths by counting islands from a single pairwise alignment of managable size.

Let us close with a general remark: While the numerics presented in the present study was restricted to position-independent scoring functions, this is not a prerequisite for the application of our theory. In fact, we expect that the asymptotic value of λ to remains1 as long as the probability conservation condition (4) is *locally satisfied* at each node of the alignment lattice. This can be readily accomplished for position-specific substitution and indel weights by generalizing our previous result (Yu & Hwa, 1999).

Acknowledgments

We thank Rolf Olsen for much help and many useful suggestions during the course of this study. This research is supported in part by the Beckman Foundation and the NSF through Grant No. DMR-9971456. TH further acknowledges the financial support of a Guggenheim Fellowship, and RB acknowledges

[3] The statistical uncertainties associated with the $K(N)$'s are much larger because the actual parameter used in the Gumbel fit was $K(N)N^2$.

a Hochschulsonderprogramm III fellowship of the DAAD. Finally, TH and YKY gratefully acknowledge the hospitality of the Center for Studies in Physics and Biology at Rockefeller University where this work was initiated.

6 Appendix A: Consistency of our approximations

As a simple consistency check of our assumption (23), we see that insertion of Eq. (24) into Eq. (15) recovers the result (17), as long as α is *length-independent.* To probe the validity of the assumption more closely, let us introduce the notation $\langle ... \rangle^*$ to be the statistical average over the extremal ensemble $Q_L^* = W_L \cdot P_0$, i.e.,

$$\langle F[\mathbf{a},\mathbf{b}]\rangle^* \equiv \sum_{\{\mathbf{a},\mathbf{b}\}} F[\mathbf{a},\mathbf{b}] W[\mathbf{a},\mathbf{b}] P_0[\mathbf{a},\mathbf{b}]. \tag{32}$$

Then, the distribution (13) can be rewritten as

$$D(h|L) = e^{-h} \langle \delta(h - \ln W_L[\mathbf{a},\mathbf{b}])\rangle^*, \tag{33}$$

while the result of our assumption, Eq. (24), can be re-written as

$$D(h|L) \approx e^{-h} \delta\left(h - \langle \ln W_L[\mathbf{a},\mathbf{b}]\rangle^*\right). \tag{34}$$

Thus, the approximation made here is to replace the random variable $\ln W_L$ by its *typical* value in the extremal ensemble, $E^*[\ln W_L] \equiv \langle \ln W_L[\mathbf{a},\mathbf{b}]\rangle^*$. Approximations of this nature are poor in cases where the distribution of $\ln W_L$ is broad, e.g., if

$$\mathrm{var}(\ln W_L) \equiv \langle \ln W_L^2[\mathbf{a},\mathbf{b}]\rangle^* - E^*[\ln W_L]^2$$

is comparable to $E^*[\ln W_L]^2$. Due to the fact that the alignment of a pair of sequences of the extremal ensemble can be split into independent pieces of length ξ as discussed in the main text, we get $\mathrm{var}(\ln W_L) \approx L/\xi$ for $L \gg \xi$ according to the central-limit theorem. Since $E^*[\ln W_L] = \alpha L$, we have

$$\mathrm{var}(\ln W_L)/E^*[\ln W_L]^2 \to 0$$

in the limit of large L. In this way, the equivalence between Eq. (33) and Eq. (34) is justified.

7 Appendix B: Z and W

In this appendix we show that the restricted local alignment weight $Z_{m,n}$ at fixed (m,n) is at large values statistically equivalent to $\overline{Z}$ as defined in Eq. (25). First, we note that due to the symmetry of $\mathcal{W}_{m',n';m,n}$ and the fact that random sequences are statistically equivalent to their reverse sequences, the statistics of $Z_{m,n}$ is identical to the statistics of $\widehat{Z}_{M-m,N-n}$ defined by reversing the sequences, i.e., by

$$\widehat{Z}_{m,n} \equiv 1 + \sum_{m'=1}^{M-m} \nu^{m'} + \sum_{n'=1}^{N-n} \nu^{n'} + \sum_{\substack{m+1 \le m' \le M \\ n+1 \le n' \le N}} \mathcal{W}_{m+1,n+1;m',n'}. \tag{35}$$

Moreover, since we assume that the distribution of $Z_{m,n}$ is translationally invariant as long as we stay far enough away from the edges $m \approx 1$ and $n \approx 1$, we can without loss of generality pick $(m,n) = (M,N)$ and study the statistics of $\widehat{Z}_{0,0}$. This statistics should at most very weakly depend on the values of $\widehat{Z}_{m,n}$ for very large m and n and we can therefore extend the summations in Eq. (35) to infinity. To summarize, we expect that the statistics of $Z_{m,n}$ for a fixed (m,n) are identical to the statistics of

$$\widetilde{Z} \equiv 1 + \sum_{m'=1}^{\infty} \nu^{m'} + \sum_{n'=1}^{\infty} \nu^{n'} + \sum_{m'=1}^{\infty} \sum_{n'=1}^{\infty} \mathcal{W}_{1,1;m',n'}. \tag{36}$$

In order to study $\widetilde{Z}$ we rewrite it as

$$\widetilde{Z} = 1 + \sum_{L=1}^{\infty} \widetilde{W}_L \tag{37}$$

with

$$\widetilde{W}_L \equiv 2\nu^L + \sum_{m'=1}^{L-1} \mathcal{W}_{1,1;m',L} + \sum_{n'=1}^{L-1} \mathcal{W}_{1,1;L,n'} + \mathcal{W}_{1,1;L,L}. \tag{38}$$

This quantity looks very similar to $\overline{W}$ as defined in Eqs. (26) and (5) and indeed we can easily convince ourselves using Eq. (3) that

$$\overline{W}_L \le \widetilde{W}_L \le \frac{1}{1-\nu} \overline{W}_L. \tag{39}$$

Thus, the $\overline{Z}$ defined in Eq. (25) bounds $\widetilde{Z}$ as

$$\overline{Z} \le \widetilde{Z} \le \frac{1}{1-\nu} \overline{Z} \tag{40}$$

and since we are only interested in the logarithms of these quantities they become statistically equivalent for large enough values of $\overline{Z}$.

Under all these transformations, sequence configurations which make $Z_{m,n}$ large for a fixed (m,n) are directly related to configurations of the (reversed !) sequences $a_m a_{m-1} \dots a_{m-L_0}$ and $b_n b_{n-1} \dots b_{n-L_0}$ which make $\overline{W}_{L_0}$ large for some L_0. Thus, these reversed sequences are drawn from the extremal ensemble $Q^*_{L_0}[a_m a_{m-1} \dots a_{m-L_0}, b_n b_{n-1} \dots b_{n-L_0}]$ derived in Eq. (22).

References

1. Altschul, S.F., Gish, W., Miller, W., Myers, E.W., and Lipman, D.J., 1990. Basic Local Alignment Search Tool. J. Mol. Biol. **215**: 403–410.
2. Altschul, S.F., 1991. Substitution Matrices from an Information Theoretic Perspective. *J. Mol. Biol.* **119**: 555–565.
3. Altschul, S.F., and Gish, W., 1996. Local Alignment Statistics. *Methods in Enzymology* **266**: 460–480.
4. Altschul, S.F., Madden, T.L., Schäffer, A.A., Zhang, J., Zhang, Z., Miller, W., and Lipman, D.J., 1997. Gapped BLAST and PSI-BLAST: a new generation of protein database search programs. *Nucleic Acids Research* **25**: 3389–3402.
5. Altschul, S.F., Bundschuh, R., Hwa, T., and Olsen, R., 2001. The estimation of statistical parameters for local alignment score distributions. *Nucleic Acids Research* **29**: 351–361.
6. Arratia, R., Morris, P., and Waterman, M.S., 1988. Stochastic scrabbles: a law of large numbers for sequence matching with scores. *J. Appl. Prob.* **25**: 106-119.
7. Bishop, M.J., and Thompson, E.A., 1986. Maximum likelihood alignment of DNA sequences. *J. Mol. Biol.* **190**: 159-165.
8. Bundschuh, R., 2000. An Analytic Approach to Significance Assessment in Local Sequence Alignment with Gaps. *RECOMB 2000.*
9. Collins, J.F., Coulson, A.F.W., and Lyall, A., 1988. The significance of protein sequence similarities. *CABIOS* **4**: 67–71.
10. Dayhoff, M.O., Schwartz, R.M., and Orcutt, B.C., 1978. A Model of Evolutionary Change in Proteins. In *Atlas of Protein Sequence and Structure*, Dayhoff M.O. and Eck, R.V., eds., **5** supp. 3: 345–358, Natl. Biomed. Res. Found.
11. Drasdo, D., Hwa, T., and Lassig, M., 1998. A Scaling Theory of Sequence Alignment with Gaps. *ISMB98*: 52-58.
12. Gumbel, E.J., 1958. *Statistics of Extremes.* New York, NY: Columbia University Press.
13. Henikoff, S., and Henikoff, J.G., 1994. Position-based Sequence Weights. *J. Mol. Biol.* **162**: 705-708.
14. Hughey, R., and Krogh, A., 1996. Hidden Markov Models for Sequence Analysis: Extension and Analysis of the Basic Method. *CABIOS* **12**: 95-107.
15. Hwa, T., and Nattermann, T., 1995. Disorder-induced depinning transition. *Phys. Rev. B* **51**: 455–469.
16. Hwa, T., and Lässig, M., 1996. Similarity Detection and Localization. *Phys. Rev. Lett.* **76**:2591–2594.
17. Karlin, S., and Altschul, S.F. 1990. Methods for assessing the statistical significance of molecular sequence features by using general scoring schemes. *Proc. Natl. Acad. Sci. USA* **87**: 2264–2268.
18. Karlin, S., and Dembo, A., 1992. Limit distributions of maximal segmental score among Markov-dependent partial sums. *Adv. Appl. Prob.* **24**: 113–140.
19. Karlin, S., and Altschul, S.F., 1993. Applications and statistics for multiple high-scoring segments in molecular sequences. *Proc. Natl. Acad. Sci. USA* **90**: 5873–5877.
20. Mott, R., 1992. Maximum likelihood estimation of the statistical distribution of Smith-Waterman local sequence similarity scores. *Bull. Math. Biol.* **54**: 59–75.
21. Needleman, S.B., and Wunsch, C.D., 1970. A general method applicable to the search for similarities in the amino acid sequence of two proteins. *J. Mol. Biol.* **48**: 443–453.

22. Olsen, R., Bundschuh, R., and Hwa, T., 1999. Rapid Assessment of Extremal Statistics for Gapped Local Alignment. *Proceedings of The Seventh International Conference on Intelligent Systems for Molecular Biology (ISMB99)*. T. Lengauer et al. eds., 211–222 (AAAI Press, Menlo Park).
23. Pearson, W.R., 1988. Improved Tools for Biological Sequence Comparison. *Proc. Natl. Acad. Sci. USA* **85**: 2444-2448.
24. Smith, T.F., and Waterman, M.S., 1981. Identification of Common Molecular Subsequences. *J. Mol. Biol.* **147**: 195-197.
25. Smith, T.F., Waterman, M.S., and Burks, C., 1985. The statistical distribution of nucleic acid similarities. *Nucleic Acids Research* **13**: 645–656.
26. Thorne, J.L., Kishino, H., and Felsenstein, J. 1991. An Evolutionary Model for Maximum Likelihood Alignment of DNA Sequences. *J. Mol. Evol.* **33**: 114-124.
27. Thorne, J.L., Kishino, H., and Felsenstein, J., 1992. Inching toward Reality: An Improved Likelihood Model of Sequence Evolution. *J. Mol. Evol.* **34**: 3-16.
28. Waterman, M.S., and Vingron, M., 1994a. Sequence Comparison Significance and Poisson Approximation. *Stat. Sci.* **9**: 367–381.
29. Waterman, M.S., and Vingron, M., 1994b. Rapid and accurate estimates of statistical significance for sequence data base searches. *Proc. Natl. Acad. Sci. U.S.A.* **91**: 4625–4628.
30. Yu, Y.-K., and Hwa, T., 1999 Statistical Significance of Probabilistic Sequence Alignment and Related Local Hidden Markov Models. *Submitted to J. Comp. Biol.*.

On the design of optimization criteria for multiple sequence alignment

Dannie Durand[1] and Martin Farach-Colton[2]

[1] Department of Biological Sciences, Carnegie Mellon University, Pittsburgh, PA 15213, USA
[2] Department of Computer Science, Rutgers University, Piscataway, NJ 08855, USA

Abstract. Multiple sequence alignment (MSA) is important in functional, structural and evolutionary studies of sequence data. While MSA construction has traditionally been an interactive process, the rapid growth of genetic sequence data has engendered a need for automated sequence analysis without human intervention. This requires more accurate methods based on rigorous mathematical models that reflect sequence biology in a realistic way. Focusing on MSA as an optimization problem, we examine the problem of unifying mathematical tractability with biological accuracy in cost function design. In particular, we consider tree alignment, which is often viewed as the most "biological" of the rigorous approaches to MSA. We point out several important pitfalls in current optimization approaches to MSA and identify characteristics for good cost function design. Design issues specific to approximation algorithms are also addressed. We hope these ideas will lead to future research on a biologically realistic and mathematically rigorous approach to MSA.

1 Introduction

Multiple sequence alignment (MSA) is a formal method for identifying shared features in a set of related nucleic acid or protein sequences, such as the sequences of a gene of interest (e.g., myoglobin) in several different species or the sequences of a family of related genes (e.g., myoglobin, the α-globins and the β-globins) in a single species. The goal is to align the sequences in such a way that biological relationships between sequence elements are revealed. MSA's are used to build evolutionary trees, extract information about the function and structure of proteins, identify patterns for data base searching and motif recognition and to design probes and primers for laboratory experiments. In some of these applications, such as evolutionary tree reconstruction, the MSA will be used as input to another program.

To state the MSA problem formally, we represent each biopolymer, P_1, $P_2, \ldots, P_k$, as a string of symbols chosen from an alphabet, Σ, where $\Sigma = \{A, C, G, T\}$ for nucleic acid sequences and Σ contains the twenty amino acids for protein sequences. Then an MSA is a set of sequences $\hat{P}_1, \hat{P}_2, \ldots, \hat{P}_k$ drawn from $\Sigma \cup \{\_\}$, where $\_$ is called a *blank*, such that P_i is the string derived by removing all blanks from $\hat{P}_i$. Additionally, each $\hat{P}_i$ should be of the same length. We will call $\hat{P}_1[j], \ldots, \hat{P}_k[j]$ the *jth column* of the MSA. The standard interpretation of blanks is that they represent mutations in the form of insertions or

```
HUMAN  MKWVTFISLL FLFSSAYSRG V--FRRDA-H KSEVAHRFKD LGEENFKALV
RABBIT MKWVTFISLL FLFSSAYSRG V--FRREA-H KSEIAHRFND VGEEHFIGLV
PIG    --WVTFISLL FLFSSAYSRG V--FRRDT-Y KSEIAHRFKD LGEQYFKGLV
CHICK  MKWVTLISFI FLFSSATSRN LQRFARDAEH KSEIAHRYND LKEETFKAVA
```

Fig. 1. A multiple alignment of four partial albumen sequences

deletions, and therefore they are sometimes called *indels*. However, the sequences $P_1, \dots, P_n$ need not be evolutionarily related, a fact which we will explore further below. As an example, an MSA of four amino acid sequence fragments, taken from human, rabbit, pig and chicken albumin, is shown in Fig. (1). Two indels have been inserted after the central valine in the first three sequences to reveal several conserved features in the alignment such as the KSE motif starting at column 31.

Columns in an MSA should share a target biological feature. The feature sought depends on the application of the alignment. For example, if the multiple alignment is used to illustrate evolutionary relationships, then the residues in each column should have a shared evolutionary history. If the multiple alignment is used to determine structure or function, then the residues in each column should have a shared structural or functional role. It is exactly this kind of feature extraction that is the goal of MSA computation, the underlying assumption of which is that the sequences of a molecule contain enough information to extract the feature.

Numerous programs are available to construct MSA's, many of them based on heuristic approximations. A traditional approach has been to use such software to generate an initial multiple alignment and then adjust that alignment "by hand". Under this paradigm, highly accurate methods are not needed. However, as sequence data bases grow in the era of whole genome sequencing, the need to generate large numbers of MSA's automatically is also growing. Methods that can construct reliable MSA's without human oversight are required. Such methods must depend on formal models that accurately reflect the biology of sequences.

A number of formal, mathematical models of MSA have been proposed and a wealth of papers elaborating on these models has appeared in the theoretical computational molecular biology literature in the last ten to fifteen years. These approaches generally formulate MSA as an *optimization problem*, by defining a *cost function* over the set of all possible MSA's and then seeking the MSA which optimizes[1] the cost. A multiple sequence alignment may be viewed as a hypothesis concerning the true relationship of a set of biopolymers, be that relationship evolutionary, structural or functional. A cost function is a method for evaluating the quality of such hypotheses and may be used to compare alternate alignments, whether generated by exact algorithms or by heuristics. The advantage of this optimization approach is that the cost function makes explicit the

[1] We will interchangeably minimize and maximize the function in our examples, as needed.

assumptions upon which the optimization is based. Because they are explicit, these assumptions are open to scrutiny and falsification.

The success of this approach depends on the assumption that it is possible to design a cost function such that the MSA with optimal cost is also the best explanation of the biological relationships between the sequences. In the face of genome scale data, research that attempts to unify biological and mathematical notions of accurate alignment is needed. The goal of the current paper is to examine the optimization approaches currently in use and identify problems that must be addressed on the path to robust, reliable MSA for automatic analysis of genomic data. In Section 2, we examine several popular cost functions that have received widespread attention in the literature, including sum-of-pairs, star and tree alignment. We focus primarily on tree alignment, the most biologically motivated of the three. We explore mathematical aspects of cost function design in Section 3. In particular, design issues related to approximation algorithms are discussed. Important biological considerations that must be addressed by any sound optimization criterion are elucidated in Section 4. Here we also specifically address the biological reasonableness of tree alignment for MSA computation. Biological validation of multiple sequence alignments is discussed in Section 5. We conclude that while optimization and approximation offers the best hope for well founded MSA computation, the current crop of optimization criteria are seriously wanting. We suggest some avenues of attack in correcting the situation.

2 Optimization criteria

The desired output of an MSA computation is the column relationships of the alignment. Thus, it is natural to focus on optimization criteria which associate a score with each column, and most criteria in the literature take this approach. We will not give a comprehensive survey of multiple sequence alignment algorithms here. Such surveys can be found in [1] and in the introduction to [2]. Statistical approaches, such as hidden Markov models to MSA are discussed in [3]. These approaches are used primarily for local multiple alignments. An experimental comparison of multiple sequence algorithms is described in [4]. The column score expresses how well matched the residues in the column are. The score of the MSA is then the sum of the scores of its columns. We seek a set of $\hat{P}$ maximizing

$$D(A) = \sum_j d(\hat{P}_1[j], \ldots, \hat{P}_k[j]),$$

for column scoring function $d(\cdot)$. The choice of the column scoring function, $d(\hat{P}_1[j], \ldots, \hat{P}_k[j])$, should reflect the application of the alignment. Residues that maximize $d(\hat{P}_1[j], \ldots, \hat{P}_k[j])$ should belong together. Note that in our formulation, the scores of columns are completely independent. However, in some formulations, the cost of a gap might depend on the length of the gap in many ways. For example, for so-called affine gap penalties, the cost of a gap of length l is $\alpha + \beta l$ for some constants α and β. This separation of the gap penalty into α, a gap initiation cost, and β, a gap extension cost, reflects the assumption

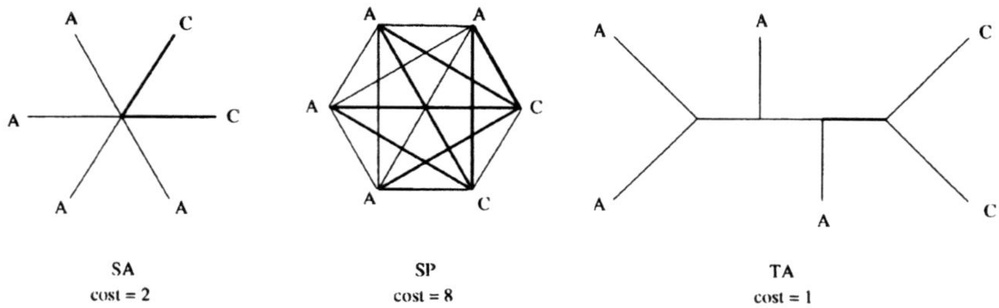

Fig. 2. Comparison of alignment costs for a multiple sequence alignment column, (A,A,A,A,C,C), under three different cost functions. Edges representing mutations are shown in bold

that whole pieces of sequence can be deleted in single events. Such gap functions typically are used for pairwise alignments and for multiple alignments that are heuristically constructed from pairwise alignments, and they will not be considered further here.

Three cost functions commonly used to evaluate each column are sum-of-pairs (SP), tree alignment (TA) and star alignment (SA), shown graphically in Fig. 2. The sum-of-pairs cost [5,6] of a column $\hat{P}_1[j], \ldots, \hat{P}_k[j]$ is the sum of the costs of all unordered pairs in the column

$$d_{SP}(\hat{P}_1[j], \ldots, \hat{P}_k[j]) = \sum_{p<q} \delta(\hat{P}_p[j], \hat{P}_q[j]),$$

for some binary cost function $\delta(x, y)$. This definition is mathematically natural but not biologically intuitive.

Tree alignment [7,8] is based on the assumption that the residues in the columns of the multiple sequence alignment share an evolutionary history and that this history can be expressed as a single tree for all columns. Under this model, a column is scored by computing the cost of the underlying tree. The score of the tree expresses the strength of our belief, under some model of evolution, that these residues are related in the manner described by this tree.

In order to use this approach, several issues must be resolved. First, a tree topology is needed. In general, the underlying tree is not known. In fact, multiple sequence alignments are generally used to estimate evolutionary trees and not vice versa. Second, it is generally assumed that every column in the alignment has the same underlying tree topology. As we shall see below, this may not always be the case. Third, in order to compute the branch costs of the tree, we need to know the ancestral sequences associated with the internal nodes. Fourth, given a tree topology with ancestral sequences at the nodes, a cost must be associated with each branch of the tree. This implies an underlying model of evolutionary change. The appropriate model will vary with the data.

Let us consider how to infer the ancestral sequences and compute the branch costs when the topology is known. The k extant sequences are associated with the k leaves of the tree. Sequences for the internal nodes are selected so that the tree is the best estimate of the true tree under some model of evolutionary change. The model in use is *maximum parsimony*, in which it is assumed that the true tree required a minimum number of evolutionary steps. Under this model, the cost of an edge (X_i, X_j) in the tree is the minimum number of mutations required to transform sequence X_i into sequence X_j. That is, internal nodes are selected such that the sum of edge costs, i.e. the total number of mutations required along the branches of the tree, is minimized. This assumption that evolution is parsimonious, is a subject of much debate.

The above approach assumes that the topology of the evolutionary tree is known, e.g. from morphological data. If the tree topology is not known, then the tree which yields the best score must be found. Since the number of trees with k leaves grows exponentially with k, this approach rapidly becomes prohibitively expensive if exhaustive enumeration is used. In general, no fast algorithm is known.

A variant on tree alignment is star alignment (SA), in which it is assumed that the underlying tree is a star. Restricting the topology makes this approach much more tractable, and it has been used as an underlying structure in many algorithms and analyses (e.g., [9,10]).

Altschul and Lipman [9] pointed out that SP, TA and SA costs for the same column can be very different, as shown in Fig. 2. SP overcounts mutations in a column because it considers that a separate mutation occurred between each pair in the column. In some sense, it is a "least parsimonious," or profligate, model of mutation. In contrast, TA has been traditionally used with the maximum parsimony criterion. For most data sets the true answer is somewhere in between.

Efficiency considerations: The MSA problem has been shown to belong to the class of *NP-compete problems* [11,12], for both sum-of-pairs and tree alignment. Problems in this class, which could be described colloquially as "officially hard problems," have the property that the only known way to obtain an optimal solution to the problem is to generate all possible solutions and compare them. (See, for example, [13], for a discussion of NP-completeness.) Since the number of possible MSA's increases exponentially with the number of sequences, the time required to find the optimal solution grows exponentially with the number of sequences as well. While no one has been able to prove that it is necessary to examine every possible solution to an NP-complete problem to find the optimum, no one has been able to solve any NP-complete problem exactly without enumerating all solutions. Since any NP-complete problem can be converted into any other NP-complete problem, a fast method to solve just one of these problems would provide a fast solution to all of them. Proving or disproving the existence of fast solutions to NP-complete problems is considered one of the most important unsolved mathematical challenges of our time. Since this problem has been the subject of intense scrutiny, without definitive resolution, by the theoretical computer science community for the last thirty years, we should not hope for a fast, exact MSA algorithm anytime soon.

In the face of this harsh reality, there are three approaches to MSA construction. For small data sets, typically eight to ten sequences of length at most 500 elements, it is possible to enumerate all alignments and select the optimal one. An optimal alignment of k sequences of length at most N can be obtained using dynamic programming in $O(2^k N^k)$ evaluations[2] of the SP cost function using $O(N^k)$ space [15]. An exact tree alignment algorithm for a given tree topology has been presented by Sankoff [7]. This requires $O(M(2N)^k)$ steps, where M is the number of internal nodes.

For larger data sets, one may turn to one of the numerous heuristic approaches that generate MSA's with no optimality guarantee. Many of these are based on the *progressive alignment* approach, in which an optimal pairwise alignment is computed for every pair of sequences in the data set. These pairwise alignments are then merged to obtain a multiple alignment. Progressive alignment was first introduced by Taylor [16] and is the approach upon which the widely used CLUSTALW program [17] is based. The various progressive alignment methods differ in the rules used to determine in what order the pairwise alignments should be merged. Since the time required to compute an optimal pairwise alignment is $O(N^2)$ and the number of pairwise alignments is $O(k^2)$, the time required to compute a progressive alignment grows quadratically with the size of the input in comparison with the exponential time required to compute an optimal MSA.

Typically, it is possible to analyze up to 5000 sequences using a progressive alignment heuristic. However, progressive alignment will not, in general, give an optimal solution and will generate different solutions depending on the order in which the pairwise alignments are merged. This is illustrated in the multiple alignment of the sequences ACTCCAT, AGTCCT and ACGTCAT in Fig. 3. Consider a sum-of-pairs cost function where the cost of an indel is $i = 2$ and the cost of a substitution is $s = 3$. The costs of the three optimal pairwise alignments

```
(1) ACTCCAT          (1,2) + (2,3)
(2) AGTCC-T          (1) A-CTCCAT
                     (2) A-GTCC-T
(2) A-GTCCT          (3) ACGTCA-T
(3) ACGTCAT
                     (1,2) + (1,3)
(1) AC-TCCAT         (1) AC-TCCAT
(3) ACGTC-AT         (2) AG-TCC-T
                     (3) ACGTC-AT
```

Fig. 3. Multiple alignment of the sequences CTCCAT, AGTCCT and ACGTCAT. Pairwise alignments these sequences are shown on the left. Two different progressive multiple alignments appear on the right, showing that the final form of the multiple sequence alignment depends on the order in which the pairwise alignments are merged

[2] O-notation is used to describe the asymptotic behavior of a program in terms of the size of it's input. Colloquially, $O(f(n))$ refers to a function whose behavior is proportional to $f(n)$ for large n. For a full treatment of O-notation, see [14].

shown in Fig. 3 are $s+i=5$, $s+i=5$ and $2i=4$, respectively. The progressive alignment algorithm requires that we select one pairwise alignment as the seed of the multiple sequence alignment. At each step of the procedure, a new pairwise alignment is selected that contains one sequence not yet included in the growing MSA. The new sequence is added to the MSA using the pairwise alignment as a guide. The juxtaposition of the elements of the sequences in partial multiple alignment from the previous iteration must remain fixed. These elements may not be rearranged later to improve the score as additional sequences are added to the MSA. This restriction is responsible for both the increase in speed and the loss in accuracy associated with the progressive alignment heuristic. Fig. 3 shows two MSA's, both using the alignment of Sequences 1 and 2 as the seed. In the first MSA, the third sequence is added by aligning it with Sequence 2. In the second, the third sequence is aligned with Sequence 1. As the figure shows, two different MSA's result. The SP cost for the first alignment is $(s+i)+(s+i)+(2s+2i)=4s+4i=20$, while the second alignment costs $(s+i)+(s+3i)+(2i)=2s+5i=16$.

As the above example shows, heuristic approaches will not in general yield an optimal solution, nor indicate whether the solution presented differs from the optimum and, if so, how. *Approximation algorithms* [18] offer a promising compromise between exact, but computationally intractable methods and the heuristics currently most often used in practice. An a-approximation is a solution to an optimization problem that is guaranteed to have a score that differs from the optimal score by a factor of at most a. Approximation algorithms for both SP and tree alignment include those in [2,10,19–21]. Notably, a polynomial time approximation scheme, yielding a $(1+\epsilon)$-approximation for arbitrarily small ϵ, has been presented by Jiang, Lawler and Wang [22,23] for tree alignment on a fixed topology. Approximation algorithms are powerful because they rigorously quantify the price to be paid by using an approximate method. However, to be practically useful, we need a better understanding of this price in *biological* terms. The work cited above has not, for the most part, investigated the biological implications of the approximation factor, a.

3 Approximations and convergence

Since any reasonable cost function for MSA will probably be NP-hard to optimize [11,12], it is natural for researchers to look for approximation algorithms which find near optimal solutions to the MSA problem. These solutions will not be meaningful unless they satisfy (at least) the following two properties.

Smoothness: The success of the optimization approach to MSA depends on finding a cost function such that the optimal solution with respect to that cost function is close enough to the true solution. The definition of "true" solution depends on the application (e.g., reconstructing phylogeny, determining structure or function), as will be discussed in the next section. So the cost function should reflect the application in some way. Unlike problems such as the Traveling Salesman Problem, where the cost function (total distance traveled) is directly related to the problem under consideration (finding the shortest tour), mathematical cost functions used in MSA often have little relationship to the solution sought.

For approximation, stricter constraints on the properties of MSA cost functions are required. Intuitively, an MSA that is close in score to the optimal MSA should also be similar biologically. We call this notion *smoothness*. Unfortunately, "biologically similar" is a vague notion. It would be useful to make this notion concrete, for approximation algorithms are only meaningful if we understand the behavior of the cost function in the neighborhood of the optimal solution. What mathematical properties does a cost function require in order to exhibit smoothness? Under tree alignment, for example, it is not generally true that MSA's with similar scores have similar biological interpretation, since it is possible for many different trees to have the same parsimony score.

Consistency: Biological wisdom holds that MSA computation should get easier as more sequences are introduced, and harder as the sequences get longer. Roos [24] and McClure, Vasi and Fitch [4] have described specific instances of this behavior in their research. By contrast, phylogeny computation is held to get harder as the number of species increases but easier as the length of the sequences describing the species increases. Our optimization criteria should capture this varied dependence of problem instance difficulty on the size of the data set if they have any hope of capturing the underlying biology.

Ideally, the maximum of our optimization criterion should coïncide with the "true" or biological alignment. A weaker notion, and one inspired by the considerations above, is that as the number of sequences being considered increases, the maximum of our optimization criterion should *converge* on the true alignment. This idea of converging on the true answer as we observe more data is basically the notion of *consistency* [25,26]. For concreteness, we define an algorithm to be *consistent* with respect to some parameter k of data size (e.g., number of species for MSA or sequence length for phylogeny) if it converges to the true solution as k goes to infinity. Similarly, a cost function is *consistent* if an algorithm that outputs its maximum is consistent. Notice that for MSA we would not expect a cost function to be consistent with respect to sequence length.

Suppose we have a cost function which is consistent. Would an approximation algorithm for the function be consistent? We consider a couple of examples. Suppose

$$g(\phi, d_k) = \begin{cases} 2 & \text{if } \phi \text{ is the true alignment for data } d_k \text{ on } k \text{ sequences;} \\ 1 & \text{otherwise.} \end{cases}$$

Now trivially, g is a consistent cost function. But a $1/2$-approximation algorithm for g would be inconsistent, because *any* alignment yields a score at least half the maximum, so a $1/2$ approximation algorithm need not converge on the true answer. On the other hand, suppose

$$h(\phi, d_k) = \begin{cases} k & \text{if } \phi \text{ is the true alignment for data } d_k \text{ on } k \text{ sequences;} \\ 1 & \text{otherwise.} \end{cases}$$

Then any constant factor approximation algorithm for h will be consistent. This is because a $1-\epsilon$ approximation algorithm for h will be forced to output the true alignment for $k > 1/(1-\epsilon)$.

So we see that consistency for approximation algorithms requires more than an understanding of the biology of MSA, but also an understanding of the entire solution space for the cost function of choice. In particular, we need to know what happens to the set of alignments that are within a factor of $1-\epsilon$ of the best alignment, and how this set changes with k. We will say that a cost function f is *ϵ-consistent* if the set of all solutions within $1-\epsilon$ of the optimal for f converges to the true solution as k goes to infinity. We say f is *strongly consistent* if f is ϵ-consistent for every constant ϵ. In our example above, function g is ϵ consistent, for any $\epsilon < 1/2$, while function h is strongly consistent.

For phylogeny, a strongly consistent cost function was proposed in [27]. It was possible to do so in the case of phylogeny because the stochastic process relating phylogenies to DNA sequences is relatively well understood. We must seek such an understanding of MSA's before we can propose meaningful optimization criteria for this problem.

4 Evaluating tree alignment as a model for MSA

How well do existing cost functions reflect biological relationships in MSA's? There is an unstated assumption that tree alignment is the gold standard, the ideal cost function, for multiple sequence alignment because the residues in each column are thought to be related by an evolutionary tree. The implication is that tree alignment is not used only because it is intractable. In this section, we present some examples that demonstrate that a tree is not always the appropriate model for multiple sequence alignment. In considering whether tree alignment is appropriate for a given set of sequences, two issues must be addressed:

- Is a tree the correct model for describing the relationship between residues in each column? A tree may not be a suitable model because the relationship between residues is functional or structural rather than historical. Even if the relationship is historical, for some data sets, no single tree will describe all columns in the alignment.
- What is the correct mutational model for scoring the branches of the tree? Tree alignment has historically been based exclusively upon the parsimony criterion. Data that does not happen to be parsimonious can favor the wrong tree model. In addition, column-oriented optimization approaches to MSA usually assume that sequence positions are independent and identically distributed. In general, these assumptions do not hold for biological sequence data.

The residues in a column to not share a common ancestor: When alignment is used to study function or structure, residues in a column do not always share a common ancestor. The goal is to align residues that share the same role.

Although functional or structural residues usually share an evolutionary history, sometimes functional or structural roles can migrate to neighboring residues.

A possible example of shifting function occurs in dihydrofolate reductase (DHFR) gene – an important chemotherapeutic target in treating cancer and various infectious diseases. In their studies on protozoan parasites, Roos and colleagues have sought to design drugs that inhibit a metabolic protein in the parasite without affecting the infected host [28,29,24]. This requires identifying regions of structural or functional importance that differ substantially between protozoan and human versions of the DHFR protein.

Early sequence alignments placed the malaria parasite DHFR residue Phe^{223} downstream of structurally conserved regions of the protein (within the linker region which joins DHFR to thymidylate synthase, forming a bifunctional protein in protozoa and plants)[30,31]. This result was puzzling because mutational studies had suggested that Phe^{223} plays an important role in drug sensitivity. Realignment using sequences from the related parasite *Toxoplasma gondii* indicated that Phe^{223} is more likely homologous to a portion of the β-sheet which comprises the enzyme backbone [24]. The *T. gondii* sequence thus provided additional information that suggested an alternative alignment. Other protozoan sequences in the alignment have substantial, and different, nucleotide biases[3]. The *T. gondii* gene, which has relatively equal nucleotide distribution links the other protozoan sequences, facilitating alignment. This is another example of the observation, also made by McClure, Vasi and Fitch [4],that MSA's are very sensitive to sequence choice.

Phe^{223} is thought to play an indirect role in enzyme activity, interacting with His^{34}, within the active site [32]. The residue that plays this stabilizing role may have changed over time: a residue in a different position may provide this stabilizing effect in certain taxa (e.g. kinetoplastid parasites such as *Leishmania* and *Trypanosoma*) [33]. In this scenario, a random mutation allows a previously inactive residue to take on the functional role played by Phe^{223} . In Roos' alignment, the residues currently thought to provide this stabilizing effect appear in the same column, but these residues in this column may not all share a common ancestor.

The tree is not unique: Another case where a tree is not an appropriate model occurs when the residues in any particular column share a common ancestor but the columns themselves have different evolutionary histories. A tree may describe any given column, but the columns taken together cannot be modeled by a single tree.

One situation where this occurs is exon shuffling. Most vertebrate genes consist of coding regions (exons) separated by DNA segments that are not translated into proteins (introns). The discovery of this intron/exon structure lead to the theory of exon shuffling, first proposed by Gilbert in 1978 [34]. This theory posits

[3] The statistical profile of the primary sequence of genes can vary substantially, resulting in variations in, for example, the percentage of *GC* nucleotides. Such differences tend to obscure similarities between related sequences.

that exons represent functionally and/or structurally important subunits of proteins, that introns occur at the boundaries of these modules and that proteins share and reüse the modules that exons encode.

The first evidence that the same exons appear in more than one gene was found in the human low-density lipoprotein (LDL) receptor gene [35,36]. The LDL receptor gene was shown to share exons with genes for epidermal growth factor, blood clotting factor IX and complementation factor C9. Since then, many such "mosaic" genes have been discovered [37–40]. In aligning mosaic genes, a different tree may be needed for each exon. If the exon boundaries are known, each exon could be aligned separately. However for many sequences, the splice sites have not been determined. Detecting such boundaries would require that the alignment already be known.

Residues with different evolutionary histories within a single gene can also occur due to horizontal gene transfer, the transfer of genetic material between species. For example, sequences similar to the Fn3 module in fibronectin have been found in bacterial proteins [41]. Fibronectin is a protein found in animals. Since Fn3 is found in both bacterial and animal proteins, one would expect to find Fn3 modules throughout the tree of life. However, Fn3 sequences have not been found in simpler eukaryotes, plants or fungi [41], suggesting a direct transfer of genetic material between bacteria and animals. Thus, in an alignment of genes containing the Fn3 sequence, one would not expect the residues in the Fn3 module to share an evolutionary history with other residues in the alignment. More than one tree is needed to model the sequence and, as in the case of exon shuffling, it may not be possible to know where the module boundaries occur. Other examples of mixing of genetic material possibly requiring more than one tree include transposition and gene conversion.

The tree is not parsimonious: Sequence data are not generally parsimonious, especially between distantly related sequences. Multiple substitution (e.g., A $\rightarrow$ C $\rightarrow$ T), coincidental substitution (e.g., A $\rightarrow$ C vs. A $\rightarrow$ G), parallel substitution (e.g., A $\rightarrow$ C vs. A $\rightarrow$ C) and back substitution (e.g., A $\rightarrow$ C $\rightarrow$ A), can all obscure the evolutionary history of a sequence.

Convergent evolution results in a situation where the parsimony criterion will lead to the wrong tree model. Residues that appear to be closely related may simply be similar due selective pressure because they perform the same function. An example of this occurs in cows and colobine monkeys, species that independently evolved foregut fermentation [42]. In order to maximize the nutritional benefits of foregut fermentation, the enzyme lysozyme had to evolve in these species to function in the acidic, pepsin-rich environment of the stomach. In other species, lysozyme is only needed in the intestines. As a result, lysozymes in cows and colobine monkeys exhibit amino acid substitutions not found in other species, suggesting, wrongly, that the two species are closely related. This would result in the wrong tree for those residues.

Tetraloops in rRNA provide another example of convergent evolution [43,44]. Tetraloops are strings of six bases that form loops at the end of helices in rRNA structures. The two end bases bond, allowing the internal four bases to form a loop. Although there are 256 possible inner loop sequences, only a small number actually occur in nature. Tetraloop sequences will tend to appear to be closely related, even when they are not.

Sequence data is not i.i.d.: Structural constraints prevent sequence data from being independent. Structural integrity depends on interactions between non-adjacent residues in the sequence. For example, α-helices are characterized by a heptad repeat, so that there are chemical interactions between every seventh residue in helical regions. Similarly, in order to maintain the structure of the RNA molecule, distant residues bind to form a structural interaction. Compensatory mutations between distant residues that form structural bonds are selectively favored.

Sequence data are not identically distributed either. Structural constraints on protein sequences result in variations in selective pressure at different positions, depending on whether they are located in α-helices, β-sheets or random coils and whether they have a role in tertiary structure or biochemical function. This fact has been recognized and exploited by researchers who developed structure-specific substitution matrices for recognizing specific secondary structures and motifs [45,46].

Additional variations in selective pressure occur at the DNA level. In protein coding regions, substitutions can be non-synonymous (resulting in an amino acid substitution in the protein coded for) or synonymous (resulting in a different codon for the same amino acid). Due to differences in selective pressure, synonymous changes are seen with more frequency than non-synonymous changes. Originally, it was thought that sequence positions could be classified as replacement sites, synonymous sites and non-coding sites and that mutation rates within each class would be relatively constant. More recently, evidence has emerged that suggests that selective pressure can vary within each class, even within a single gene or intron [47,48]

5 Biological validation of MSA

As new optimization approaches to MSA construction are proposed, how do we determine if the resulting alignments are biologically sound? Structural and functional information has been used to validate multiple sequence alignments. In a comprehensive study of twelve different multiple alignment programs, McClure, Vasi and Fitch [4] measured algorithm performance by computing a numerical score based on the ability to find known motifs in four different data sets, for which supporting evidence from structural or mutational studies was available. Barton and Sternberg [49] have also used structural alignments to validate sequence-based alignments.

Structural information has also been used guide the computation of MSA's. Some of the earliest work in structure-based alignments was presented by Lesk and Chothia [50], who used superposition of secondary structures to align globin sequences. Today, there are several common approaches to structural alignment, as surveyed in [51]. First, one may associate a secondary structure type (α-helix, β-sheet or random coil) with each residue and then impose the additional constraint that only residues associated with the same type of secondary structure may be aligned (for example, [52,24]). This type of structural alignment is often done manually. Second, structural alignments may be performed by minimizing the root mean square distance between the aligned α-carbons in the backbone (see [53,54], for an example). A third approach is to minimize the difference between the distance matrices of the two proteins, where a distance matrix represents the distance between every pair of α-carbons in the protein (e.g., [55,56]). Structural alignments can also be evaluated by comparing the contact maps of the aligned proteins [57]. A contact map is a matrix describing the interactions of the amino acid side chains within the protein.

The use of structural information to guide or validate alignments is only possible for data sets where structural information is available. While the fraction of biopolymers for which structural information is available is small, the majority of newly sequenced biopolymers turn out to be members of families for which some structure is known. Another consideration is whether a structural approach to MSA is useful in all cases. For example, as discussed in Section 4, residues that share a structural or functional role do not always share an evolutionary history.

6 Conclusion

Optimization is an extremely effective tool for attacking many computational problems. In the case of problems from computational biology, one must be careful that the optimization criterion used closely follows some specific biological model of how the data relate to the underlying structure sought, e.g. a phylogeny or an MSA. In this vein, tree alignment has been touted as a "biological" approach to the MSA problem. However, as argued in Section 4, the definition of tree alignment (i.e. a single tree under a parsimonious scoring scheme) renders it meaningless in many biological settings. Furthermore, we argue in Section 3 that approximation algorithms for tree and other alignment criteria do not exploit biologically intuitive notions of convergence.

None of this need be the case. We are not pointing out fundamental shortcomings of the optimization/approximation approach in computational biology. Indeed, we believe that this approach offers the best hope for a principled attack on MSA and other problems. However, the strength of this approach, most notably a transparent association between underlying assumptions and computational methods, is rendered meaningless if the underlying assumptions are not biological. We therefore call for a more focused search for a biologically grounded model for MSA computation. Our paper points out many pitfalls of any such search, and suggests new directions for improvement. It should be considered a call to action.

Acknowledgements

The authors wish to thank Sergei Agulnik, Maja Bucan, Dan Gusfield, Laura Landweber, Eugene Myers, Dalit Naor, R. Ravi, David Roos, Ilya Ruvinsky, Mona Singh and, especially, the late Chris Overton for helpful discussions. D.D. was supported by NSF Grants BIR-94-13215 A01 and BIR-94-12594. M.F-C. was supported by NSF Career Development Award CCR-95-01942, NSF Grant BIR-94-12594, an Alfred P. Sloan Research Fellowship and NATO Grant 96-0215.

References

1. S. Chan, A. Wong, D. Chiu: Bulletin of Mathematical Biology **54**, 563–598 (1992)
2. P. Pevsner: Journal of Applied Mathematics **52**, 1763–1779 (1992)
3. R. Durbin, S. Eddy, A. Krogh, G. Mitchison: *Biological Sequence Analysis: Probabilistic Models of Proteins and Nucleic Acids* (Cambridge University Press, 1999)
4. M. McClure, T. Vasi, W. Fitch: Mol. Biol. Evol. **11**, 571 – 592 (1994)
5. D.J. Bacon, W.F. Anderson: Journal of Molecular Biology **191**, 153–161 (1986)
6. M. Murata, J.S. Richardson, J.L. Sussman: Proc.Natl.Acad.Sci. USA **82**, 3073–3077 (1985)
7. D. Sankoff: Journal of Applied Mathematics **28**, 443–453 (1975)
8. D. Sankoff, R.J. Cedergren: Simultaneous Comparison of Three or More Sequences Related by a Tree, in *Timewarps, Edits and Macromolecules: The Theory and Practise of Sequence Comparison* (Addison-Wesley, Reading, MA, 1983), pp. 253–258
9. S. Altschul, D. Lipman: Journal of Applied Mathematics **49**(1), 197–209 (1989)
10. D. Gusfield: Bulletin of Mathematical Biology **55**, 141–154 (1993)
11. E. Sweedyk, T. Warnow: (1992), Manuscript
12. L. Wang, T. Jiang: Journal of Computational Biology **1**(4), 337–348 (1994)
13. M.R. Garey, D.S. Johnson: *Computers and Intractability: A Guide to the Theory of NP-Completeness* (W. H. Freeman and Company, 1979)
14. T.H. Cormen, C.E. Leiserson, R.L. Rivest: *Introduction to Algorithms* (MIT Press/McGraw-Hill, 1990)
15. J.C. Setubal, J. Meidanis: *Introduction to Computational Molecular Biology* (PWS Publishing Company, Boston, 1997)
16. W.R. Taylor: CABIOS **3**(2), 81–7 (1987)
17. J.D. Thompson, D.G. Higgins, T.J. Gibson: NAR **22**(22), 4673–80 (1994)
18. D. Hochbaum: *Approximation Algorithms for NP-hard Croblems* (PWS Publishing Company, Boston, 1997)
19. V. Bafna, E.L. Lawler, P. Pevzner: Approximation Algorithms for Multiple Sequence Alignment, in *5th Ann. Symp. On Pattern Combinatorial Matching* Vol. 807 (1994), pp. 43–53
20. R. Ravi, J.D. Kececioglu: Approximation algorithms for multiple sequence alignment under a fixed evolutionary tree, in *6th Ann. Symp. On Pattern Combinatorial Matching, Springer Verlag Lecture notes in Computer Science* (1995)
21. L. Wang, D. Gusfield: Improved Approximation Algorithms for Tree Alignment, in *7th Ann. Symp. On Pattern Combinatorial Matching* Vol. 1075 (1996), pp. 220–33
22. J. Jiang, L. Wang, E. Lawler: Algorithmica **16**, 302–15 (1996)

23. T. Jiang, E. Lawler, L. Wang: Aligning sequences via an evolutionary tree: Complexity and approximation, in *Proceedings of the Symposium on the Theoretical Aspects of Computer Science* (1994), pp. 760–769
24. D. Roos: J. Biol. Chem. **268**, 6269–6280 (1993)
25. J. Cavender: Mathematical Biosciences **40**, 271–280 (1978)
26. J. Felsenstein: Syst. Zool. **22**, 240–249 (1978)
27. M. Farach, S. Kannan: Efficient algorithms for inverting evolution, in *Proceedings of the Symposium on the Theoretical Aspects of Computer Science* (1996)
28. R.G. Donald, D.S. Roos: Proc.Natl.Acad.Sci. USA **90**, 11 703–11 707 (1993)
29. R.G.K. Donald, D.S. Roos: Molec. Biochem. Parasitol. **63**, 243–253 (1994)
30. J. Hyde: Pharmacol Ther **48**(1), 45–59 (1990)
31. M. Tanaka, H.M. Gu, D.J. Bzik, W.B. Li, J.W. Inselburg: Mol Biochem Parasitol **39**, 127–134 (1990)
32. M. Reynolds, D. Carter, M. Schumacher, D.S. Roos: Personal communication
33. D.S. Roos: Personal communication
34. W. Gilbert: Nature **271**, 501 (1978)
35. T.C. Sudhof, J.L. Goldstein, M.S. Brown, D.W. Russell: Science **228**, 815–822 (1985)
36. T.C. Sudhof, D.W. Russell, J.L. Goldstein, M.S. Brown, R. Sanchez-Pescador, G.I. Bell: Science **228**, 893–895 (1985)
37. R.L. Dorit, W. Gilbert: Curr Opin Genet Dev **1**, 464–469 (1991)
38. M.D. Adams *et al.*: Science **287**(5461), 2185–9 (2000)
39. International Human Genome Sequencing Consortium: Nature **409**(682), 860–921 (2001)
40. J.C. Venter *et al.*: Science **291**(5507), 1304–51 (2001)
41. P. Bork, R.F. Doolittle: Proc.Natl.Acad.Sci. USA **89**, 8990–8994 (1992)
42. C.B. Stewart, A.C. Wilson: Cold Spring Harbor Symposium on Quantitative Biology **52**, 891–899 (1987)
43. R. Gutell, N. Larsen, C. Woese: Microbiological Reviews **58**(1), 10–26 (1994)
44. C.R. Woese, S. Winker, R.R. Gutell: Proc.Natl.Acad.Sci. USA **87**, 8467–8471 (1990)
45. R. Luthy, A.D. McLachlan, D. Eisenberg: Proteins **10**, 229–239 (1991)
46. P. Mehta, J. Heringa, P. Argos: Protein Science **4**, 2517–2525 (1995)
47. M. Kreitman, R.R. Hudson: Genetics **127**, 565–582 (1991)
48. S.W. Schaeffer, C.F. Aquadro: Genetics **117**, 61–73 (1987)
49. G. Barton, M. Sternberg: Protein Engineering **1**, 89–94 (1987)
50. A. Lesk, C. Chothia: Journal of Molecular Biology **136**, 225–270 (1980)
51. A. Godzik: Protein Science **5**, 1325–1338 (1996)
52. A. Aevarsson: Journal of Molecular Evolution **41**, 1096 – 1104 (1995)
53. A. Valencia, M. Kjeldgaard, E.F. Pai, C. Sander: Proc.Natl.Acad.Sci. USA **88**, 5443–5447 (1991)
54. G. Vriend, C. Sander: Proteins **11**(1), 52–58 (1991)
55. L. Holm, C. Sander: Journal of Molecular Biology (1993)
56. S. Pascarella, P. Argos: Protein Engineering **5**, 121–37 (1992)
57. A. Godzik, J. Skolnick, A. Kolinski: Protein Engineering **6**(8), 801–10 (1993)

Red queen dynamics and the evolution of translational redundancy and degeneracy

David C. Krakauer[1], Vincent A.A. Jansen[2], and Martin Nowak[1]

[1] Institute for Advanced Study,
Princeton NJ 08540, USA
[2] School of Biological Sciences, Royal Holloway, University of London,
Egham, Surrey, TW20 0EX, UK

Abstract. We explore adaptive theories for the diversity of protein translation based on the genetic code viewed as a primitive immune system. Immunity is acquired through a genetic mechanism of non-recognition of parasite genomes. Modifying the set of codons bound by tRNA anticodon molecules or changing the specificity of binding, reduces the replication rate of translational parasites such as viruses. Changing the binding specifity can be thought of in terms of varying degrees of redundancy and degeneracy. Redundancy in the genetic code is commonly attributed to using a four base triplet mechanism to encode the 20 amino acids. This has been referred to as synonym redundancy. There are however at least a further two forms of redundancy associated with the code and one source of degeneracy. A first form of redundancy arises from decoding all 61 possible sense codons using fewer than 61 anticodons. Such a strategy involves reduced binding specificity. A second source of redundancy is present in the multiplicity of copies of each unique tRNA (tRNA copy redundancy). Degeneracy arises when different anticodons become associated with a single amino acid to increase specificity. Variation in these strategies across taxa ensures that the translational machinery is diverse whereas the code remains approximately constant. We construct a red queen theory for translational diversity: a theory in which host translational strategies – as defined by the degree of redundancy or degeneracy of anticodons – are constantly shifting through time to evade parasitism but where neither parasite nor host gains a systematic advantage.

1 Introduction to the code

We start this chapter with some neccessary background to the genetic translation machinery. We introduce the concepts of redundancy and degeneracy and illustrate their variability in the way of a contrast to the relative taxonomic uniformity of the genetic code. This information is then used to explore how parasites such as viruses might influence the evolution of translation, and how host translational preferences feed-back to influence parasite codon usage.

The eukaryotic genome ranges in size from around 10Mb to over 100,000Mb. Among the larger genomes only some small fraction of DNA, around 3 percent, encodes proteins. The remainder is either non-coding or extragenic. Among viruses and bacteria almost 100 percent of their RNA or DNA is used to encode proteins. This difference derives in large part from replication time pressures – present in small parasites and sometimes organelles – that are not experienced

	T	C	A	G
T	TTT Phe (F)	TCT Ser (S)	TAT Tyr (Y)	TGT Cys (C)
	TTC "	TCC "	TAC	TGC
	TTA Leu (L)	TCA "	TAA **Ter**	TGA **Ter**
	TTG "	TCG "	TAG **Ter**	TGG Trp (W)
C	CTT Leu (L)	CCT Pro (P)	CAT His (H)	CGT Arg (R)
	CTC "	CCC "	CAC "	CGC "
	CTA "	CCA "	CAA Gln (Q)	CGA "
	CTG "	CCG "	CAG "	CGG "
A	ATT Ile (I)	ACT Thr (T)	AAT Asn (N)	AGT Ser (S)
	ATC "	ACC "	AAC "	AGC "
	ATA "	ACA "	AAA Lys (K)	AGA Arg (R)
	ATG Met (M)	ACG "	AAG "	AGG "
G	GTT Val (V)	GCT Ala (A)	GAT Asp (D)	GGT Gly (G)
	GTC "	GCC "	GAC "	GGC "
	GTA "	GCA "	GAA Glu (E)	GGA "
	GTG "	GCG "	GAG "	GGG "

Fig. 1. The canonical genetic code. Each codon (triplet of three nucleotides) is associated with an amino acid. When two or more codons are associated with each amino acid this gives rise to synonym, codon redundancy in the code. The mechanism producing codon redundancy is not apparent from the code table and involves redundancy and degeneracy in the anticodon (See Fig. 2)

by larger organisms. In extreme cases this pressure for genomic reduction has lead to the loss of those genes encoding the replicative machinery. This promotes parasitism whereby a second species provides the translational proteins ensuring that replication of the reduced genome can proceed. The clearest example of this approach is the virus life cycle.

The mechanisms of protein synthesis via translation remain largely the same across all taxa. Each amino acid building block constituting a protein is associated with one or more unique 'triplet' codons of three nucleotides of DNA or RNA. Each codon encodes an amino acid through an adaptor molecule carrying an amino acid, the transfer RNA (tRNA). The tRNA binds to the codon with a complementary anticodon according to the so called Watson-Crick base paring rules. These state that the purines Adenine and Guanine bind with the pyrimidines Thymine, Cytosine and Uracil. The mapping from codon to amino acid is many to one and this gives rise to synonym redundancy in the genetic code (Fig. 1). One mechanism for ensuring that each of the 64 possible codons is associated with at least one of the 20 possible amino acids is through 'wobble' of the first or third nucleotide base of the codon. This arises as the first and third sites of the anticodon bind more weakly than the second. As a consequence, in many triplets the third base of a codon is by and large 'silent', reducing the specificity of the anticodon and thereby expanding the set of

Adpative explanations for the redundant structure of the genetic code tend to focus on the non-random assignment of amino acids to codons. In other words, these theories address why it is certain codons encode certain amino acids and why these associations remain so widespread. Mutational hypotheses observe that the twenty sense codon phenotypes (encoding 20 amino acids), are assigned to 61 codons so as to minimize the effects of mutations to the DNA on the amino acid sequence [1,2]. The code is therefore seen as error buffering, ensuring that

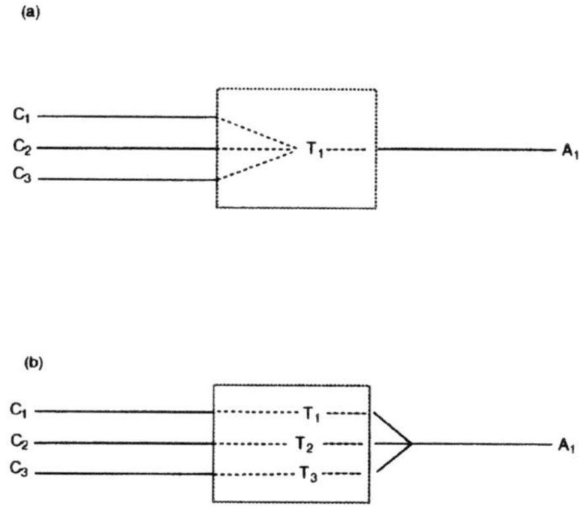

Fig. 2. Anticodon redundancy (a) and anticodon degeneracy (b). Anticodon redundancy requires that each anticodon (T_i) recognizes several different codons (C_i) and maps these onto a single amino acid (A_1) . Degeneracy involves specific anticodons binding to specific codons where each anticodon is associated with the same amino acid.

biochemically similar amino acids are encoded by sets of mutationally-related codons. A good example of this correspondence is that amino acids with a U at the second codon position are all hydrophobic while those with an A are all hydrophilic. These adaptive theories can help to explain how the otherwise arbitrary codon assignments in the genetic code show so little variability across distantly related species. One can further distinguish anticodon redundancy and anticodon degeneracy, both of which are far more variable.

1.1 Anticodon redundancy and degeneracy

Redundancy will be taken to refer to cases where populations of identical elements (anticodons) can process alternative inputs (different codons) to produce a single outcome (translation into an amino acid). Whereas degeneracy involves non-identical elements (different anticodons) processing alternative inputs (different codons) to produce a single outcome (translation into an amino acid). These two strategies are illustrated in Fig. 2. The total number of tRNAs can vary theoretically between 22, in which case each anticodon must recognizes on average 3 codons, or 61 in which case each anticodon binds strictly to its complementary codon (setting aside the 3 stop codons), leading to two or more tRNAs carrying the same amino acid (isoaccepting tRNAs). Using the above terminology, employing a large number of specific anticodons (61 for example) to encode the 20 amino acids is a strategy of degeneracy as non-identical anticodons map onto a common set of amino acids. By employing a smaller number of anticodons than codons, the system becomes redundant, as the same anticodon can bind to more than one codon.

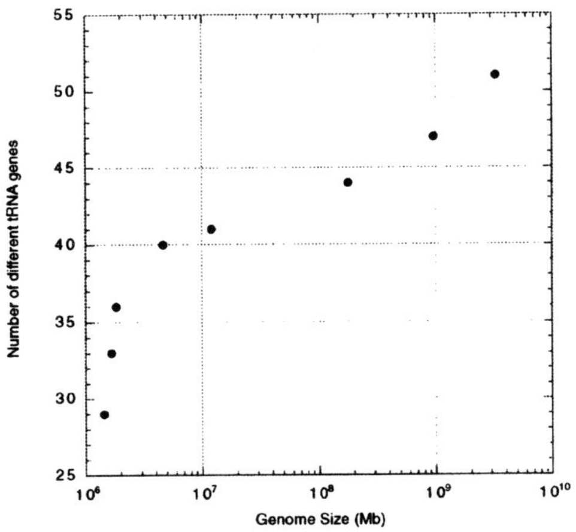

Fig. 3. The relationship between genome size and tRNA gene numbers for 3 eukaryotic and 5 prokaryotic species. One observes a strong influence of genome size on the degree of anticodon redundancy and degeneracy. Small genomes tend towards redundancy, whereas large genomes favor degeneracy.

The degree of redundancy at one extreme and degeneracy at the other, more or less spans the theoretical range. Mitochondria decode all codons with around 22 unique tRNAs not counting copies of the same anticodon. In order to do so mitochondria have evolved special redundant mechanisms of translation and have zero degeneracy. Each mitochondrial tRNA is used for a set of two or four redundant codons. Those tRNAs interacting with a four-member set, employ a U at the 5' terminus of the anticodon able to interact with any of the four bases of the codon. In contrast, *Homo sapiens* carry around 51 unique tRNAs; *C. elegans* around 47; *Drosophila melanogaster* 44; *Saccharomyces cerevisiae* 41; *Haemophilus influenza* 36; *Helicobacter pylori* 33; and *Borrelia burgdoferi* 29. In Fig. 3 we have plotted the number of different tRNA genes as a function of genome size for a number of prokaryotic and eukaryotic species. One finds that genome size is a strong determinant or translational strategy. Smaller genomes favor redundancy, whereas larger genomes favor degeneracy.

Anticodon redundancy arises through modification of the standard Watson-Crick base pairing rules, whereas degenerate anticodons arise by ensuring that the rules are interpreted strictly: 61 tRNAs are required to bind all 20 amino acids. In both cases, each ot the tRNAs is charged with one amino acid by one of twenty tRNA synthetase enzymes. Anticodon redundancy and anticodon degeneracy influence the fidelity of translation, the rate of replication, and code modifiability. A redundant strategy allows genomes to carry fewer unique tRNA molecules and solves the problem of error buffering by not distinguishing among similar codons. However, this involves a potential cost in terms of reduced binding specificity and frequent mismatch errors [3]. With redundant strategies, codon recognition depends on modifying the specificity of the anticodon. A degenerate strategy requires more unique tRNAs and an attendant increase in the genome size. It also provides greater specificity of binding, reducing translation errors by reducing the incidence of near-cognate codon readings leading to greater translational accuracy. In this case, error buffering must be solved at the level of charging of the related tRNAs with amino acids by the synthetase enzymes. Using a degenerate strategy, the set of bound codons can be modified by expanding or contracting the set of anticodons.

Another important source of redundancy present in the code is the number of tRNA gene copies associated with each unique anticodon.

Thus for a given amino acid there can be one or more tRNA gene copies, the numbers of which influence the efficiency of translation. Increasing the number of duplicates of a given tRNA has been shown to significantly amplify the rate of translation of highly expressed genes [4–6]

2 Translational parasitism

In this chapter we explore the results of assuming that the code evolved and persists in an environment in which there is a constant risk of translational parasitism. This takes the form of short sequences of RNA or DNA which make use of a host's protein synthesis machinery. A contemporary example is a virus, an obligate parasite incapable of self-replication without host factors. Increased resistance against infection can be achieved by biasing the anti-codons of the host tRNA pool away from the pathogen codon set by reducing anticodon redundancy or degeneracy and by modifying the copy number of a required tRNA anticodon. These would be effective as translation of highly expressed genes is observed to require large numbers of complementary anti-codons in the host [7–9]. Modulating the tRNA anti-codons and tRNA availabilities, restricts virus genes access to protein synthesis machinery. The translational mechanism serves as a biochemical recognition system, analogous to other strategies for host defense [10]. A fully redundant or degenerate code becomes more remarkable in light of parasitism, as parasites with different codon biases are all able to infect a fully redundant or fully degenerate host genome. A less degenerate code - one utilizing a subset of the 61 sense codons found in the universal genetic code, combined with no anticodon redundancy - would be immune from all those parasites not sporting a complementary codon table, and thereby restrict a parasite's host range. To explore these ideas we have developed a number of models in which parasites and their hosts can vary the degree of effective synonym redundancy in their genomes (how many synonymous codons are employed), and in which hosts can vary the specificity and number of their tRNA molecules (degeneracy). Such models have obvious parallels with 'red queen' models for the evolution of sex [11]. The red queen model describes those situations in which each improvement in one species is matched by an equivalent improvement in a competitor. The net result is that while strategies constantly change, relative fitness remains constant. In this chapter we explore the coevolution of host translation and parasite codon usage in order to understand: (1) what the consequences of different degrees of redundancy and degeneracy are on virus codon usage and replication, (2) how changes in translation strategies at the cellular level are influenced by virus replication at the population level, and (3) review the experimental data relating to molecular mechanisms that might modify host translation and provide evidence of shifting parasite codon usage.

2.1 Translation, redundancy and degeneracy

To quantify redundancy and degeneracy and to explore the consequences for the translation of a cell's own genome and that of its parasites we develop a model for the translation process of RNA into proteins. We assume that there are m different codons, m anticodons and n amino acids. The $m \times 1$ vector $\vec{c}$ specifies the abundances of each of the m codons in the RNA strain. The process of translation involves matching these codons with tRNA anticodons. The abundance of anticodon i in the cell is given by elements v_i of the anticodon vector $\vec{v}$. The binding rate of anticodon i with codon j is given by element w_{ij} of the binding matrix $\mathbf{W}$.

The total rate of binding to codon j is given by $\sum_i^m v_i w_{ij}$, hence the average time it takes for a codon to be matched is given by

$$f_j = (\sum_i v_i w_{ij})^{-1}. \tag{1}$$

It also follows that codon j is matched with anticodon i with probability

$$u_{ij} = v_i w_{ij} f_j. \tag{2}$$

Because each codon will be matched with an anticodon we have $\sum_{i=1}^m u_{ij} = 1$. The $m \times m$ matrix $\mathbf{U}$ has the probabilities u_{ij} as elements. Each tRNA anticodon is associated with an amino acid. The association between anticodon and amino acid is described by the $n \times m$ binary (i.e. containing only zeros and ones) matrix $\mathbf{A}$. The elements of $\mathbf{A}$ are denoted a_{ij}. Each column of $\mathbf{A}$ consist of $n-1$ zeros and only a single one as there is a unique association between tRNA and amino acid. In case there is degeneracy rows of $\mathbf{A}$ can contain many ones. The abundance in amino acids after translation is given by $\mathbf{AU}\vec{c}$.

If $m > n$ the anticodons can be redundant and/or degenerate. Binary $\mathbf{U}$ matrices are degenerate with no redundancy. Matrices in which any element $u_{ij} < 1$, have redundancy in the ith anticodon. We can state this more formally by appreciating that redundancy is a measure of the number of states that a given anticodon can occupy, and thereby define the average redundancy as,

$$S = -\frac{1}{m}\sum_{i=1}^{m}\sum_{j=1}^{m} u_{ij} \log_2 u_{ij}.$$

The value S is a binary measure of the number of codons bound by anticodons, and is therefore maximal when most redundant. Degeneracy described how many different anticodons code for identical amino acids. We can derive a similar degeneracy measure as for redundancy from the matrix **A**,

$$H = -\frac{1}{n}\sum_{i=1}^{n}\sum_{j=1}^{m}\frac{a_{ij}}{\sum_{j=1}^{m}a_{ij}}\log_2\frac{a_{ij}}{\sum_{j=1}^{m}a_{ij}}.$$

Degeneracy is maximal when many anticodons are associated with a single amino acid.

The total translation time is given by the sum of the time it takes to initiate the translation process, ϵ, and the total time to match all codons

$$\tau = \epsilon + \sum_{j}^{m}(c_j f_j)^{\mu} \tag{3}$$

The parameter μ determines the degree of competition ($\mu > 1$) or cooperativity ($\mu < 1$) among cognate tRNAs for the A site of the ribosome. For values of $\mu < 1$ binding the ribosome becomes increasingly rate limiting. The rate of of translation is given by $q = 1/\tau$.

Because the codon abundance of the parasite can differ from that of the host, the host can to a certain extent control the replication of parasites in its cells. Our model allows us to infer the consequences of redundancy and degeneracy on the translation rate of parasite genomes.

We now distinguish between parasite and host codon usage. Let the $m \times 1$ vector $\vec{c_p}$ contain the abundances of codons in the parasite genome. The first step in translation is the transcription into mRNA, which is a possible source of error. Another source of error is mutation during replication. The rate of mutation of parasite codons during replication is determined by the matrix **B** where elements b_{ij} specify the probability of codon i mutating into codon j. Thus the effective number of codons of type i is determined by the product of the mutation matrix and the codon abundance vector, $\vec{\pi} = \mathbf{B}\vec{c_p}$. The total time needed for translation of a parasite l in a host k is given by τ_{lk}. Within a host cell τ_{lk} is derived from the degree of complementarity between the parasite and host genomes and is derived form the mean time for translation of all parasite codons:

$$\tau_{lk} = \epsilon + \sum_{j}^{m}(\pi_j f_j)^{\mu}$$

The rate of translation of parasite l in a host k is $q_{lk} = 1/\tau_{lk}$ A parasite is incapable of infecting a host when $q_{lk} = 0$. This will result when $\sum_i f_i \pi_i$ tends to infinity, i.e. when the parasite has a set of codons which find no match in the host's set of anticodons. Consider the translation of a pair of codons into a single amino acid. We assume that $\mathbf{B} = \mathbf{I}$, $\vec{v} = \{\alpha, 1-\alpha\}$, $\vec{c_p} = \{\gamma, 1-\gamma\}$, $\vec{c_h} = \{\beta, 1-\beta\}$, and

$$\mathbf{W} = \begin{pmatrix} 1-d & d \\ d & 1-d \end{pmatrix}. \tag{4}$$

The α value determines the usage of anticodons, the γ value determines the usage of parasite codons, the β value determines the usage of host codons, and the d value the binding constants. Given these assumptions, the rate of translation of a single amino acid of a parasite l in a host k, is given by

$$q_{lk} = \left(\frac{1-\gamma}{1-(d+\alpha-2d\alpha)} + \frac{\gamma}{d+\alpha-2d\alpha} + \epsilon\right)^{-1}. \tag{5}$$

The parasite is able to modify only the value of γ, its codon usage, specifying the relative frequency of synonymous codons. We can ask what value of γ maximizes the rate of translation provided fixed host anticodon abundances $\vec{v}$ and binding affinities $\mathbf{W}$. One finds that this depends on a simple parameter grouping, $d + \alpha - 2d\alpha = \begin{cases} > 1/2 & \text{then } \gamma = 1 \\ < 1/2 & \text{then } \gamma = 0 \end{cases}$ Space is thus divided into four quadrants with two outcomes. This tells us that small biases in the host affinity for codons, translates into the exclusive use of the favored codons by the parasite. The same reasoning should hold for the translation of host messages – all else being equal the host can optimize its translation process by only using the codon that gives rise to the highest translation rate. Alternatively the host can adjust the affinity matrix or anticodon abundance to optimize translation. However, codon usage also determines the error rate during transcription into mRNA of its nuclear DNA. A more extensive discussion of translation efficiency is provided in [4] and [12] where transcript elongation is also considered.

These assumptions serve as a simple mechanistic background for the dynamical models that follow. We place the translational kinetics into the larger context of parasite-host population dynamics. This allows us to explore the ways in which population level selection feeds back to modify parasite codon usage and host translational strategies.

2.2 Population dynamics

For expositional reasons consider the translation into a single amino acid a pair of synonymous codons. The abundance of hosts is given by x and the abundance of parasites by y. Assuming that parasites infect hosts with an efficiency proportional to translation rates, and that host replication rates are proportional to the translation of their own genes, we can write down the differential equations,

$$\begin{aligned} \dot{x} &= x(r - yq(d,\alpha,\gamma)) \\ \dot{y} &= y(xq(d,\alpha,\gamma) - \delta) \end{aligned} \tag{6}$$

The equilibrium abundances of the parasites and hosts are given by:

$$\hat{x} = \delta/q(d,\alpha,\gamma) \qquad \text{and} \qquad \hat{y} = r/q(d,\alpha,\gamma).$$

Thus any increase in the rate of translation leads to a concomitant reduction in the abundances of both parasite and host. A reduction in the rate of translation of host genes, leads to a reduction in the abundances of parasites without any impact on the host abundance. We might expect over the course of evolution translation within the host to become less and less efficient. This follows from competition among hosts whereby translation is reduced in order to exclude any competitors that require higher levels of translation in order to persist.

2.3 Evolutionary dynamics of single parasite and host

In the previous paragraphs we speculated on the evolutionary pressures effecting codon usage in host and parasite employing optimization arguments. Using the population dynamics of host and parasites we can make these arguments precise by deriving the evolutionary dynamics of redundancy. We will do this by considering the fate of mutants of the host which have different redundancies. For simplicity we will assume that a change in redundancy is only achieved through a change in the parameter d. In this way we hold the level of degeneracy fixed. For the sake of understanding the population dynamics we shall only consider the redundancy value. We will denote the mutant parameter with a superscripted star, hence the mutant affinity is given by d^*. Similarly, we will consider mutant parasites that have a different codon usage, γ^*. The invasion rate of such a mutant when it is rare, relative to the other resident strains, is a measure of its fitness [13]. If a mutant host is rare it will predominantly be infected by a resident parasite. Similarly, mutant parasites will predominantly infect resident hosts. The dynamics of the resident hosts and parasites are approximated by the system (6) given above as long as the mutants are rare, and the average densities are given by $\hat{x}$ and $\hat{y}$. The dynamics of rare mutant hosts, x^* and parasites y^* is therefore given by

$$
\begin{aligned}
\dot{x}^* &= x^*(r - \hat{y}q(d^*, \gamma)) \\
\dot{y}^* &= y^*(\hat{x}q(d, \gamma^*) - \delta)
\end{aligned}
$$

The growth rate of a mutant host when rare is thus given by $r(1 - \frac{q(d^*,\gamma)}{q(d,\gamma)})$ and the growth rate of a mutant parasite when rare, $\delta(\frac{q(d,\gamma^*)}{q(d,\gamma)} - 1)$. These expressions show that the mutant parasite can always invade if it has a higher value of q, whereas the mutant host can invade if it has a lower value of q than the resident host. These expressions also show that the host that causes the highest parasite density cannot be replaced by any other hosts. The parasite that induces the lowest host density cannot be replaced by other parasites. Host evolution thus maximizes parasite density while parasite evolution minimizes host density. This is an example of 'spite', evolution favors those traits that create the worst possible environments for competitors. This is also known as the pessimisation principle [14].

In creating these worst of possible worlds, the hosts are selected to minimize q while the parasites are selected to maximize it. Host and parasites thus have opposing evolutionary interests. This raises the question as to what translation rate will be finally established and the nature of the evolutionary dynamics of codon usage and redundancy.

To reconstruct the evolutionary dynamics of d and γ we will follow [15] who derive the evolutionary dynamics of a trait as the product of the mutation rate times the number of individuals in which a mutation can occur times the selection differential. This yields the following system of differential equations:

$$\dot{d} = -\mu_d \hat{x}\,\hat{y}\,\frac{\partial q}{\partial d}$$
$$\dot{\gamma} = \mu_\gamma\,\hat{x}\,\hat{y}\,\frac{\partial q}{\partial \gamma}$$

The parameters μ_d and μ_γ represent the mutation rates of resp. d and γ. To restrict the evolutionary dynamics to the unit square these parameters take the value zero if mutation is outside the unit interval.

To analyze the evolutionary dynamics we will locate the equilibrium of this system. The isocline for γ is given by $d = 1/2$, the isocline for d by $\gamma = (d + \alpha - 2d\alpha)^2(1 - 2(d + \alpha - 2d\alpha)(1 - (d + \alpha - 2d\alpha)))^{-1}$. The isoclines intersect at $d = 1/2$, $\gamma = 1/2$ at which point the equilibrium of this dynamical system can be found. If the affinity of the codon-anticodon bond is the same for both pairs and if the parasite uses both codons this state cannot be altered by coevolution. Note that there is maximal redundancy for these parameter values.

This does not mean that the evolutionary dynamics proceed to this equilibrium. Simulations show that the dynamics spiral away from this point (Fig. 4). The host and the parasites continuously alter their codon usage and redundancy

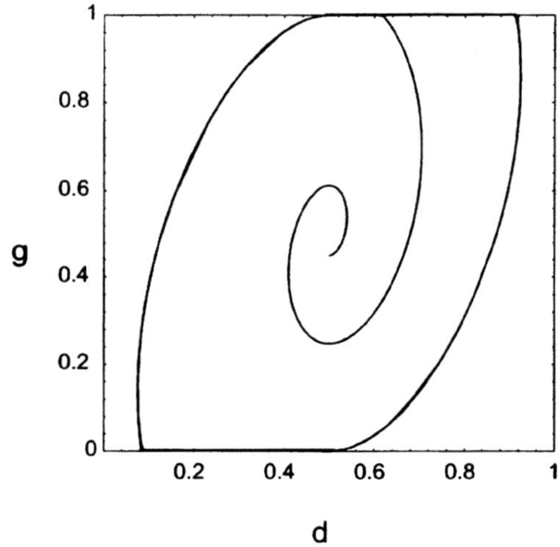

Fig. 4. The oscillatory (red queen) dynamics of monomorphic host and parasite populations. The dynamics are depicted as a phase portrait for the parameters d and γ. One observes cycles in the prefered parameter values, with an increase in redundancy, there is a concomitant decrease in parasite codon usage and vice versa. Parameter values are: $\epsilon = 1$, $\alpha = \beta = 0.25$, $r = \delta = 0.1$ $\mu_d = 0.25$, $\mu_\gamma = 1$

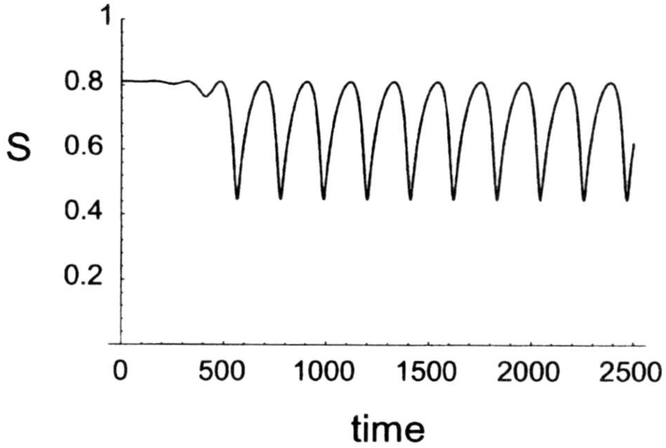

Fig. 5. The redundancy quantified over time illustrating red queen dynamics of monomorphic host and parasite populations. Parameters as in Fig. 2

to make life worse for their competitors. In doing so they modify the translation rate in opposite directions. This results in a never ending arms race in which both partners have to keep running in order to stand still: the evolutionary dynamics of the red queen. If the parasite's codon usage should remain constant, the host would modify its codon usage such that it employed a single, non-matching codon and redundancy would be minimized. However, the cyclic evolutionary dynamics cause the redundancy to fluctuate between extreme values (Fig. 5).

2.4 Evolutionary dynamics of two pairs of host and parasites

In the previous section it was tacitly assumed that the host and parasite populations are monomorphic. There is no *a priori* reason why this should be the case. In fact, more than one pair of hosts and parasites can coexist. The simplest example being two hosts, both with minimal redundancy achieved using two different codons and a pair of parasites specializing on these hosts.

To investigate the coevolutionary dynamics of two pairs of hosts and parasites we need to consider four different traits. In line with the previous section we have chosen the parameter d_1 and d_2, which denote the binding rates between codon and anticodon for host 1 and 2, and γ_1 and γ_2 which represent the codon abundances of parasites and 1 and 2. Let the densities of host i be given by x_i and the density of parasite i by y_i. The resident dynamics of the four host and parasites is given by:

$$\begin{aligned}
\dot{x_1} &= x_1(r - y_1 q_{11} - y_2 q_{12}) \\
\dot{x_2} &= x_2(r - y_1 q_{21} - y_2 q_{22}) \\
\dot{y_1} &= y_1(x_1 q_{11} + x_2 q_{21} - \delta) \\
\dot{y_2} &= y_2(x_1 q_{12} + x_2 q_{22} - \delta)
\end{aligned}$$

with $q_{ij} = q(d_i, \gamma_j)$. The equilibrium values are given by:

$$\hat{x}_1 = \frac{\delta(q(d_2,\gamma_2) - q(d_2,\gamma_1))}{q(d_1,\gamma_1)q(d_2,\gamma_2) - q(d_2,\gamma_1)q(d_1,\gamma_2)}$$

$$\hat{x}_2 = \frac{\delta(q(d_1,\gamma_1) - q(d_1,\gamma_2))}{q(d_1,\gamma_1)q(d_2,\gamma_2) - q(d_2,\gamma_1)q(d_1,\gamma_2)}$$

$$\hat{y}_1 = \frac{r(q(d_2,\gamma_2) - q(d_1,\gamma_2))}{q(d_1,\gamma_1)q(d_2,\gamma_2) - q(d_2,\gamma_1)q(d_1,\gamma_2)}$$

$$\hat{y}_2 = \frac{r(q(d_1,\gamma_1) - q(d_2,\gamma_1))}{q(d_1,\gamma_1)q(d_2,\gamma_2) - q(d_2,\gamma_1)q(d_1,\gamma_2)}$$

Also in this system the long term averages of the densities converge to the equilibrium values.

The dynamics of rare mutant hosts x_1^* and x_2^* and parasites y_1^* and y_2^* with respective traits d_1^*, d_2^*, γ_1^* and γ_2^* is given by:

$$\begin{aligned}
\dot{x_1^*} &= x_1^*(r - \hat{y}_1 q(d_1^*,\gamma_1) - \hat{y}_2 q(d_1^*,\gamma_2)) \\
\dot{x_2^*} &= x_2^*(r - \hat{y}_1 q(d_2^*,\gamma_1) - \hat{y}_2 q(d_2^*,\gamma_2)) \\
\dot{y_1^*} &= y_1^*(\hat{x}_1 q(d_1,\gamma_1^*) + \hat{x}_2 q(d_2,\gamma_1^*) - \delta) \\
\dot{y_2^*} &= y_2^*(\hat{x}_1 q(d_1,\gamma_2^*) + \hat{x}_2 q(d_2,\gamma_2^*) - \delta)
\end{aligned}$$

Following the same reasoning as before we can derive the evolutionary dynamics as a system of four coupled ordinary differential equations of the form:

$$\begin{aligned}
\dot{d}_1 &= -\mu_d \hat{x}_1 \left(\hat{y}_1 \left.\frac{\partial q(d,\gamma_1)}{\partial d}\right|_{d=d_1} + \hat{y}_2 \left.\frac{\partial q(d,\gamma_2)}{\partial d}\right|_{d=d_1} \right) \\
\dot{d}_2 &= -\mu_d \hat{x}_2 \left(\hat{y}_1 \left.\frac{\partial q(d,\gamma_1)}{\partial d}\right|_{d=d_2} + \hat{y}_2 \left.\frac{\partial q(d,\gamma_2)}{\partial d}\right|_{d=d_2} \right) \\
\dot{\gamma}_1 &= \mu_\gamma \hat{y}_1 \left(\hat{x}_1 \left.\frac{\partial q(d_1,\gamma)}{\partial \gamma}\right|_{\gamma=\gamma_1} + \hat{x}_2 \left.\frac{\partial q(d_2,\gamma)}{\partial \gamma}\right|_{\gamma=\gamma_1} \right) \\
\dot{\gamma}_2 &= \mu_\gamma \hat{y}_2 \left(\hat{x}_1 \left.\frac{\partial q(d_1,\gamma)}{\partial \gamma}\right|_{\gamma=\gamma_2} + \hat{x}_2 \left.\frac{\partial q(d_2,\gamma)}{\partial \gamma}\right|_{\gamma=\gamma_2} \right)
\end{aligned}$$

The values of all evolutionary parameters are restricted to the unit interval. Simulations show that the evolutionary cycling disappears when the host and parasite populations are dimorphic (Fig. 6). Whereas in the monomorphic case the redundancy fluctuated, here the system evolves towards minimal redundancy. In the evolutionary equilibrium the hosts both use a single, but different anticodon.

Once the hosts have evolved to minimal redundancy a third host type with maximal redundancy can appear. In the present model this host can never invade the system because it will be attacked by both parasites. If, however, this host would derive a further advantage from having more than one anticodon

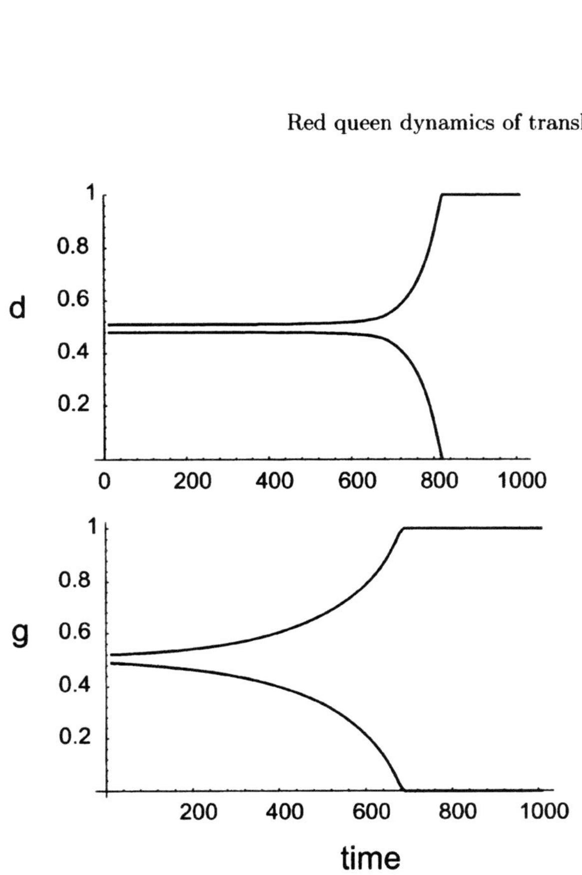

Fig. 6. The dynamics of dimorphic host and parasite populations. Both parasites (γ parameter) and host (d parameter) evolve towards minimal redundancy with non-overlapping codon usage. Populations are initially homogeneous. Over the course of evolution the populations diverge into two specialized solutions. Parameters as in Fig. 2

available, for instance an increased r as a result of a lower mutational load, it is possible that it invades and replaces the two non-redundant hosts already present. In this way increased redundancy can evolve. This is therefore consistent with mutational theories for the evolution of genetic code redundancy [1,2]. The detailed modelling and analysis of the evolutionary dynamics lies beyond the scope of this chapter.

3 Adaptive modification of tRNA base composition

The models illustrate how changes in anticodon redundancy and degeneracy are expected to modify parasite replication rate, and how parasite-host population dynamics feedback to influence translation. In biological systems this could be achieved in one of two ways: (1) changing the abundance and anticodon-usage of all tRNAs in the germ line or (2) facultatively modifying individual tRNAs in infected cells or there level of expression. The first solution is problematic as it would interfere with the translation of host genes, and presumably, it could only occur at a fraction of the rate of change of parasite codon usage. The second strategy would be preferable as it could target individual cells, and operate over very short – molecular – time scales.

In this section we review evidence for molecular mechanisms that modify degeneracy and redundancy within cells. Transfer RNAs possess a large number of modified bases. These are chemical modifications of one of the four standard RNA bases, A, G, C and U. Modification can be modest in which case a single methyl group is added to a base, or more dramatic, when the purine ring is restructured. The result of modification is to change the binding affinity of the anticodon for the codon, thereby increasing or reducing adherence to canonical Watson-Crick base pairing rules. Furthermore, gene expression of each tRNA can be modified when a gene requires a minimum concentration of cognate tRNAs in order to be expressed.

In papillomavirus, codon bias has been observed to suppress the production of capsid protein. Upregulating tRNA expression can restore this protein [16] In *E. coli* uridine in the wobble position of tRNAGlu and tRNALys is modified to reduce the possibility of misreading near-cognate codons and reduces the range of cognate codon readings. Modification causes GAG to be translated more slowly and GAA more quickly than the average codon [17]. Modification of tRNA bases can alter the incidence of retroviral ribosomal frameshift mutations. In particular a modified quinine in the wobble position increases frameshifting at one frameshift site codon and reduces frameshifting at another site [18]. In the starfish mitochondrial genome there is no tRNA anticodon able to translate all four serine codons. This deficiency is overcome by having a 7-methylguanosine at the wobble position, expanding the isoaccepting class such that the tRNA can form bases with all four nucleotides [19]. Structural studies of tRNA with a queuosine modified tRNAasp show that this modifies base pairing. Unmodified tRNAasp form stable association with GAC but unstable association with GAU. The modified form exhibits no bias for either of these codons and is therefore thought to regulate protein synthesis under different codon biases[20]. Wildtype tRNALeu(UUR) found in mitochondria possess a modified base at the wobble position. Mutants possess an unmodified base, and this is associated with the muscle wasting MELAS disease phenotype. It is thought that mutants lacking the modification are unable to decode their cognate codons efficiently [26].

Thus widespread modification provides an epigenetic means of increasing variability in host translation machinery, without having to change genetically tRNA copy numbers or codon usage. The facultative nature of the response suggests that modification need not be systemic but rather be regulated by a context dependent mechanism. Thus infected cells could undergo anticodon modification as a means of handicapping the highly expressed translational parasite, without impairing translation in uninfected cells.

4 Long term trends in codon usage

In previous sections we introduced models for thinking about translation kinetics and evolutionary dynamics. We then provided some evidence for changes in the anticodon leading to changes in anticodon specificity. We now explore some time series data from flu infection collected over several years to test whether we observe systematic variation in parasite codon usage, as is predicted to follow from translation selection.

It is worth emphasizing that codon bias is believed to result from a number of processes including mutational bias [4], the demands made by DNA folding and stability [22] and translation selection [21]. The importance of translation selection is illustrated by studies such as those of codon usage of highly expressed genes in C. cerevisiae [7] and its positive correlation with tRNA abundance [23] the higher the expression rate of a codon the larger number of tRNA copies with a corresponding anticodon.

Unlike mutational theories of codon usage, parasite-host coevolution is frequency-dependent. We therefore expect cycles in codon usage as hosts attempt to reduce the efficiency of infection, while parasites track any changes in host codon usage so as to maximize replication efficiency through translation selection. Thus unlike structural or mutational theories for codon bias which predict constant within species bias, parasite theories predict systematic variation in codon bias, correlated with parasite load.

Sequence data required to test these ideas remains scarce. An example of the sort of data required are provided by the [24] study which collected sequence data from the HA1 domain of human influenza A virus genes over the course of 12 years (1984-1996). The HA1 domain is under strong selection pressure as it encodes part of the hemaggluutinin molecule - -an important epitope for humoral recognition. Modification of the amino acids in the HA lead to immune escape. Silent mutations should have no influence on the ability of the virus to evade antibody detection. According to the logic of our models, codon usage is another way in which the virus interacts with the host for access to the the cellular ribosomes. As a proxy for codon usage we have analyzed the effective codon number [25] and the synonymous GC content at the third site in the HA1 gene over the course of the 12 year study. The effective codon number varies from 20, where there is extreme bias and one codon is used for each amino acid, to 61 where alternative synonymous codons are used with equal probability. In Fig. 7 one observes that there is a statistically significant decline in both the effective number of codons ($y = 67 - .13x, r^2 = .35, p < 0.001$) and the synonymous, third site GC content ($y = -3 + .04x, r^2 = .6, p < 0.001$) in the HA1 gene. The change in GC content is insufficient to explain the decline in codon usage. A further analysis of the variance through time, shows that there is a significant increase in the variance of the effective number of codons ($y = 0.5 - 0.01x, r^2 = .28, p < 0.001$), whereas variance in the silent GC content does not deviate significantly from zero. According to the translation selection mechanism, such a pattern would reflect a shift in the mean virus population codon usage towards reduced degeneracy, whereas individual viruses become more specialized in their codon

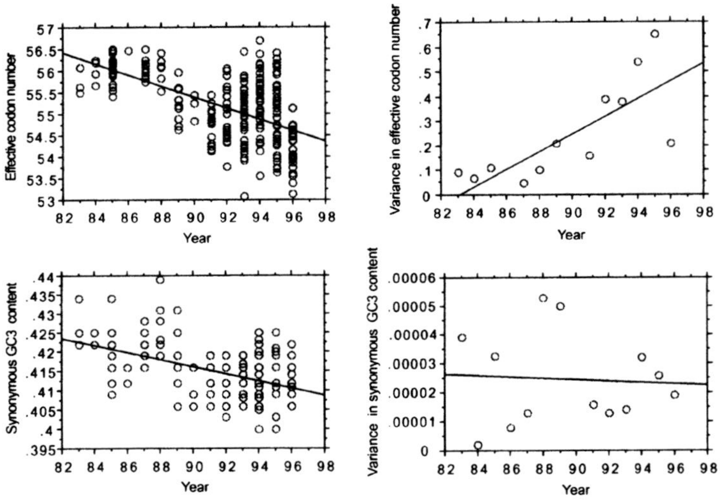

Fig. 7. Long term trends in codon usage of HA1 domain of human influenza

usage, promoting population heterogeneity. These trends do not seem to reflect neutral variation in the nucleotide pool.

5 Summary and conclusions

The mechanisms that organisms employ to translate proteins remain highly conserved across taxa. However, at the level of the anticodon-codon recognition, there is a great deal of variation both within and between species. This variation is reflected in codon usage, tRNA copy number, and anticodon redundancy and degeneracy. In this chapter, degeneracy and redundancy in the translational machinery of the genetic code has been explored in relation to infection with parasites utilizing host protein synthesis pathways. It is suggested that features of the translational machinery can be understood as adaptations to reduce the costs of infection. Assuming that codon usage influences parasite replication rate, a monomorphic host population continually modifies its tRNAs so as to reduce parasite translation efficiency. Mechanisms which facilitate epigenetic modification of the anticodon, thereby modifying codon-anticodon binding specificity, are likely to have evolved to bring the rate of host evolution in line with the rapid genetic evolution of parasites.

When populations contain more than one host type (vary in their anticodon redundancy and degeneracy), then each host and parasite pair evolve towards low levels of redundancy whereas the population as a whole becomes polymorphic in codon usage. Thus each parasite becomes a translational specialist on the host tRNA background.

The frequency-dependent nature of fitness suggest that the compass of isoaccepting tRNAs should fluctuate through time as parasites track host codon usage, and hosts modify their usage to escape infection. Time series data on codon usage can address this prediction. Data on the flu HA1 domain was analyzed, and shows evidence of a systematic reduction in virus codon usage and a concomitant increase in codon-usage heterogeneity at the population level. Thus there is some data consistent with the idea that host translational mechanisms impose a selection pressure on parasite codon-usage.

Acknowledgements

DCK and MAN thank the Alfred P Sloan Foundation, The Ambrose Monell Foundation, The Florence Gould Foundation, and the J. Seward Johnson Trust. VAAJ gratefully acknowledges support of The Wellcome Trust.

References

1. Haig, D., Hurst, L. D. A quantitative measure of error minimization in the genetic code. J Mol Evol **33**, 412-7 (1991)
2. Freeland, S. J., Hurst, L. D. 1998. The genetic code is one in a million. J Mol Evol **47**, 238-48 (1998)
3. Knight, R. D., Freeland, S. J., Landweber, L. F. Selection, history and chemistry: the three faces of the genetic code [see comments]. Trends Biochem Sci **24**, 241-7 (1999)
4. Bulmer, M. The selection-mutation-drift theory of synonymous codon usage. Genetics **129**, 897-907 (1991)
5. Dong, H., Nilsson, L., Kurland, C. G. Gratuitous overexpression of genes in Escherichia coli leads to growth inhibition and ribosome destruction. J Bacteriol **177**, 1497-504 (1995)
6. Sharp, P. M., Stenico, M., Peden, J. F., Lloyd, A. T. Codon usage: mutational bias, translational selection, or both? Biochem Soc Trans **21**, 835-41 (1993)
7. Bennetzen, J. L., Hall, B. D. Codon selection in yeast. J Biol Chem **257**, 3026-31 (1982)
8. Kurland, C. G. Codon bias and gene expression. FEBS Lett **285**, 165-9 (1991)
9. Lammertyn, E., Van Mellaert, L., Bijnens, A. P., Joris, B., Anne, J. Codon adjustment to maximize heterologous gene expression in Streptomyces lividans can lead to decreased mRNA stability and protein yield. Mol Gen Genet **250**, 223-9 (1996)
10. Frank, S. A. Recognition and polymorphism in host-parasite genetics. Philos Trans R Soc Lond B Biol Sci **346**, 283-93 (1994)
11. Hamilton, W.D., Axelrod, R. Tanese, R. Sexual selection as an adaptation to resist parasites (a review). Proc, Natl. Acad. Sci. USA **87**, 3566-3573 (1990)
12. Xia, X. How optimized is the translational machinery in Escherichia coli, Salmonella typhimurium and Saccharomyces cerevisiae? Genetics **149**,37-44 (1998)
13. Metz J.A.J., Geritz S.A.H., Meszéna G., Jacobs F.J.A., van Heerwaarden JS: Adaptive Dynamics: A Geometrical Study of the Consequences of Nearly Faithful Reproduction. IIASA Working Paper WP-95-099. In: *van Strien SJ, Verduyn Lunel SM (eds.) Stochastic and Spatial Structures of Dynamical Systems*, Proceedings of the Royal Dutch Academy of Science (KNAW Verhandelingen), North Holland, Amsterdam, pp.183-231 (1996).

14. Mylius, S.D., Diekmann, O. On evolutionarily stable life histories, optimization and the need to be specific about density dependence Oikos **74** 218-224 (1995)
15. Dieckmann U, Law R: The Dynamical Theory of Coevolution: A Derivation from Stochastic Ecological Processes. Journal of Mathematical Biology **34**, 579-612. (1996)
16. Zhou, J., Liu, W. J., Peng, S. W., Sun, X. Y., Frazer, I. Papillomavirus capsid protein expression level depends on the match between codon usage and tRNA availability. J Virol **73**, 4972-82 (1999)
17. Kruger, M. K., Pederson, S., Hagervall, T. G., Soreson, M. A. The modification of the wobble base of tRNAGlu modulates the translation rate of glutamic acid codon in vivo. J. Mol. Evol **284**, 621-631 (1998)
18. Carlson, B. A., Kwon, S. Y., Chamorro, M., Oroszlan, S., Hatfield, D. L., Lee, B. J. Transfer RNA modification status influences retroviral ribosomal frameshifting. Virology **255**, 2-8 (1999)
19. Matsuyama, S., Ueda, T., Crain, P. F., McCloskey, J. A., Watanabe, K. A novel wobble rule found in starfish mitochondria. Presence of 7-methylguanosine at the anticodon wobble position expands decoding capability of tRNA. J Biol Chem **273**, 3363-8 (1998)
20. Morris, R. C., Brown, K. G., Elliott, M. S. The effect of queuosine on tRNA structure and function. J Biomol Struct Dyn **16**, 757-74 (1999)
21. Sharp, P. M., Averof, M., Lloyd, A. T., Matassi, G., Peden, J. F. DNA sequence evolution: the sounds of silence. Philos Trans R Soc Lond B Biol Sci **349**, 241-7 (1995)
22. Karlin, S., Mrazek, J. What drives codon choices in human genes? J Mol Biol **262**, 459-72 (1996)
23. Percudani, R., Pavesi, A., Ottonello, S. Transfer RNA gene redundancy and translational selection in Saccharomyces cerevisiae. J Mol Biol **268**, 322-30 (1997)
24. Fitch, W. M., Bush, R. M., Bender, C. A., Cox, N. J. Long term trends in the evolution of H(3) HA1 human influenza type A. Proc Natl Acad Sci U S A **94**, 7712-8 (1997)
25. Wright, F. The 'effective number of codons' used in a gene. Gene **87**, 23-9 (1990)
26. Wright, F. Defect in the modification of anticodon wobble base of mutant mitochondrial tRNAs in MELAS mitochondrial encephalomyopathy. **87**, 23-9 (1990)

A testable genotype-phenotype map: modeling evolution of RNA molecules

Peter Schuster[1,2]

[1] Universität Wien, Institute für Theoretische Chemie und Molekulare Strukturbiologie, Währingerstraße 17, A-1090 Wien, Austria
[2] Santa Fe Institute, 1399 Hyde Park Road, Santa Fe NM 87501, USA

Abstract. Recent experiments and progress in modelling evolution *in silico* converge towards a coherent view of Darwinian evolution in molecular systems. Conventional population genetics and quasi-species theory model evolution in genotype space and properties of phenotypes enter evolutionary dynamics as parameters only. RNA evolution *in vitro* is an appropriate basis for the development of a new and comprehensive model of evolution, which is focussed on the phenotype and its fitness relevant properties. Relation between genotypes and phenotypes are described by mappings from genotype space onto a space of phenotypes. These mappings are many-to-one and thus give ample room for neutrality. The RNA model reduces genotype-phenotype relations to a mapping from sequences into secondary structures with minimal free energies and allows to derive otherwise inaccessible quantitative results. RNA sequences that fold into the same structure form neutral networks in genotype space, which determine the course of evolution. Neutral networks are embedded in sets of compatible sequences. Intersections of these sets represent regions in sequence space where single molecules can form two or more structures. Continuity and discontinuity in evolution are defined through straightforward interpretation of computer simulations of RNA optimization. *In silico* evolution provides insight into the accessibility of phenotypes and demonstrate the constructive role of random genetic drift in the search for phenotypes of higher fitness. New experimental data, among them the results of genome research, will present a solid basis for test and further development of the model for phenotype evolution.

1 Evolution experiments *in vitro*

The first experimental studies on evolution of molecules in the test tube were performed by Sol Spiegelman and his group [1]. They investigated populations in the range of 10^{12} to 10^{15} RNA molecules in a medium that sustains replication. Consumed materials were replaced and the increase in RNA concentration was compensated by serial transfer of small samples into fresh replication medium. Spiegelman observed evolution through mutation and selection in a cell-free assay, just as it follows from Darwin's principle. Later, in the nineteen-eighties, the mechanism of RNA replication was studied by the methods of chemical reaction kinetics [2]. The astonishing efficiency of organismic evolution in the creation of highly specific adaptations provoked the idea to use evolutionary techniques in biotechnology, in particular for the design of biomolecules with predefined properties [3]. Experiments along this line were successfully initiated by Manfred Eigen and his group in Göttingen, Germany, and later on extended by Larry

Gold in Boulder, Colorado, Gerald Joyce in La Jolla, and Jack Szostak, Boston, Massachussetts (For reviews see the volume [4] and the reviews [5,6]). Research on the application of evolutionary principles to the production of molecules of biomedical and pharmacological interest became a rapidly growing discipline in its own right.

One reason for the success of evolutionary techniques is related to the large numbers of molecules ($\approx 10^{15}$), which can be tested in a single selection cycle. Another important advantage of evolutionary design of biomolecules consists in the dispensability of knowledge on molecular structures. Molecules are optimized for function, and not for structure. Rational design of molecules (See, for example, [7]), on the other hand, is obviously dependent on detailed insights into structures and relations between changes in sequences and their consequences for structures. An analogy to organismic biology might be illustrative: Evolution of molecules in the test tube and evolutionary design are in the same mutual relation to each other as natural and artificial selection are. Evolution without intervention by the experimenter leads to optimization of replication rate or fitness, respectively. Animal breeders and evolutionary bio-technologists pick out the variants that fit best a predefined purpose and discard the rest (irrespective of their reproductive success). Almost all applications of evolutionary techniques follow a scheme consisting of amplification, diversification, and selection as sketched in fig.1. As characteristic examples we mention here two techniques which were applied in the optimization of RNA molecules: (i) the SELEX (**s**ystematic **e**volution of **l**igands by **ex**ponential enrichment) method, and (ii) the production of catalytically active RNA molecules called ribozymes (**ribo**nucleotide en**zymes**) whose activities are monitored by means of chemical tags.

In SELEX experiments RNA molecules, called aptamers, are produced, which recognize arbitrarily chosen targets through specific binding. The target molecules are immobilized through covalent chemical bonds to the solid-state phase of a chromatographic column. Solutions containing initial populations of up to 10^{15} RNA molecules with different sequences as obtained by random chemical synthesis are applied to the column. Molecules, which are retained on the column, are eluted and used for further selection rounds. Sequence diversity is usually produced by replication with artificially high mutation rates. The solvent is modified during a series of selection cycles in such a way that successively larger binding constants are required to be retained on the column. Commonly, twenty to thirty selection cycles are sufficient to isolate optimally binding aptamers. In favorable cases, like the aptamers binding to the antibiotic *tobramycin*, strong binding constants in the nanomolar range were achieved [6].

Design of ribozymes with novel catalytic functions started from reprogramming naturally occurring catalytic RNA molecules to accept unnatural substrates [8,9]: A specific RNA cleaving ribozyme, a class I (self-splicing) intron, was modified through variation and selection until it operated efficiently on DNA. The evolutionary path of such a transformation in the catalytic activity has been recorded in detail [10]. The essential problem in the evolutionary de-

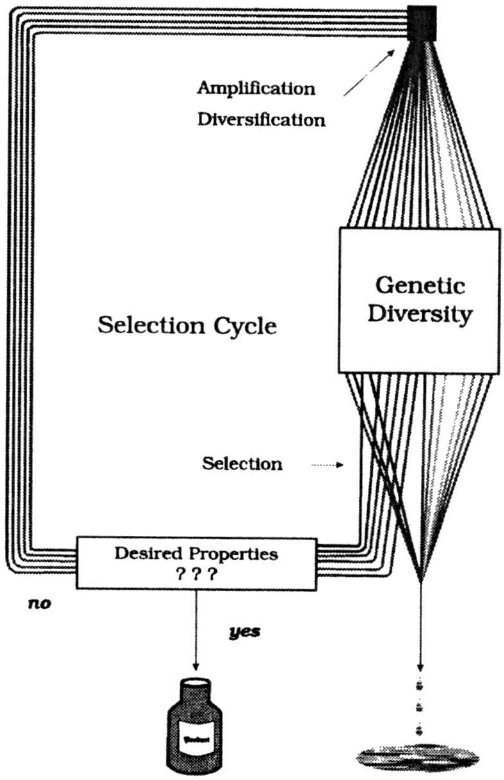

Fig. 1. Evolutionary design of biopolymers. Properties and catalytic functions of biomolecules are optimized iteratively through selection cycles. Each cycle consists of three different phases: (i) amplification, (ii) diversification by replication with problem adjusted high error rates (or random synthesis), (iii) and selection. Amplification and diversification are carried out by well established methods in molecular biology. Examples are the polymerase-chain-reaction (PCR) and the self-sustained sequence replication reaction (3SR-reaction), both of which can be carried out with enhanced mutation rates. Selection still requires ingenious concepts. Two examples, the SELEX method and chemical tagging are discussed in the text.

sign of new catalysts is the availability of an appropriate analytical tool for the detection of the activity. The technique of chemical tagging uses a covalently attached detectable marker, which is, for example, cleaved off in the molecules with the desired catalytic activity. Inactive molecules are unable to split off the tag. They can be detected by the presence of the tag after reaction, and are easily excluded from further selection cycles. Ribozymes with new catalytic activities were selected from pools of random RNA sequences (See, for example, [11]).

2 Molecular evolution including phenotypes

Population genetics became the mathematical basis of the synthetic theory [12] and is still seen by many biologists as the current frame for understanding evolution. It focuses on frequencies of genes or genomes in populations and describes their temporal changes caused by selection of variants of higher fitness or, rarely, by random drift. The basic concept was developed in the nineteen-thirties by the three scholars Ronald Fisher [13], John Haldane, and Sewall Wright, at a time when the molecular natures of genes and proteins were not yet known. Population genetics saw a major extension by Motoo Kimura [14] who based his work on the assumption that adaptive mutations are rare, most mutants are selectively neutral, and the predominant role of selection is the elimination of deleterious variants. Kimura's view was strongly supported by data obtained from comparative sequence analysis of proteins and nucleic acids [15] which became the basis of current molecular phylogeny [16,17]: Genotypes are continuously changing also during epochs of phenotypic stasis and they do so with an approximately constant rate per year and site. This constancy gave rise to the notion of a molecular clock setting the pace for evolution (See, however, [18]). Despite overwhelming indirect support for neutral evolution from molecular data, the first direct evidence came only last year from experiments on bacterial evolution under the controlled conditions of a serial transfer experiment [19]: Within epochs of phenotypic stasis the changes in genotypes occur at rates which are as high as, if not higher than, those recorded during adaptive periods. Populations genetics, although successful in its own rights, suffers from two major shortcomings when confronted with present-day molecular biology: (i) Mutation is considered as some external event, which is not part of the regularly considered dynamics, and (ii) the phenotype is represented only by its fitness value and other evolutionarily relevant properties like mutation rates, which are assigned as parameters to the corresponding genotype.

In 1971, almost at the time of Spiegelman's *in vitro* evolution experiments with RNA, Eigen published his seminal work on self-organization of biological macromolecules [20]. His concept unites evolutionary dynamics with knowledge from molecular biology. Replication and mutation are seen as parallel chemical reactions and evolution is visualized as a process in an abstract space of genotypes, called sequence space. Every RNA or DNA sequence is a point in sequence space, $\mathcal{I}$, and the Hamming distance [21], d^h_{ij},[1] induces a metric in this space. The temporal development of the distribution of genotypes in (sufficiently large) populations is described by the selection-mutation equation:

$$\frac{d\xi_i}{dt} = \dot{\xi}_i = \xi_i \Big(Q_{ii} a_i \; - \; \Phi(t) \Big) \; + \; \sum\nolimits_{j=1, j \neq i}^{n} Q_{ij} a_j \xi_j \; ; \; i = 1, \ldots n \; . \qquad (1)$$

[1] The Hamming distance between two end-to-end aligned sequences I_i and I_j, d^h_{ij}, is defined as the number of positions, in which the two sequences differ. If changes in genotypes are restricted to point mutations, the move set of evolution is restricted to differences of Hamming distance one in the sequences.

The variables $\xi_i(t)$ describe the frequencies of individual genotypes I_i. The constraint $\Phi(t) = \sum_{i=1}^{n} a_i \xi_i$ takes care of normalization of genotype frequencies $\sum_{i=1}^{n} \xi_i$. The square matrix $Q = \{Q_{ij}; i,j = 1,\dots,n\}$ contains replication accuracies Q_{ii} as diagonal elements and mutation rates as off-diagonal terms.[2] Although the use of differential equations to describe an inherently stochastic phenomenon like mutation may be seriously questioned, the concept of Eigen and coworkers, called **quasi-species** theory [22,23], passed successfully the experimental test [24]. At sufficiently accurate replication, that means low enough mutation rates, populations modelled by eq.(1) approach stationary mutant distributions, called quasi-species, which are centered around a most frequent genotype, the master sequence (fig.2). The major result of quasi-species theory is the existence of a sharp evolutionarily relevant threshold for mutation rates: At rates above the threshold value, populations do not approach stationary states but drift randomly through sequence space and genetic information is lost. Evolution is confined to mutation rates between a lower and an upper limit: The lower limit is given by the maximal accuracy of the replication machinery and the upper limit is set by the maximal sustainable fraction of error copies determined by the error threshold. Quasi-species theory was applied quantitatively not only to *in vitro* evolution of RNA [24] but also to virus evolution [25,26]. In rough agreement with the concept of molecular quasi-species, Jan Drake found remarkable constancy of mutation rates per genome replication within families of organisms, in particular with simple RNA viruses, retro-viruses, DNA-based microbes and even higher eukaryotes [27–30], respectively.

Although quasi-species theory gave otherwise inaccessible insights into the molecular mechanism of evolution and solved the mutation problem, it still suffers from the lack of an explicit consideration of phenotypes: Fitness relevant properties of phenotypes appear only as parameters of genotypes in the differential equations, for example a_i and Q_{ij} in eq.(1). Any comprehensive theory of evolution, however, has to handle the phenotype as an integral part of the model. Mathematics, in principle, has formal tools which are capable of such an integration. The relation between genotypes and phenotypes may be considered as a mapping from genotype space $\mathcal{I}$ into phenotype space $\mathcal{S}$:

$$\psi : \{\mathcal{I}; d^h_{ij}\} \Rightarrow \{\mathcal{S}; d^s_{ij}\} \quad \text{or} \quad S_k = \psi(I_.) \ . \qquad (2)$$

Writing $I_.$ instead of a fully defined subscript is meant to express the fact that counting in sequence space is different from counting in shape space and, as we shall see later, many sequences $I_.$ may form the same phenotype S_k.

Sequence space in eq.(2) is opposed by phenotype space $\mathcal{S}$, for which we assume the existence of a distance between phenotypes d^s_{ij}.[3] Eq.(2) implies that a unique phenotype is assigned to every genotype, whereas the inverse is commonly not true: Many genotypes may be mapped onto the same phenotype, and

[2] The element Q_{ij} is the frequency with which the genotype I_i is obtained as an error-copy of I_j, this means through a mutation of $I_j \rightarrow I_i$.

[3] It is not hard to define a metric in phenotype space, but it is anything but trivial to find an evolutionarily relevant measure of distance between phenotypes.

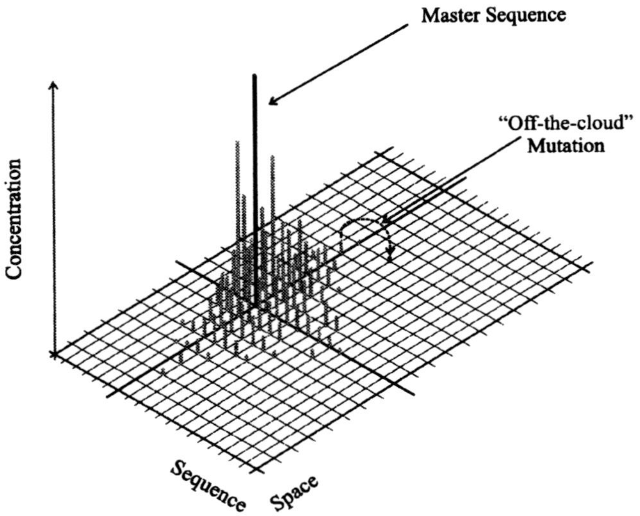

Fig. 2. Population structure of molecular quasi-species in sequence space. The quasi-species is an ordered stationary distribution of polynucleotide sequences (RNA or DNA) in sequence space $\mathcal{I}$. The fittest genotype or master sequence is surrounded in sequence space by a "cloud" of closely related sequences. Relatedness of sequences to the master is expressed by the number of point mutations, which are required to produce them as mutants of the master genotype. Precisely, the quasi-species is defined as the stable stationary solution of eq.(1) [22,23]. Such a solution exists only if the mutation rate lies below a maximal sustainable value, called the error threshold. Above the critical error rate all genotypes have the same probability to be present in the stationary population. This is tantamount to a uniform distribution, which, however, can never be realized in nature or *in vitro* since the numbers of possible nucleic acid sequences (4^{ℓ}) exceed the numbers of individuals by many orders of magnitude even in the largest populations. The actual behavior is then determined by incorrect replication leading to a breakdown of inheritance, and populations drift randomly through sequence space. The figure shows also an "off-the-cloud" mutation, which creates a genotype that has not been present in the population before. Such mutations represent true "innovations" and will be discussed in section 5. We remark that the sequence space $\mathcal{I}$, which is sketched as a two-dimensional object for the purpose of illustration, is of dimension ℓ in reality (where ℓ is the sequence or genome length).

then the map is not invertible. In other words, knowing the phenotype is not sufficient to derive the genotype, from which it had been developed. In essence, this is the mathematical formulation of phenotypic neutrality. A word of precaution is required here: The unique assignment of phenotypes to genotypes is an approximation in biology. Phenotypes are not exclusively determined by genotypes, since environmental factors and epigenetic effects are also relevant. Accordingly, the concept developed here is based on two assumptions: (i) Environments do

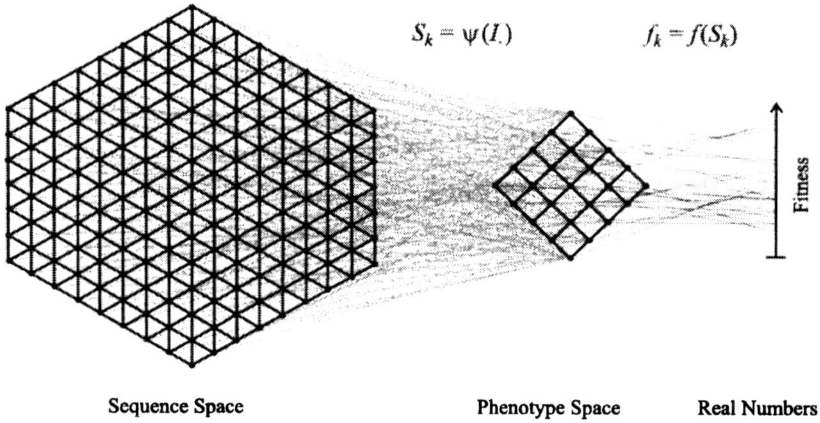

Fig. 3. Mapping genotypes onto phenotypes and into fitness values. The sketch shows a map from sequence (or genotype) space onto phenotype space, as described by eq.(2), and further into the real numbers ($\mathbb{R}^1$) resulting in fitness values assigned in two steps to the individual genotypes. The second map corresponds to eq.(3) and is a landscape, which could also be illustrated by a sketch with fitness plotted on the vertical axis. Both mappings are usually many-to-one and thus non-invertible. As in fig.2, the high-dimensional objects, sequence and phenotype space are sketched by two-dimensional illustrations.

not interfere with uniqueness of phenotype assignment, and (ii) mutation caused changes in epigenetics are negligible.

In order to complete evolutionary dynamics the fitness relevant properties have to be extracted from the phenotype. Again we may use the notion of a map for this purpose:

$$f : \{\mathcal{S}; d_{ij}^h\} \Rightarrow \mathbb{R}^1 \quad \text{or} \quad f_k = f(S_k) , \tag{3}$$

The mapping (3) assigns a real number to every phenotype S_k and is called a landscape.[4] As in the previous case there is room for neutrality: Many phenotypes may have the same fitness and then, the map is non-invertible. Realistic fitness landscapes are complex objects with a very large number of local peaks and steep valleys (See section 3). For the mathematically interested reader we mention that recent attempts to develop a theoretical frame for classification and analysis of such landscapes were successful [31–34].

Now we are already in a position to distinguish formally the concepts which neglect explicit consideration of phenotypes from more elaborate ones that deal with them. The first class of models use direct random [35–38] or non-random (for example single- and double-peak and other simple landscapes [39,40]) model

[4] The expression "landscape" is a generalization of the notion used in common-sense or geography for the representation of a three-dimensional relief on Earth as a mapping from two dimensions (longitude, latitude) into the real numbers (altitude).

assignments of fitness values to genotypes

$$g: \{\mathcal{I}; d^h_{ij}\} \Rightarrow \mathbb{R}^1 \ \text{ or } \ g_k = g(I_k) \ . \tag{4}$$

Models of the second class use a two-step relation with the phenotype as intermediate:

$$f: \{\mathcal{I}; d^h_{ij}\} \Rightarrow \{\mathcal{S}; d^s_{ij}\} \Rightarrow \mathbb{R}^1 \ \text{ or } \ f_k = f(S_k) = f\big(\psi(I_.)\big) \ .$$

The fitness values are a result of two mappings, (2) and (3), which are applied consecutively (fig.3). The importance of the two-step-approach lies in the fact that it can be analyzed in case of simple but realistic systems and it can be used to fit experimental data (See sections 3 and 7).

What remains to be done in order to complete the integration of the phenotype into evolutionary dynamics is to combine the genotype-phenotype-fitness map with replication-mutation dynamics as, for example, expressed by eq.(1):

$$\dot{\xi}_i = \xi_i \Big(Q_{ii}\, a\big(\psi(I_i)\big) - \Phi(t) \Big) + \sum_{j=1, j\neq i}^{n} Q_{ij}\, a\big(\psi(I_j)\big)\, \xi_j \ ; \ i = 1, \dots n \, . \tag{5}$$

It is straightforward to interpret this equation: Whenever a new genotype, say I_ℓ, is produced by mutation, it is converted into a phenotype $S_\ell = \psi(I_\ell)$ which has a replication rate parameter of $a_\ell = a\big(\psi(I_\ell)\big)$.[5] The sequence of events addressed in eq.(5) is sketched in figure 4. Replication-mutation dynamics need not meet the criteria for applicability of differential equations, but then the same sequence of events would provide the input for the stochastic process.

Consideration of environmental changes is also possible by means of time dependent mappings: $f_k = f\big(S_k, \mathcal{E}[t]\big) = f\big(\psi(I_k, \mathcal{E}[t]), \mathcal{E}[t]\big)$ where $\mathcal{E}[t]$ symbolizes the environmental influence at time t. A necessary condition for the applicability of the conventional analytical tools to separate evolutionary dynamics and environmental change is that the two processes occur on different time-scales. Namely, changes due to mutation and selection have to be fast compared to fitness changes induced by environment in order to justify the landscape concept. Alternatively, environmental changes that are much faster than mutation-selection dynamics are effective only in terms of their averages.

What can we actually gain from the proposed concept to integrate the phenotype into evolutionary dynamics? Provided suitable experimental data are available they can be readily fit into the formalism (See section 7). Another feature of the two mappings, ψ and f, is built upon empirical accessibility of genotype-phenotype relations and fitness values of phenotypes. Models are conceivable that are sufficiently simple for mathematical analysis and exploration through algorithms but, at the same time, sufficiently realistic to encapsulate relevant features (See, for example, section 3). Instead of assigning fitness values to

[5] In eq. (5) we made the tacit assumption that the dependence of mutation frequencies Q_{ij} on changes in the phenotype is negligible. This need not be fulfilled but then the incorporation of the phenotype into the mutation matrix is straightforward: $Q_{ij} = Q\big(\psi(I_j)\big)$.

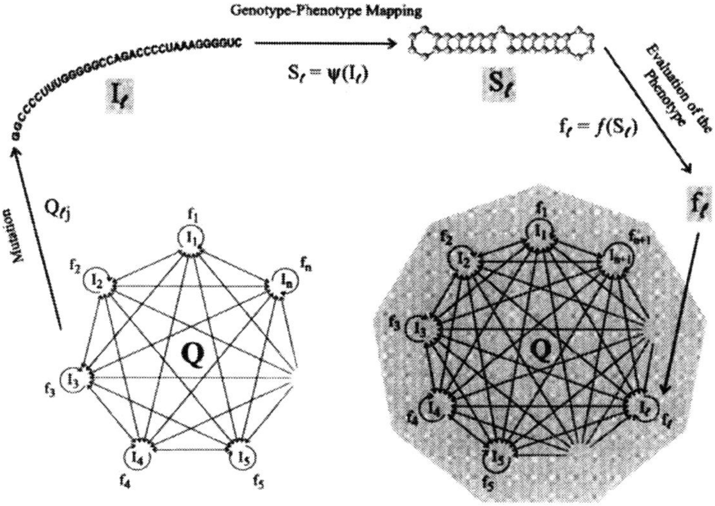

Fig. 4. Integration of phenotypes into replication mutation dynamics. The sketch shows the formation of a new variant, I_ℓ, through mutation of a genotype present in the population (see also the "off-the-cloud" mutation in fig.2). The symbol **Q** stands for the mutation matrix $Q \doteq \{Q_{ij}\}$ which determines the individual mutation frequencies. The polynucleotide sequence is processed through the mapping (2) to yield the corresponding phenotype S_ℓ. The phenotype, in turn, is evaluated by the mapping (3), which returns fitness relevant properties in quantitative terms. These values appear in the dynamical system as parameters of the new species.

genotypes either at random or on the basis of biologically unclear model assumptions, phenotypes are built from genotypes by a set of predefined rules which reflect important features of real systems. Mathematical analysis will reveal the generic properties of mappings which are often or almost always fulfilled. These regularities can be found, therefore, in all kinds of genotype-phenotype maps. In other words, if there are generic features of genotype-phenotype mappings, the usefulness of the concept depends on the success to find a suitable model and an experimental system to which it applies, but then generalizations will be possible.

3 The RNA model

In vitro evolution of RNA molecules reduces evolutionary dynamics to a simple experimental model which can be used as a test case for the model of evolution presented in the previous section 2. RNA is especially well suited for this purpose because genotype and phenotype are two features of the same molecule, the sequence and the molecular structure, respectively. The genotype-phenotype map in the RNA model boils down to the relation between sequences and structures. To elucidate this relation is still a hard but properly formulated and fairly

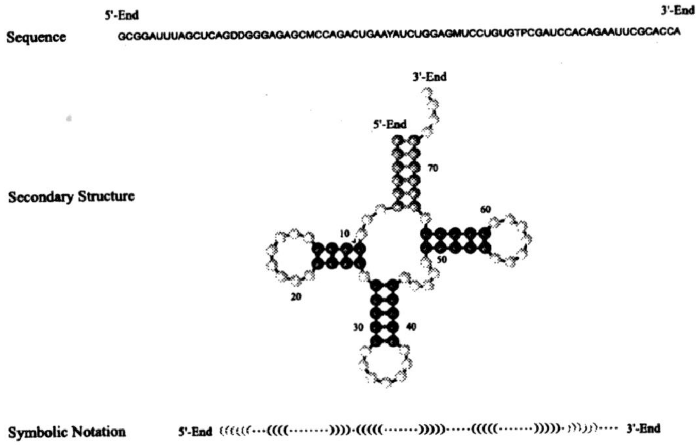

Fig. 5. RNA secondary structures. The nucleotide sequence of $tRNA^{phe}$ (shown in the upper string) is presented together with the secondary structure of minimal free energy and the symbolic notation (lower string). The sequence contains several modified nucleotides (**D**,**M**,**P**,**T**,**Y**) in addition to the conventional bases (**A**,**U**,**G**,**C**). Individual nucleotides in the secondary structure are shown as light grey (single bases), dark grey and black pearls (base pairs). The tRNA structure is a clover-leaf with three hairpin loops (adjacent stacks are shown in black) and a closing stack (shown in dark grey) which completes the central multiloop. The symbolic notation assigns one of the three symbols '·', '(', and ')' to each nucleotide depending on its binding state, "unpaired", "paired downstream", and "paired upstream", respectively. Downstream and upstream refer here to the conventional direction in polynucleotide sequences going from the 5'-end to the 3'-end. The three stacks of the hairpin loops (black) are embraced by the base pairs of the closing stack (grey).

understood standard problem in structural biology. Instead of predicting the structure for a given sequence we shall be concerned here with mapping the entire sequence space onto an RNA shape space, which is tantamount to the phenotype space in the RNA model (For example ψ in fig.3).

Despite progress in structural biology of RNA [50], full three-dimensional RNA structures are still too complex to be efficiently modelled. In addition, computation of full structures is time consuming and not suitable for large scale studies. Biochemists, however, have developed a coarse-grained version of RNA structure called secondary structure [51][6] which is physically meaningful: (i) a major portion of the free energy of structure formation is already captured by the secondary structure, (ii) the secondary structure represents (almost always) an experimentally observable intermediate in the folding process, (iii) the secondary structure is conserved in phylogenetic evolution, and (iv) despite the simplicity of the model assumptions RNA secondary structure predictions are

[6] In essence, an RNA secondary structure is a listing of Watson-Crick (**AU**,**GC**) and **GU** base pairs that can be drawn as a planar graph, which is free of knots and pseudoknots (fig.5).

Table 1. Various strategies applied to study sequence-structure maps of RNA

	Method	Advantage	Disadvantage	Ref.
Mathematical model	Random graph theory	Analytical expressions	Limited validity of model assumptions	[41]
Exhaustive folding and enumeration	Folding algorithm and handling of large samples up to 10^{10} objects	Exact results	Limited to short chain lengths: **GC**, $\ell \leq 32$ **AUGC**, $\ell \leq 16$	[42,43]
Statistical evaluation	Inverse folding or random walks in sequence space	Applicability to longer sequences	Limited accuracy due to statistics	[44,45]
Simulation of evolutionary dynamics	Chemical kinetics of replication and mutation	Evolutionary relevance	Restriction to small parts of sequence space	[46–49]

fairly reliable and reach often scores of 70 % on the basis of individual nucleotides. RNA secondary structures are readily computed by highly efficient algorithms [52,53], and they are easily encoded in strings of three-symbols: '·', '(', and ')' for "unpaired base", "base pair opening", and "base pair closure", respectively (See fig.5). The property of being encodable in strings and the modular building principle[7] are indeed the ultimate reasons why, in contrast to other notions of structure, RNA secondary structures are accessible to mathematical analysis (See table 1).

Strategies to investigate sequence – secondary structure mappings of RNA are summarized in table 1: (i) construction of mathematical models, (ii) exhaustive folding of sequence space and enumeration, (iii) statistical evaluation of random samples, and (iv) simulation of evolutionary processes. The most straightforward strategy consists in folding all sequences into structures and determining properties by complete enumeration [42,43,56,57]. Exhaustive folding is limited by the number of objects that can be handled, classified, stored, and retrieved in reasonable times by standard computational equipment. This limit lies between 10^9 and 10^{10} genotypes and implies that exhaustive studies of sequence spaces over a four-letter alphabet (for example the natural alphabet, **AUGC**; $\kappa = 4$) are confined to chain lengths $\ell \leq 16$; those over two-letter alphabets (**AU** or **GC**; $\kappa = 2$) can hardly exceed $\ell = 32$.[8]

[7] The modular building principle of RNA secondary structures implies that the structure can be partitioned into substructures, which contribute additively to (extensive) properties of the entire molecule. This is, for example, true for (free) energies and it represents the basis of the applicability of dynamic programming in the computation of structures with minimal free energies [54,55].

[8] Alphabets with less than four letters are indeed relevant for the design of RNA molecules. The three-letter alphabet **AUG**, for example, is sufficient for the formation of efficient ribozymes [58].

The simplicity of the base pairing logic allows to apply combinatorial methods and to calculate the maximum number of acceptable RNA secondary structures by means of recursions [55,59]. In addition, it was possible to derive a closed expression for these numbers

$$N_S(\ell) \approx s(\ell) = 1.4848 \times \ell^{-3/2}(1.84892)^{\ell} ,$$

which is asymptotically correct in the limit of long chains, $\lim \ell \to \infty$. The numbers of structures grow with a smaller exponent than the numbers of sequences, $N_\kappa(\ell) = \kappa^\ell$, and the mapping from RNA sequence space onto the secondary structures is many-to-one therefore.

The results obtained from studies on RNA sequence-structure maps can be casted in four statements:

(i) **More sequences than structures**. The numbers of sequences exceed the numbers of acceptable structures by many orders of magnitude (See, for example, the computed data in [57]).

(ii) **Few common and many rare structures**. Exhaustive folding of entire sequence spaces revealed widely different frequencies for the individual structures. The frequency distribution is strongly biassed towards the low frequency end: We have many rare structures opposed by relatively few frequent or common ones. In the limit of long chains almost all sequences fold into a negligibly small fraction of all structures [42,43].

(iii) **Common structures are found almost everywhere in sequence space**. In a sphere of radius r_{cov} around any randomly chosen reference sequence we have for every common structure at least one sequence, which forms it as its minimum free energy conformation. This **shape space covering** principle [45] yields a covering radius for common structures which is substantially smaller than the dimension of sequence space: $r_{\text{cov}} \ll 2\ell$.

(iv) **Neutral networks of common structures extend over whole sequence space**. The set of all sequences folding into a given structure S_k, $G_k = \psi^{-1}(S_k) \doteq \{I_j | \psi(I_j) = S_k\}$ is converted into a graph, called the neutral network $\mathcal{G}_k$, by connecting all pairs of sequences in G_k with Hamming distance one by en edge. Neutral networks of common sequences are connected and span almost the entire sequence space.

The last statement (iv) is particularly relevant for evolutionary dynamics and requires some elaboration.

The set of all sequences folding into a given structure S_k, the neutral set $G_k = \psi^{-1}(S_k) \doteq \{I_j | \psi(I_j) = S_k\}$, is the pre-image of the structure S_k in sequence space. The corresponding mapping is one-to-one, $G_k \Leftrightarrow S_k$. Neutral networks $\mathcal{G}_k$ are modelled by random graph theory in order to describe their properties by mathematical expressions. The nodes, $|G_k|$ in number, are assigned to the graph at random. Characteristic quantities of graphs or networks are the local connectivities which in case of $\mathcal{G}_k$ count the number of neutral neighbors at individual nodes: $\lambda_k(I_j)$, $j = 1, \ldots, |G_k|$. The average taken over the whole network, $\bar{\lambda}_k = \frac{1}{|G_k|} \sum_{j \in G_k} \lambda_k(I_j)$, is the parameter which determines global network properties. Networks are connected (that means they consist of a single component) if $\bar{\lambda}$ exceeds a threshold value λ_{cr}. Below threshold random graph theory predicts scattered networks with many components and one dominant "giant component". The threshold value is readily computed from the size of the genetic alphabet, κ:

$$\mathcal{G}_k \text{ is } \begin{cases} \text{connected} & \bar{\lambda}_k > \lambda_{cr} = 1 - \kappa^{-\frac{1}{\kappa-1}} , \\ \text{disconnected} & \bar{\lambda}_k < \lambda_{cr} = 1 - \kappa^{-\frac{1}{\kappa-1}} . \end{cases} \tag{6}$$

Examples for both cases are sketched in fig.6. The existence of a sharp threshold separating connected and partitioned networks reminds of a percolation phenomenon.

Neutral networks were extensively studied in **GC** sequence spaces of small chain lengths [42,43]. In addition to connected networks and networks with a single giant component, as predicted by the random graph model, networks with two and four equal largest components, and three-component networks with a ratio 1:2:1 in component size were observed. All three deviations from generic behavior could be explained by the peculiarities of secondary structure formation. This special case is a nice demonstration of the usefulness of mathematical models: Specific and generic features are easily separated by an inspection of the rules applied in secondary structure formation since straightforward molecular interpretations of deviations from generic behavior can be given.

The existence of extended and connected neutral networks in RNA sequence space was proven by an elegant experiment recently published by Erik Schultes and David Bartel [60]. They designed an RNA sequence which forms two known structures (of chain length $\ell = 88$) with different catalytic activities, an RNA ligase evolved in the laboratory [61] and a natural cleavage ribozyme isolated from hepatitis delta virus RNA [62]. The two structures have no base pair in common. Folding the synthesized chimeric sequence into structures yielded indeed both activities, although they were substantially weaker than those of the reference ribozymes, the ligase and the cleavage ribozyme, respectively. Only two or three selected point mutations or base pair exchanges are required, however, to reach full catalytic efficiency. Still, the two optimzed RNA molecules have a Hamming distance around forty from their reference sequences. Then, Schultes and Bartel [60] explored further the mutational neighborhoods and found neutral paths of Hamming distance about 40, by preparing and analyzing series of RNA sequences, in which neighboring sequences differ in a single base or base pair only. Without interruption these neutral paths lead from the RNA with both catalytic activities to the two reference ribozymes.

4 Computer simulation of RNA evolution

In order to explore RNA structure space by means of an evolutionary process a kind of flow reactor was implemented for evolution *in silico* [63,64]. Populations of one thousand to one hundred thousand RNA molecules are subjected to replication and mutation in the constant environment of a flow reactor. In the reactor

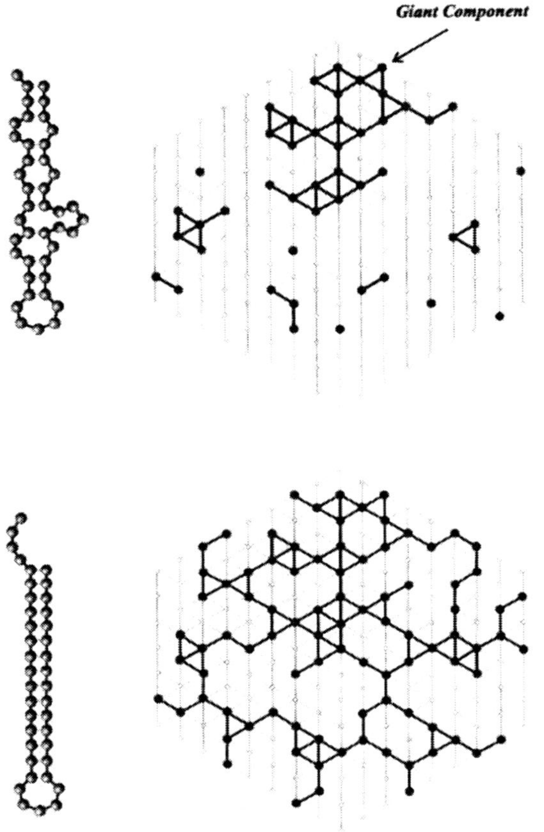

Fig. 6. Neutral networks in sequence space. The pre-image of the structure in the lower part of the figure is a connected neutral network spanning whole sequence space. Networks of this class are typical for frequent structures. The upper part of the figure shows an example of a disconnected network which consists of one giant component and many small islands. Connectivity is determined by the mean fraction of neutral neighbors ($\bar{\lambda}$) in sequence space.

consumed materials are continuously renewed and molecules produced in excess are removed. The computer implementation is based on an algorithm developed for the simulation of chemical reactions and published by Daniel Gillespie in the nineteen-seventies [65,66]. More recent computer simulations investigated the conservation of RNA phenotypes in neutral evolution [48,67] and studied optimization dynamics in the task to match a predefined target structure, S_τ, which was chosen to be the clover-leaf of tRNA^{phe} [47,46]. Progress of evolution is monitored by the average distance between the population and the target structure, $\overline{d^s_\tau}(t) = \frac{1}{N}\sum_{j=1}^{N} d^s_{j\tau}$ taken at time t (grey trajectory in fig.7). Individual runs

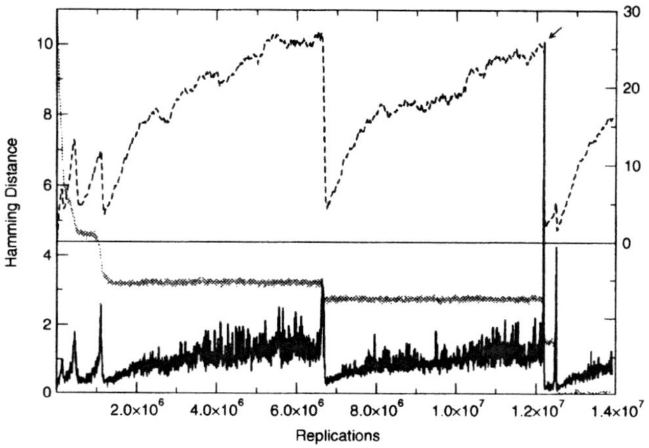

Fig. 7. Variability in genotype space during punctuated evolution. Shown are the results of a typical simulation of RNA optimization towards a tRNA target with population size $N = 3000$ and mutation rate $p = 0.001$ per site and replication: (i) the trace of the underlying trajectory recording the average distance from target, $\overline{d^s_{\mathcal{T}}}(t)$ (grey, left ordinate scaled by 0.22, or full length is 50), and (ii) plots of measures of the progress of evolution in genotype space, $\overline{d^h_{\mathbf{P}}}(t)$ and $\Delta d^h_{\mathbf{CP}}(t, \Delta t)$ with $\Delta t = 8\,000$ replications, against time expressed as the total number of replications performed until t. The upper plot shows the mean Hamming distance within the population ($\overline{d^h_{\mathbf{P}}}(t)$; dotted line, right ordinate) which is a measure of genotype diversity. The lower curve presents the Hamming between the centers of the population at times t and $t + \Delta t$ ($\Delta d^h_{\mathbf{CP}}$; full line, left ordinate), and records the drift velocity of the population center. The arrow indicates a sharp peak of $\Delta d^h_{\mathbf{CP}}$ at the end of the second long plateau reaching a Hamming distance of about ten.

were analyzed in terms of relay series, which are uninterrupted time ordered series of phenotypes leading from the initial structure to the target molecule. The length of a relay series, expressed as the number of structures it contains, n_{rl} (table 2), is one characteristic quantity of the optimization process. Typical trajectories consist of an initial phase of fast approach towards target followed by punctuated optimization: Long quasi-stationary epochs of phenotypic stasis are interrupted by short periods of adaptations and, eventually, the populations reaches the target. The number of steps or discontinuous transitions, n_{dt}, is an other characteristic of evolutionary optimization (table 2). In order to measure the effort that is required to reach the goal we recorded total numbers of replications.

Individual trajectories are not reproducible. The mean number of replications shows vast scatter and increases slowly with population size. As expected the

Table 2. Statistics of evolutionary trajectories. Twenty trajectories of *in silico* evolution towards a tRNA target were recorded for each value of population size N [68]. Shown are, the numbers of replications, the lengths of the relay series, n_{tr}, the numbers of discontinuous or major transitions, n_{dt}, and the mean structure distance between the population at the end of the fast adaptive initial period and the target shape, $\overline{d^s_\tau}(t_{\mathrm{in}})$, being the distance which is bridged by the punctuated part of the trajectory.

Population Size N	Number of Replications	Number of Relay Steps n_{tr}	Number of Transitions n_{dt}	Initial Phase $\overline{d^s_\tau}(t_{\mathrm{in}})$
1 000	$(7.8 \pm 7.4) \times 10^7$	114.1 ± 88.5	6.5 ± 1.7	17.6 ± 2.3
2 000	$(1.13 \pm 1.50) \times 10^8$	62.8 ± 25.6	6.5 ± 1.7	18.5 ± 2.3
3 000	$(1.97 \pm 4.00) \times 10^8$	49.1 ± 23.5	6.3 ± 2.2	17.6 ± 2.4
10 000	$(1.50 \pm 1.07) \times 10^8$	37.1 ± 11.0	5.7 ± 1.3	16.5 ± 1.0
20 000	$(1.82 \pm 1.05) \times 10^8$	29.2 ± 5.0	5.7 ± 1.3	16.5 ± 1.0

relative standard deviation is smaller in larger populations. Runs recorded under identical conditions[9] yielded relay series with different structures and substantial variation in lengths (table 2). The expectation value of n_{tr} decreases strongly and the widths of the probability distribution becomes smaller with increasing population size N. The number of discontinuous transitions, n_{dt}, however, stays remarkably constant within the range of population size variation. Another quantity which is fairly insensitive to the variation in population site reported here is the fraction of optimization score in the initial period versus the epochal scenario. In order to be able to interpret these results we have to find an evolutionarily relevant measure of nearness in phenotype space. Before doing so, however, we shall consider evolution in genotype space.

Populations may be characterized by their weighted center in sequence space, **CP**,[10] and the width of the distribution expressed by the mean Hamming distance $\overline{d^h_{\mathbf{P}}}$. The drift velocity of the population center in sequence space, $\Delta d^h_{\mathbf{CP}}$ and the mean Hamming distance are plotted as a function of time in fig.7. They reveal the course of evolution on the plateaus of phenotypic stasis. Genetic diversity increases during quasi-stationary epochs and reaches a maximum at the end of the plateau. Selection that sets on at the edge of the plateau causes a sudden reduction of diversity. In other words, the population passes through a bottle-neck, which is defined by the mutation that terminates the period of phenotypic stasis, and the fitter phenotype it had produced. The population center,

[9] Identical means here that all parameters and the initial conditions are the same except the seeds of the random number generator.

[10] The weighted center of a population is a string of ℓ vectors, **CP**$= (\mathbf{v}_1, \mathbf{v}_2, \ldots, \mathbf{v}_\ell)$, each of which gives the average nucleotide distribution at the corresponding position: $\mathbf{v}_1 = (a_i, u_i, g_i, c_i)$ means a_i**A**$+u_i$**U**$+g_i$**G**$+c_i$**C** at position 'i'. It is related to the consensus sequence which carries at every position the nucleotide with the largest weight, **A** at position i if and only if $a_i = \max\{a_i, u_i, g_i, c_i\}$, etc.

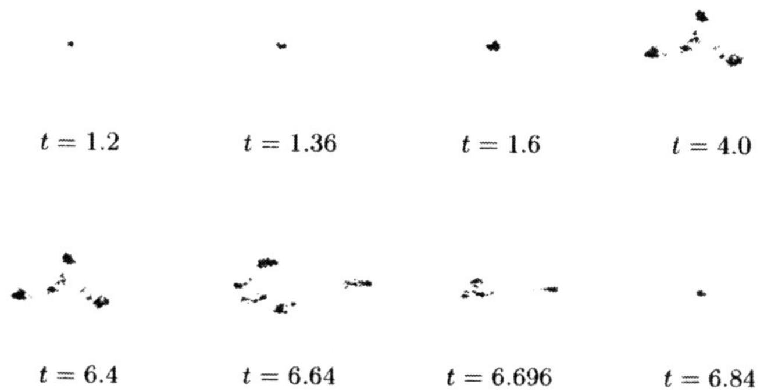

Fig. 8. Spreading of populations during quasi-stationary epochs. The individual images are snapshots of the distribution of genotypes recorded during the simulation experiment presented in fig.7 at times corresponding to 1.2, 1.36, 1.6, 4.0, 6.4, 6.64, 6.696, and 6.84×10^6 replications. In order to visualize population spreading, genotype distributions were transformed to principal axes and individual sequences were projected onto the plane spanned by the two largest eigenvectors. Along the series we observe an important and characteristic feature of neutral evolution: The populations break up in smaller sub-clusters which diffuse radially away from the center of the distribution (See also the model on neutral evolution discussed in [48,69]). Whenever an innovation with increase in fitness happens in one of the sub-clusters this sub-cluster takes over and all other sub-clusters die out.

on the other hand, drifts slowly during stasis and makes a sudden jump and the end of the epoch.

Both observations are readily interpreted by visualizing the population and its evolution in sequence space (fig.8). For this goal the population is projected onto the most instructive plane in sequence space.[11] After passing the bottleneck at the end of a quasi-stationary epoch the population enters the next fitness plateau and spreads in sequence space. At first the population expands as a coherent cloud and then, when the population is too small for further expansion, it breaks up into individual clones which continue to separate until the end of the static epoch. During the entire period the center of the population drifts slowly (with a velocity of Hamming distance one per 8000 replications). The expansion of the population is reflected by the increase in mean Hamming distance (fig.7), which settles down at a rather large value of more than 20 provided the period of stasis is sufficiently long. The epoch ends when a favorable mutant is produced (by a single error mutation) in one of the clones. This variant plays the role of a “founder molecule”: It is selected because of higher fitness and becomes the center of the population during the next epoch. The sequence distribution of the population narrows down since all clones die out except the one, in which the advantageous mutation had occurred. The jump of the population from one

[11] The genotype distribution in sequence space is transformed to principal axes and projected onto the plane spanned by the two largest eigenvectors.

center to the next manifests itself in a spike in the migration velocity (see fig.7), which is also a reliable diagnostic for discontinuous transitions.

5 Transitions, continuity, and statistical neighborhoods

The previously defined distance between pairs of RNA secondary structures, d^s_{ij}, is easy to obtain but, unfortunately, it tells nothing about evolutionary relatedness of structures. Since point mutations are the elementary moves of evolution in sequence space, accessibility of molecular phenotypes has to be considered in genotype space. The pre-images of structures are the neutral networks and thus neighborhood relations have to be expressed in terms of closeness of the corresponding graphs in sequence space. All possible single point mutations of all sequences forming structure S_j, $(\kappa - 1)\,\ell\,|G_j|$ in number, form the boundary set B_j, which after folding yields a set of neighboring structures Σ_j. In general, the neighboring structures have very different frequencies. The frequent ones are readily accessible whereas there will be many rare ones, which can hardly be reached through a search based on point mutations. The question is now, how to define an evolutionarily justified borderline between accessible and non-accessible structures (which, inevitably, will be of probabilistic nature). Assume, the neutral networks of two structures, S_j and S_k, were $\mathcal{G}_j$ and $\mathcal{G}_k$, and had γ_{kj} contacts of Hamming distance one. The relevant quantity is the probability to reach a node of $\mathcal{G}_k$ from a node of $\mathcal{G}_j$ by a single point mutation, which is easily obtained by dividing through the number of all possible point mutations of all sequences forming S_j:

$$\varrho(S_k; S_j) \;=\; \frac{\gamma_{kj}}{|B_j|} \;=\; \frac{\gamma_{kj}}{(\kappa - 1)\,\ell\,|G_j|}\;. \tag{7}$$

This probability is not symmetric, $\varrho(S_k; S_j) \neq \varrho(S_j; S_k)$: Although the numbers of Hamming distance one contacts between two neutral sets are, of course, symmetric, that is $\gamma_{kj} = \gamma_{jk}$, the sizes of the networks will in general be different, $|G_j| \neq |G_k$. Examples of cases where one structure is a substantially more frequent neighbor of another structure than vice versa are easily found with RNA molecules [46].

The distribution of structures in the one-error neighborhood of a large neutral network corresponding to frequent structure S_j exhibits characteristic regularities that can be seen best in a log frequency – log rank plot [46]: The reference structure has the highest probability (which, in essence, is expressed by $\bar{\lambda}_k$), it is followed by a relatively small number of neighbors which are approximately equally frequent and clearly separated from the less frequent ones. This less frequent tail fulfils a power law in the sense of Zipf's law [70]. Although accessibility of phenotypes is population size dependent (see next section), it is straightforward to draw a borderline at the rank where the distribution becomes approaches a straight line in the log – log plot. Then, only the structures which have a probability of occurrence above the frequency ε corresponding to this borderline form the set of statistical neighbors of S_j:

$$\Psi_\varepsilon(S_j) = \{S_k \in \Sigma_j \,|\, \varrho(S_k;S_j) \geq \varepsilon\} \ . \tag{8}$$

This statistical definition of nearness in phenotype space has weaker properties than a distance measure: In general, it does neither fulfil the triangle inequality nor is it symmetric as follows from eq.(7). Attempts have been made to use eq.(8) to induce a topology or some weaker structure of pre-topological nature in phenotype space [71–73], which still have to show their usefulness in concrete applications. Nevertheless, the nature of transitions between structures in the relay series of an evolutionary trajectory are readily classified by the concept of statistical nearness.

The concept of the statistical neighborhood has been applied to classify transitions observed in the relay series of RNA optimization [46,47]. Transitions $S_j \to S_k$ occur readily when S_k is a statistical neighbor of S_j. The two neutral networks have sufficiently many contacts to guarantee a probability of S_k in the boundary of S_j above threshold, $\varrho(S_k;S_j) > \varepsilon$. Hence, the transition is possible from almost everywhere on the network of S_j and occurs readily. It was named *continuous*, therefore. Continuous transitions are observed not only in adaptive phases but also during periods of constant mean fitness of the population, then between closely related structures of identical fitness values. *Discontinuous* transitions, on the other hand, are characterized by a probability below the threshold, $\varrho(S_k;S_j) < \varepsilon$. This implies that $S_j \to S_k$ is possible only from relatively few positions on the neutral network and the population has to search through random drift before it finds a proper place for the transition. By inspection of the relay series we see two scenarios realized during quasi-stationary epochs: (i) Constant phenotype and changing genotpyes leading to a spreading population which eventually reaches a position on the neutral network that allows for a discontinuous transition with gain in fitness, and (ii) a group of closely related phenotypes of identical fitness, which come and go in random manner until a point is reached, from which a discontinuous transition occurs that is accompanied by gain in fitness.[12]

The probabilities of continuous and discontinuous transitions see an independent confirmation through the interpretation of molecular structures. Continuous transitions involve closing or opening of single base pairs, which occur frequently on point mutations. Another typical continuous transition is represented by the opening of the terminal stack in the clover-leaf structure of tRNAs (fig.5, the upper, dark grey base paired region in the tRNA) leading to a structure with three hairpin loops joined by unpaired stretches. The terminal stack closing the multiloop is marginally stable and can be easily opened by single point mutations, which may occur in several positions. Discontinuous transitions, in contrary, involve simultaneous formation and cleavage of several base pairs. Suitable sequences have to meet several requirements and this reduces substantially the

[12] Rare, but regularly observable, "silent" discontinuous transitions occur also within periods of stationary mean fitness. Then, the transition leads to a distant phenotype of the same fitness.

probability of occurrence. Typical transitions of this kind are shifts of one or two entire stacks or the closure of the terminal stack in tRNAs [46].

6 An exercise in probabilities

The simulation data in table 2 indicate that only the total number of steps in the relay series shows appreciable population size effects. Whereas this number varies approximately by a factor three, the number of discontinuous transitions stays remarkably constant. The number of transitions in the relay series scales roughly with $1/N$ and extrapolation to infinite population size yields a number that is approximately the number of discontinuous transitions. Straightforward conclusion suggests that the numbers of continuous transitions are substantially reduced with increasing population sizes in the range investigated.

In order to derive population size effects on the relay series we calculate the probability that a newly formed neutral variant stays sufficiently long in the population. This can be derived from first passage times leading from single copies to extinction, $T_{0,1}$ [68]. Since we are interested in the neutral case, all quantities are computed for single replication events. The probability of mutation per replication is mutation rate times chain length, $P_{\mathrm{mut}} = p \cdot \ell$, where p is the error rate per site and replication and ℓ the chain length of the molecule. Here we make the assumption of uniform error rates which implies that error rates do not depend on position. From the general probability of mutation we obtain the probability of a specific mutation between two genotypes I_i and I_j (of identical chain length),

$$\mathrm{Prob}\{I_i \to I_j\} \;=\; \frac{N_i}{N} \cdot \frac{p \cdot \ell}{\ell \cdot (\kappa - 1)} \;=\; \xi_i \, \frac{p}{\kappa - 1} \;=\; \xi_i \, P_{ij} \; ,$$

with N_i being the number of genotypes of class I_i in a population of size $N = \sum_{i=1}^{n} N_i$, ξ_i being its relative frequency, and κ the size of the nucleotide alphabet. The transition matrix, $\mathbf{P} = \{P_{ij}; i,j = 1,\ldots,n\}$, is symmetric as a consequence of the uniform error rate assumption.

The relative frequency of phenotype S_j is expressed by $\eta_j = \sum_{i=1}^{n_j} \xi_i^{(j)}$ [67] where we assumed that all genotypes $I_i^{(j)}$ $(i = 1,\ldots,n_k)$ form the same phenotype. In order to investigate transition dynamics between phenotypes we compute first the probability of a mutation from phenotype S_j to phenotype S_k:

$$\mathrm{Prob}\{S_j \to S_k\} \;=\; \frac{M_j}{N} \cdot p \cdot \ell \cdot \varrho(S_k; S_j) \;=\; \eta_j \cdot p \cdot \ell \cdot \varrho(S_k; S_j) \;=\; \eta_j \, \Pi_{jk} \, ,$$

where $M_j = \sum_{i=1}^{n_j} N_i^{(j)}$ (with $\sum_{j=1}^{m} M_j = N$) and $\varrho(S_k; S_j)$ is the frequency of occurrence of S_k in the one-error neighborhood of phenotype S_j according to eq.(8). We remark that the transition matrix for mutations between phenotypes, $\mathbf{\Pi} = \{\Pi_{ij}; i,j = 1,\ldots,n\}$, is in general not symmetric, $\Pi_{jk} \neq \Pi_{kj}$, because the two networks need not have identical sizes. Mutation dynamics is readily extended to several neutral precursors of S_k:

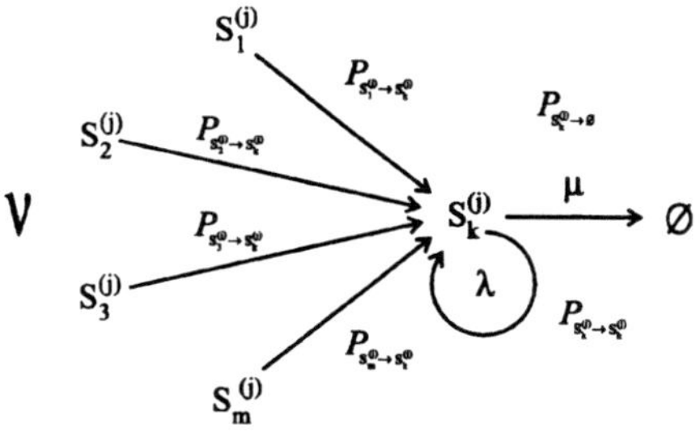

All phenotypes belong to the same class, $S^{(j)} = \{S_i^{(j)}; i = 1, 2, \ldots, m\}$, are inter-converted through continuous transitions. As in previous cases [67] all relative frequencies of phenotypes within the same class are lumped together, $\eta^{(j)} = \sum_{i=1}^{m} \eta_i^{(j)}$, and a mean value is applied for the frequency of occurrence, $\bar{\varrho}^{(j)} = \sum_{i=1}^{m} \eta_i^{(j)} \varrho(S_k^{(j)}; S_i^{(j)})/\eta^{(j)}$. The process shown in the insert above is approximated by a birth an death process with immigration in the numbers of phenotypes $M_k = |S_k^{(j)}|$ between two reflecting barriers [74] at $M_k = 0$ and $M_k = N$, and with $\sum_{i=1}^{m_j} M_i = N$ and the rate parameters $\lambda_n = n \cdot \lambda + \nu$ and $\mu_n = n \cdot \mu$ ($\bar{\lambda}(S_k^{(j)})$ being the degree of neutrality of $\mathcal{G}_k$),

$$\lambda = (1-p) \cdot \frac{1}{N} + p \cdot \ell \cdot \left(\bar{\lambda}(S_k^{(j)}) - \bar{\varrho}^{(j)}\right) \cdot \frac{1}{N} \, ,$$

$$\nu = p \cdot \ell \cdot \bar{\varrho}^{(j)} \, , \quad \text{and} \quad \mu = \frac{1}{N} \, .$$

The expectation value of the first passage time representing the mean life time of a newly formed phenotype S_k is readily computed

$$< T_{0,1} > = \frac{1}{\mu} \sum_{n=1}^{N} \frac{(n - 1 + \frac{\nu}{\lambda})_{n-1}}{n! \, \left(\frac{\mu}{\lambda}\right)^{n-1}} \, . \tag{9}$$

Two different scenarios are easily recognized: (i) small population sizes with $\mu \gg \nu$ and short first passage times $< T_{0,1} > \approx N \ln N$, and (ii) large population sizes with $\mu \ll \nu$ and long first passage times $< T_{0,1} > \geq (2\pi N \nu)^{-1/2} \left(\frac{e}{\nu}\right)^N$. The population size that separates the two regimes is defined by $\mu(N_{\mathrm{sat}}) = \nu$,

$$N_{\mathrm{sat}} = \frac{1}{p \cdot \ell \cdot \bar{\varrho}^{(j)}} \, , \tag{10}$$

and called saturation size because the probabilities of continuous transition are "saturated" there and they disappear from the relay series. Application to quasi-stationary epochs with clover-leaf structures yields a saturation size of a few thousand molecules in full agreement with the simulations.

The definition of the probability saturation size allows also to extrapolate to discontinuous transitions. In this case we calculate extremely large population sizes of 10^{40} and more, implying that these major transitions will persist and cause punctuation even in the largest sustainable populations (See, e.g., the experiments with bacteria with populations sizes of about 10^8 [75]).

7 Molecular evolution in the post-genomic age

The results reported here for RNA sequence-to-structure mappings seem to be readily extensible to proteins. Results on model proteins [76,77] revealed few common and many rare structures, and studies based on empirical potentials gave strong indications on the existence of neutral networks in protein sequence space [78]. There are, however, substantial differences between protein and RNA landscapes, which are apparently caused by the lack of a simple pairing logic for proteins: shape space covering as found with RNA secondary structures does not seem to hold for proteins. In addition, many polypeptide sequences do not give rise to soluble proteins with defined structures and this fact leads to holes in protein foldability landscapes [79–82]. More RNA and protein data will come up in the near future and they will allow for more detailed comparisons. Presumably, some generic features of biopolymer genotype-phenotype maps can then be extracted and generalized to more complex cases.

Viral evolution is the next logical step in analyzing and modelling more complex relations between genotypes and phenotypes. A great number of viral genome sequences were completed already some time ago. It remains now to explore the corresponding phenotypes. A viral phenotype comprises the entire life cycle in the host cell and its control mechanism which is commonly encoded in the molecular structures of the virus. Many data on the unfolding of viral genotypes are already available. For example, structures of non-translated regulatory sequences (IRES) have been determined by sequence comparisons within families of RNA viruses. More results on virus regulation will be available within the next few years.

Evolution experiments with bacterial populations under controlled conditions show many similarities to evolution of molecules [19,75] and we may anticipate even more analogies to pop up when further molecular details of bacterial evolution had been explored. Sequencing of further bacterial genomes adds little to understanding genotype-phenotype relations in bacteria. What is required in this case, is information on the influence of mutation on cellular metabolism whose exploration constitutes the primary goal of functional genomics. Current theoretical approaches along these lines are at the very beginning, but fast progress can be expected in the near future. The next major experimental step in modelling phenotypes is the collection and interpretation of gene expression data derived from DNA chips.

From analyzing and modelling prokaryotic evolution to an understanding of multicellular organisms is still a very long way to go and we do not know yet whether or not generic properties of simple genotype-phenotype mappings can be used as models. Progress, however, is fast and molecular genetics of development is witnessing a true explosion of new insights into the enormous complexity of higher organisms. At the current state of our knowledge it seems certainly premature to generalize the concepts developed here to evolution of plants and animals but several features of organismic evolution are already reflected by simple molecular systems [83].

Acknowledgements

Many fruitful discussions with Manfred Eigen, Christoph Flamm, Walter Fontana, Ivo Hofacker, and Peter Stadler are gratefully acknowledged. Andreas Wernitznig kindly provided computer plots on *in silico* evolution experiments in the flow reactor. Financial support of the work presented here was provided by the Austrian *Fonds zur Förderung der wissenschaftlichen Forschung* (Projects P-13093, and P-13887), by the *Jubiläumsfonds der Österreichischen Nationalbank* (Project 7813), by the Commission of the European Union (Project PL-970189), and by the Santa Fe Institute.

References

1. S. Spiegelman. An approach to the experimental analysis of precellular evolution. *Quart. Rev. Biophys.*, 4:213–253, 1971.
2. C. K. Biebricher and M. Eigen. Kinetics of RNA replication by Qβ replicase. In E. Domingo, J. J. Holland, and P. Ahlquist, editors, *RNA Genetics I: RNA-directed Virus Replication.*, pages 1–21. Plenum Publishing Corporation, Boca Raton, Florida, 1987.
3. M. Eigen and W. C. Gardiner. Evolutionary molecular engineering based on RNA replication. *Pure Appl. Chem.*, 56:967–978, 1984.
4. A. Watts and G. Schwarz, editors. *Evolutionary Biotechnology – From Theory to Experiment*, volume 66/2-3 of *Biophysical Chemistry*. Elesvier, Amsterdam, 1997.
5. L. Gold, C. Tuerk, P. Allen, J. Binkley, D. Brown, L. Green, S. MacDougal, D. Schneider, D. Tasset, and S. R. Eddy. RNA: The shape of things to come. In R. F. Gesteland and J. F. Atkins, editors, *The RNA World*, pages 497–509. Cold Spring Harbor Laboratory Press, Plainview, NY, 1993.
6. D. S. Wilson and J. W. Szostak. *In vitro* selection of fuctional nucleic acids. *Annu. Rev. Biochem.*, 68:611–647, 1999.
7. B. I. Dahiyat, C. A. Sarisky, and S. L. Mayo. *De Novo* protein design: Towards fully automated sequence selection. *J. Mol. Biol.*, 273:789–796, 1997.
8. A. A. Beaudry and G. F. Joyce. Directed evolution of an RNA enzyme. *Science*, 257:635–641, 1992.
9. R. R. Breaker. *In vitro* selection of catalytic polynucleotides. *Chem. Rev.*, 97:371–390, 1997.
10. N. Lehman and G. F. Joyce. Evolution *in vitro*: Analysis of a lineage of ribozymes. *Current Biology*, 3:723–734, 1993.
11. D. P. Bartel and J. W. Szostak. Isolation of new ribozymes from a large pool of random sequences. *Science*, 261:1411–1418, 1993.
12. E. Mayr and W. B. Provine, editors. *The Evolutionary Synthesis. Perspectives of the Unification of Biology.* Harvard University Press, Cambridge, MA, 1980.
13. R. A. Fisher. *The Genetical Theory of Natural Selection.* Oxford University Press, Oxford, UK, 1930.

14. M. Kimura. *The Neutral Theory of Molecular Evolution.* Cambridge University Press, Cambridge, UK, 1983.
15. J. L. King and T. H. Jukes. Non-Darwinian evolution: Random fixation of selectively neutral variants. *Science*, 164:788–798, 1969.
16. W.-H. Li and D. Graur. *Fundamentals of Molecular Evolution.* Sinauer Associates, Sunderland, MA, 1991.
17. L. H. Caporale, editor. *Molecular Strategies in Biological Evolution*, volume 870 of *Annals New York Acad. Sci.* New York Academy of Sciences, New York, 1999.
18. F. J. Ayala. Vagaries of the molecular clock. *Proc. Natl. Acad. Sci. USA*, 94:7776–7783, 1997.
19. D. Papadopoulos, D. Schneider, J. Meier-Eiss, W. Arber, R. E. Lenski, and M. Blot. Genomic evolution during a 10 000-generation experiment with bacteria. *Proc. Natl. Acad. Sci. USA*, 96:3807–3812, 1999.
20. M. Eigen. Selforganization of matter and the evolution of biological macromolecules. *Naturwissenschaften*, 58:465–523, 1971.
21. R. W. Hamming. Error detecting and error correcting codes. *Bell Syst. Tech. J.*, 29:147–160, 1950.
22. M. Eigen and P. Schuster. The hypercycle. A principle of natural self-organization. Part A: Emergence of the hypercycle. *Naturwissenschaften*, 64:541–565, 1977.
23. M. Eigen, J. McCaskill, and P. Schuster. The molecular quasispecies. *Adv. Chem. Phys.*, 75:149 – 263, 1989.
24. C. K. Biebricher and W. C. Gardiner. Molecular evolution of RNA *in vitro*. *Biophys. Chem.*, 66:179–192, 1997.
25. E. Domingo. Biological significance of viral quasispecies. *Viral Hepatitis Reviews*, 2:247–261, 1996.
26. E. Domingo and J. J. Holland. RNA virus mutations and fitness for survival. *Annu. Rev. Microbiol.*, 51:151–178, 1997.
27. J. W. Drake. A constant rate of spontaneous mutation in DNA-based microbes. *Proc. Natl. Acad. Sci.*, 88:7160–7164, 1991.
28. J. W. Drake. Rates of spontaneous mutation among RNA viruses. *Proc. Natl. Acad. Sci.*, 90:4171–4175, 1993.
29. J. W. Drake, B. Charlesworth, D. Charlesworth, and J. F. Crow. Rates of spontaneous mutation. *Genetics*, 148:1667–1686, 1998.
30. J. W. Drake. The distribution of rates of spontaneous mutation over viruses, prokaryotes, and eukaryotes. *Annals of the New York Academy of Sciences*, 870:100–107, 1999.
31. P. F. Stadler. Landscapes and their correlation functions. *J. Math. Chem.*, 20:1–45, 1996.
32. P. F. Stadler. Fitness landscapes arising from the sequence-structure maps of biopolymers. *J. Mol. Struct. (Theochem)*, 463:7–19, 1999.
33. P. F. Stadler, R. Seitz, and G. P. Wagner. Population dependent fourier decomposition of fitness landscapes over recombination spaces: Evolvability of complex characters. *Bull. Math. Biol.*, 62:399–428, 2000.
34. C. M. Reidys and P. F. Stadler. Combinatorial landscapes. *SIAM Rev.*, 2001. In press.
35. B. Derrida. Random-energy model: An exactly solvable model of disordered systems. *Phys. Rev. B*, 24:2613–2626, 1981.
36. C. Amitrano, L. Peliti, and M. Saber. Population dynamics in a spin-glass model of chemical evolution. *J. Mol. Evol.*, 29:513–525, 1989.
37. P. Tarazona. Error threshold for molecular quasispecies as phase transitions: From simple landscapes to spin-glass models. *Phys. Rev. A*, 45:6038–6050, 1992.

38. S. A. Kauffman. *The Origins of Order. Self-Organization and Selection in Evolution.* Oxford University Press, Oxford, UK, 1993.
39. J. Swetina and P. Schuster. Self-replication with errors - A model for polynucleotide replication. *Biophys. Chem.*, 16:329–345, 1982.
40. P. Schuster and J. Swetina. Stationary mutant distribution and evolutionary optimization. *Bull. Math. Biol.*, 50:635–660, 1988.
41. C. Reidys, P. F. Stadler, and P. Schuster. Generic properties of combinatory maps. Neutral networks of RNA secondary structure. *Bull. Math. Biol.*, 59:339–397, 1997.
42. W. Grüner, R. Giegerich, D. Strothmann, C. Reidys, J. Weber, I. L. Hofacker, P. F. Stadler, and P. Schuster. Analysis of RNA sequence structure maps by exhaustive enumeration. I. Neutral networks. *Mh.Chem.*, 127:355–374, 1996.
43. W. Grüner, R. Giegerich, D. Strothmann, C. Reidys, J. Weber, I. L. Hofacker, P. F. Stadler, and P. Schuster. Analysis of RNA sequence structure maps by exhaustive enumeration. II. Structure of neutral networks and shape space covering. *Mh.Chem.*, 127:375–389, 1996.
44. W. Fontana, P. F. Stadler, E. G. Bornberg-Bauer, T. Griesmacher, I. L. Hofacker, M. Tacker, P. Tarazona, E. D. Weinberger, and P. Schuster. RNA folding and combinatory landscapes. *Phys. Rev. E*, 47:2083–2099, 1993.
45. P. Schuster, W. Fontana, P. F. Stadler, and I. L. Hofacker. From sequences to shapes and back: A case study in RNA secondary structures. *Proc. Roy. Soc. London B*, 255:279–284, 1994.
46. W. Fontana and P. Schuster. Shaping space: The possible and the attainable in RNA genotype-phenotype mapping. *J. Theor. Biol.*, 194:491–515, 1998.
47. W. Fontana and P. Schuster. Continuity in evolution: On the nature of transitions. *Science*, 280:1451–1455, 1998.
48. M. A. Huynen, P. F.Stadler, and W. Fontana. Smoothness within ruggedness: The role of neutrality in adaptation. *Proc. Natl. Acad. Sci. USA*, 93:397–401, 1996.
49. E. van Nimwegen, J. P. Crutchfield, and M. Mitchell. Finite populations induce metastability in evolutionary search. *Phys. Lett. A*, 229:144–150, 1997.
50. R. T. Batey, R. P. Rambo, and J. A. Doudna. Tertiary motifs in structure and folding of RNA. *Angew. Chem. Int. Ed.*, 38:2326–2343, 1999.
51. P. G. Higgs. RNA secondary structures –Physical and computational aspects. *Quart. Rev. Biophys.*, 33:199–253, 2000.
52. M. Zuker and P. Stiegler. Optimal computer folding of larger RNA sequences using thermodynamics and auxiliary information. *Nucleic Acids Research*, 9:133–148, 1981.
53. I. L. Hofacker, W. Fontana, P. F. Stadler, L. S. Bonhoeffer, M. Tacker, and P. Schuster. Fast folding and comparison of RNA secondary structures. *Mh. Chem.*, 125:167–188, 1994.
54. R. Nussinov and A. B. Jacobson. Fast algorithm for predicting the secondary structure of single-stranded RNA. *Proc. Natl. Acad. Sci.*, 77:6309–6313, 1980.
55. M. S. Waterman and T. H. Byers. A dynamic programming algorithm to find all solutions in a neighborhood of the optimum. *Math. Biosc.*, 77:179–188, 1985.
56. P. Schuster and P. F. Stadler. Discrete models of biopolymers. In M. J. C. Crabbe, M. Drew, and A. Konopka, editors, *Handbook of Computational Chemistry.* Marcel Dekker, New York, 2000. In press.
57. P. Schuster. Molecular insight into the evolution of phenotypes. In J. P. Crutchfield and P. Schuster, editors, *Evolutionary Dynamics – Exploring the Interplay of Accident, Selection, Neutrality, and Function.* Oxford University Press, New York, 2001. In press.

58. J. Rogers and G. F. Joyce. A ribozyme that lacks cytidine. *Nature*, 402:323–325, 1999.
59. I. L. Hofacker, P. Schuster, and P. F. Stadler. Combinatorics of RNA secondary structures. *Discr. Appl. Math.*, 89:177–207, 1999.
60. E. A. Schultes and D. P. Bartel. One sequence, two ribozymes: Implications for the emergence of new ribozyme folds. *Science*, 289:448–452, 2000.
61. E. H. Ekland, J. W. Szostak, and D. P. Bartel. Structurally complex and highly active RNA ligases derived from random RNA sequences. *Science*, 269:364–370, 1995.
62. A. T. Perrotta and M. D. Been. A toggle duplex in hepatitis delta virus self-cleaving RNA that stabilizes an inactive and a salt-dependent pro-active ribozyme conformation. *J. Mol. Biol.*, 279:361–373, 1998.
63. W. Fontana and P. Schuster. A computer model of evolutionary optimization. *Biophys. Chem.*, 26:123–147, 1987.
64. W. Fontana, W. Schnabl, and P. Schuster. Physical aspects of evolutionary optimization and adaptation. *Phys. Rev. A*, 40:3301–3321, 1989.
65. D. T. Gillespie. A general method for numerically simulating the stochastic time evolution of coupled chemical reactions. *J. Comp. Phys.*, 22:403–434, 1976.
66. D. T. Gillespie. Exact stochastic simulation of coupled chemical reactions. *J. Phys. Chem.*, 81:2340–2361, 1977.
67. C. M. Reidys, C. V. Forst, and P. Schuster. Replication and mutation on neutral networks. *J. Math. Biol.*, 63:57–94, 2001.
68. P. Schuster and A. Wernitznig. Stochastic dynamics of neutral evolution: RNA in the flow reactor, 2001. Preprint.
69. B. Derrida and L. Peliti. Evolution in a flat fitness landscape. *Bull. Math. Biol.*, 53:355–382, 1991.
70. G. Zipf. *Human Behaviour and the Principle of Least Effort.* Addison-Wesley, Reading, MA, 1949.
71. J. Cupal, P. Schuster, and P. F. Stadler. Topology in phenotype space. In *Computer Science in Biology*, GCB'99 Proceedings, pages 9–15. Univ. Bielefeld, Hanover, DE, 1999.
72. J. Cupal, S. Kopp, and P. F. Stadler. RNA shape space topology. *Artificial Life*, 6:3–23, 2000.
73. B. M. Stadler, P. F. Stadler, G. P. Wagner, and W. Fontana. The topology of the possible: Formal spaces underlying patterns of evolutionary change. *J. Theor. Biol.*, 2000. Submitted.
74. N. S. Goel and N. Richter-Dyn. *Stochastic Models in Biology.* Academic Press, New York, 1974.
75. S. F. Elena, V. S. Cooper, and R. E. Lenski. Punctuated evolution caused by selection of rare beneficial mutants. *Science*, 272:1802–1804, 1996.
76. H. Li, R. Helling, C. Tang, and N. Wingreen. Emergence of preferred structures in a simple model of protein folding. *Science*, 273:666–669, 1996.
77. S. Govindarajan and R. A. Goldstein. Why are some protein structures so common? *Proc. Natl. Acad. Sci., USA*, 93:3341–3345, 1996.
78. A. Babajide, I. L. Hofacker, M. J. Sippl, and P. F. Stadler. Neutral networks in protein space: A computational study based on knowledge-based potentials of mean force. *Folding and Design*, 2:261–269, 1997.
79. S. Govindarajan and R. A. Goldstein. The foldability landscape of model proteins. *Biopolymers*, 42:427–438, 1997.
80. S. Govindarajan and R. A. Goldstein. Evolution of model proteins on a foldability landscape. *Proteins*, 29:461–466, 1997.

81. S. Gavrilets. Evolution and speciation on holey adaptive landscapes. *Trends in Ecology and Evolution*, 12:307–312, 1997.
82. S. Gavrilets. Evolution and speciation in a hyperspace: The roles of neutrality, selection, mutation, and random drift. In J. P. Crutchfield and P. Schuster, editors, *Evolutionary Dynamics – Exploring the Interplay of Accident, Selection, Neutrality, and Function.* Oxford University Press, New York, 2001. In press.
83. L. W. Ancel and W. Fontana. Plasticity, evolvability, and modularity in RNA. *J. Exp. Zool. (Mol. Dev. Evol.)*, 288:242–283, 2000.

Evolutionary perspectives on protein structure, stability, and functionality

Richard A. Goldstein

Biophysics Research Division and Department of Chemistry, University of Michigan, Ann Arbor, MI 48109-1055

Abstract. Proteins are the result of a long process of evolution. It is due to this process that they have developed properties rather different from those of random strings of amino acids. If we wish to understand the properties of proteins, we need to understand the underlying process of Darwinian evolution, and how its stochastic nature interacts with the underlying fitness landscape. In this review, I describe some of the underlying theory of evolution. I then discuss how these theories can help us understand the structure, thermodynamics, and functioning of naturally-occurring proteins.

Introduction

A number of different perspectives have recently changed the way that we understand evolution and its effect on biological macromolecules. The first perspective is an increased appreciation for the relationship between evolution and observed molecular properties. This has resulted from a greater availability of sequence data encouraging evolutionary studies at the molecular level, combined with increased sensitivity to the evolutionary heritage encoded in these molecules.

A second perspective was the understanding of the role of chance in evolution. Specifically, with the rise of the neutral theory we can see how adaptation is only one aspect of evolutionary change and that this change, especially at the molecular level, may represent random movement among approximately equally-fit genotypes. Biological systems are characterized by a many-to-few mapping from genotype to phenotype, with selection occurring primarily at the phenotype level. This results in large "neutral networks" of equivalent genotypes. Although mutations between these sequences may not cause appreciable changes in fitness, they can have a dramatic impact on the large-scale evolutionary process including at the phenotype level. This perspective also highlighted the need to understand the genotype-to-phenotype mapping and the nature of the fitness landscape in order to model the evolutionary process.

A third perspective was the introduction of the theory of reaction dynamics into evolutionary biology. Eigen's incorporation of the chemical-engineering theory of flow reactors allowed quantitative analysis of the behavior of populations of self-replicating molecules. In addition to allowing us to approach evolution in terms of creation of information, and demonstrating how life could have evolved from initial concentrations of RNA, these theories showed that we cannot think of a species as a homogeneous set of identical genotypes but rather as a collection of similar but different genotypes. The intrinsic variability of the genotype in the population can have a dramatic effect on the evolutionary dynamics.

In this paper, we will first briefly review these new perspectives, first discussing classical gene dynamics in infinite systems, then the neutral theory of evolution, and finally some of the consequences of Eigen's theories as developed by him and others. We will then discuss how these perspectives can help to explain some of the observed properties of proteins including the distribution of observed structures and their thermal properties. While a number of investigators have looked at this phenomena, I will necessarily concentrate on theoretical modeling performed by my group over the past few years.

Principles of molecular evolution

Classical theory of gene dynamics. In order to understand how the process of evolutionary change can influence the observed properties of proteins, we first need to consider the process of evolutionary change itself. The dominant philosophy is of course Darwin's theory of random variation followed by natural selection.

First a few definitions. (This material is covered in depth in a number of books[1–3].) A **locus** refers to a location of a given gene in the genome. Different variations of the genes are called **alleles**. Let us consider that there is a wild-type allele, that is, one version of the gene that describes every member of the population. A different version of the gene, a different allele, can result from random variation due to errors in conservation or replication. There are three possible fates of this gene. The individual or the descendents of the individual can fail to reproduce in which case this different allele is removed from the population. Alternatively, this mutant allele can spread throughout the population so that it now becomes the new "wild-type". This process is called **fixation**. Or these gene can achieve a certain representation in the population, neither dying out nor replacing the wild type. We then say that this locus is **polymorphic**. Let us see how these alternatives are possible.

Consider a diploid organism, that is, with two copies of each gene, one of which is transferred to the next generation. We will call the wild-type allele A and the mutant B. There are then three diploid genotypes possible, AA, AB, and BB. An organism is called a **homozygote** if both alleles are the same (AA or BB in this example); otherwise it is a **heterozygote**. If the fraction of the gene A in the population is p and the fraction of gene B in the population is q (so that $q = 1-p$), then at equilibrium where these ratios are maintained and where mating is random the population will have genotype AA with frequency p^2, genotype AB with frequency $2pq$, and genotype BB with frequency q^2– the so-called **Hardy-Weinberg equilibrium**. Often these ratios will not be maintained and the Hardy-Weinberg equilibrium will not hold. This is the case when these different genotypes correspond to different genotypes with different fitnesses, where the fitness measures the relative probability of contributing to the next generation. Let us consider the relative fitnesses of these three genotypes as ω_{AA}, ω_{AB}, and ω_{BB}, respectively. Often we consider fitnesses relative to the wild type AA, so $\omega_{AA} = 1$, $\omega_{AB} = 1 + s_{AB}$, and $\omega_{BB} = 1 + s_{BB}$. For an infinite population and

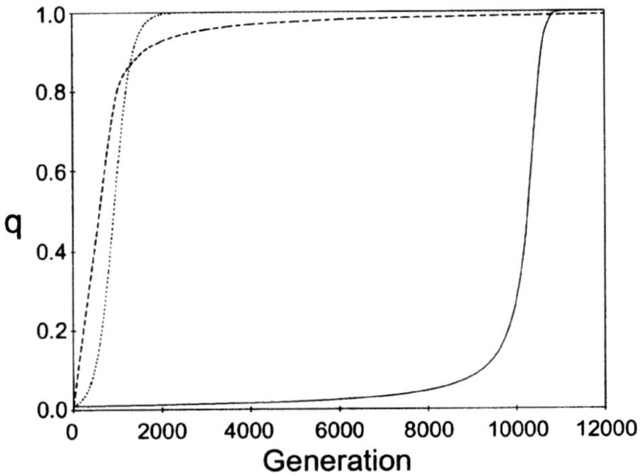

Fig. 1. Gradual fixation of a advantageous mutation starting with an initial frequency of 0.01. The relative fitnesses of the various genotypes are $s_{BB} = 0.01$ and either $s_{AB} = 0.00$ (solid line, corresponding to a recessive mutation), $s_{AB} = 0.005$ (dotted line), or $s_{AB} = 0.01$ (dashed line, corresponding to a dominant mutation). The recessive mutation requires longer for fixation due to the slow buildup of BB homozygotes with a competitive advantage. Conversely, it is difficult to achieve total fixation of a dominant advantageous mutation because of the relative fitness of the heterozygotes.

current values of p and q we can calculate how much of each allele will be present in the next generation. The fraction of allele B in the next generation, $\acute{q}$ is given by

$$\acute{q} = q + \frac{pq\left[p\left(\omega_{AB} - \omega_{AA}\right) + q\left(\omega_{BB} - \omega_{AB}\right)\right]}{p^2\omega_{AA} + 2pq\omega_{AB} + q^2\omega_{BB}} \qquad (1)$$

Figures 1 and 2 show two different examples. In figure 1, we consider the case where the new mutant gene is advantageous with $s_{BB} = 0.01$. The new allele B is fixed in the population with probability 1 with dynamics that depend upon the heterozygote fitness s_{AB}. In figure 2 we have what is called **overdominant selection** where the heterozygote AB has the highest fitness: in this case $s_{AB} = 0.02$ and $s_{BB} = 0.01$. (The classical example of overdominance is the mutation for sickle-cell anemia, where the mutant homozygote (BB) is lethal, yet the heterozygote (AB) has increased resistance to malaria.) The result is relaxation to a constant steady-state population of some B for any initial population.

Finite populations and the neutral theory. The eventual fate of a mutant gene is deterministic if the population is infinite: there are differential equations that can be solved to give the behavior described above. In reality, populations are quantized and finite. This can have a strong impact on the fate of various alleles. For instance, imagine that neither allele has a selective advantage so that the fitnesses of all three possible genotypes are equal. In this case, the

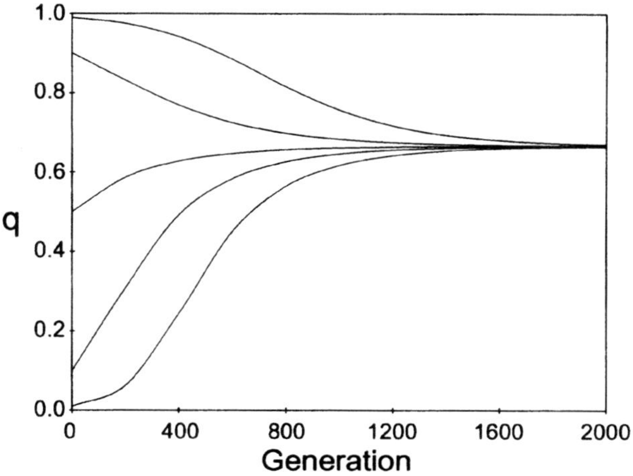

Fig. 2. Population dynamics with overdominant selection where the heterozygote has increased fitness relative to either homozygote, for a range of initial gene frequencies. The relative fitnesses of the various genotypes are $s_{AB} = 0.02$ and $s_{BB} = 0.01$.

population dynamics of infinite populations would result in the proportion of the two alleles remaining constant in a Hardy-Weinberg equilibrium. In a finite population eventually either the mutant allele would be eliminated or achieve fixation, at least if additional copies of the mutant allele are not produced by further mutations. This is because random fluctuations in the allele fractions would occur; with a finite probability that any allele frequency would decrease to zero from where it cannot recover. The discreteness of the population is key to this process, as there have to be an integral number of copies of each allele. Consider a single copy of a new allele in a population of size N (so that $q = 1/2N$, with the $2N$ coming from the fact that the individuals are diploid). In the simplest case the heterozygote has the average fitness of the two homozygotes so we can write $\omega_{AA} = 1$, $\omega_{AB} = 1 + s$, and $\omega_{BB} = 1 + 2s$. Kimura derived the probability of eventual fixation of B [4]

$$P_{\text{fixation}} = \frac{1 - e^{-(2N_e s)/N}}{1 - e^{-4N_e s}} \tag{2}$$

where N_e is the effective population size, that is, the population that is actually reproducing at any one time. (For human populations, $N_e \sim N/3$ [5]). Ignoring the difference between N and N_e results in curves of P_{fixation} as a function of s for different population sizes as shown in figure 3. As would be expected, for a neutral mutation ($s = 0$) the probability of eventual fixation represents the initial fraction of the population, $P_{\text{fixation}} = 1/2N$. In fact, all mutations with values of $|s| < 1/2N$ have approximately probability $1/2N$ of fixation; these mutations are essentially neutral. For larger values of s, the probability of fixation becomes $P_{\text{fixation}} = 2s$. For finite populations there is a chance of a negative mutation

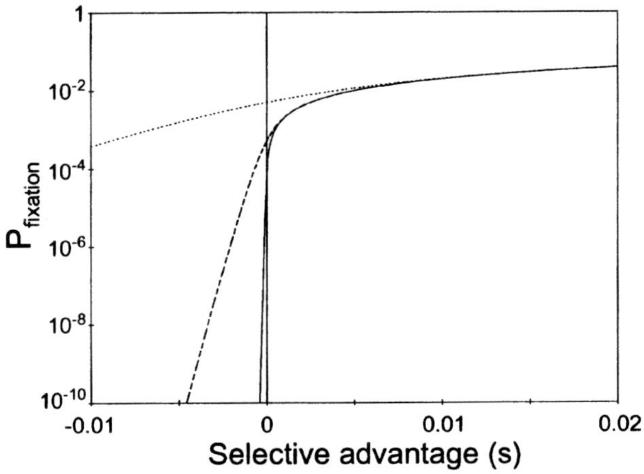

Fig. 3. Probability of fixation of a mutation with initial frequency $1/2N$ as a function of s for various population sizes: $N = 100$ (dotted line), $N = 1000$ (dashed line), and $N = 10000$ (solid line). The difference between population size and effective population size is ignored.

becoming fixed in the population, just as there is a chance of a positive mutation being removed.

In classical evolutionary theory, the process of evolutionary change involved chance advantageous mutations that became fixed, what is called **adaptive** evolution. Kimura and Jukes and King proposed the **neutral theory**, which postulated that the vast majority possible mutations are either deleterious or neutral ($|s| < 1/2N$) [4,6]. As the deleterious mutations will generally be removed from the population by purifying selection, most observed substitutions would be neutral or slightly deleterious. This was used to explain four observations. One observation had to do with the large variation in genotypes in a typical population. It was observed that many genes were **polymorphic**, that is, multiple alleles exist in the population. In the classical theory, polymorphism could result from overdominant selection (as shown in figure 2) or from frequency-dependent selection – where there was an advantage to being different from others in the population. In these cases there was selective pressure towards polymorphism. According to the neutral theory, polymorphism was generally a temporary result of a nearly-neutral mutation that had not yet been either eliminated nor fixed. Fixation and elimination times for neutral mutations are quite long, about $4N$ generations. Under these conditions it would be natural to have a large amount of polymorphism in the population. Kimura claimed that the amount of polymorphism in observed populations was too large to be explained by positive adaptation.

A second observation was that any particular gene often tends to evolve at a roughly constant rate in different organisms, that there is a **molecular clock**. (This is not to say that different genes evolve at the same rate – there are quite

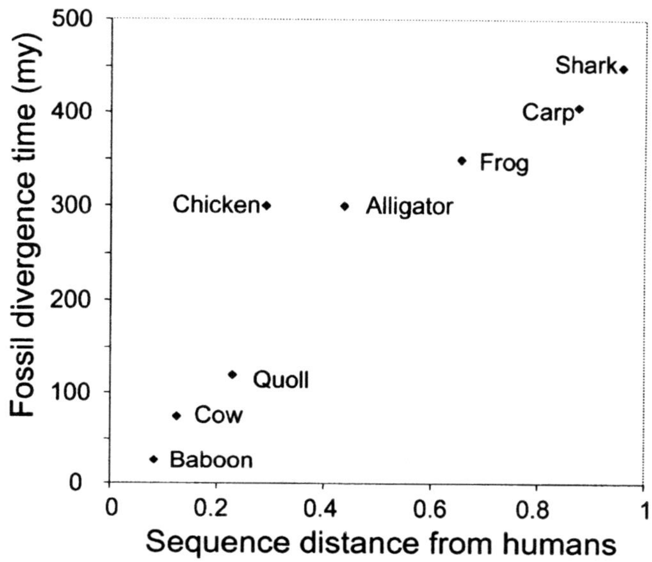

Fig. 4. Evidence for a molecular clock. Plot shows the relative time of divergence from man according to the fossil record compared with the sequence divergence. Adapted from [2].

large variations in the rate of substitution of different genes.) Figure 4, for instance, shows the relationship between the time of divergence of various species from humans according to the fossil record compared with the dissimilarity in the sequences of alpha-Hemoglobin; the near straight-line behavior across such different organisms (with the exception of chicken) is striking. This molecular clock is a natural result of the neutral theory. Neutral mutations in a population should arise at a rate proportional to the number of genes in the population, $2\mu N$. The probability of fixation is $1/2N$. Multiplying these two factors together, the total rate of introduction and fixation of neutral mutations should be μ, independent of the population size. This would explain the constant rate of genetic evolutionary change. Conversely, for adaptive change the fixation probability is $2s$, resulting in a total rate of introduction and fixation of $4\mu Ns$, proportional to the population size. This would be incompatible with the molecular clock hypothesis.

A third observation was that important genes evolve slower than less-important genes, which evolve slower than regions of the genome that do not seem to serve any purpose. If most mutations were either neutral or disadvantageous, the importance of the gene will correlate with the likelihood that changing the gene will be deleterious. As a result, mutations in less-significant regions of the genome will have a smaller probability of being removed by purifying selection, and thus more chance at fixation.

The last observation has come with the rise of genetic manipulation. Many substitutions at the DNA level do not change the amino acids of expressed proteins – these are called **synonymous substitutions**. It is likely that the vast majority of these substitutions are essentially neutral. Additionally, is now clear how plastic the resulting amino acid sequences are, that it is not difficult to find many amino acid substitutions that results in proteins with seemingly identical properties [7]. Many of the changes that we can make in the lab seem to be neutral, at least within the accuracy and the context of the experiments.

The neutral theory does not downplay the role of adaptive change. Obviously the characteristics of living creatures show that we are highly adapted to our surroundings. The argument is that adaptive changes, though important, are extremely rare and that most substitutions are either neutral or deleterious. As will be discussed below, one reason for the presence of neutral change at the genetic level is the many-to-one mapping of genotype to phenotype.

One additional concept introduced by Gould is the idea of the **spandrel** [8]. According to Gould, certain features of an organism might arise from reasons having nothing to do with selection, either through random neutral drift or as an unavoidable consequence of some other modification. The organism might then be able to adapt this feature for adaptive purpose. Accordingly, it is dangerous to explain the existence of this feature as arising as a positive adaptation towards its eventual purpose, what he called the "Panglossian paradigm" after the character in Voltaire's *Candide* who sees everything as optimal in this "best of all possible worlds".

Is the neutral hypothesis correct? This is still a topic of much debate. The degree of polymorphism seems to be somewhat between the rate predicted by adaptionists and neutralists. While there seems to be some degree of constancy to the rate of evolution of each gene across different evolutionary lines, the molecular clock seems to run somewhat erratically. Neutralists claim that these irregularities can be explained by accounting for different rates of mutation, different generational times, and some adaptive bursts. Adaptionists can also explain why important regions of the genome evolve slower than less-important regions: genetic changes in the less-important regions of the genome result in smaller changes in fitness, and smaller changes in fitness are more to be advantageous than a larger changes [9]. While it is true that it is not difficult to make genetic changes that have no observable effect on the phenotype or on the organisms chance for survival and reproduction, effects too small to be seen in the lab may still have a large impact on the evolutionary process. Even synonymous substitutions in the DNA code that have no effect on the amino acid sequence of the expressed proteins can affect the probability of survival by affecting the rate of protein transcription.

Eigen's theory of quasi-species. Kimura's neutral theory includes aspects of random change due to the finite size of populations, resulting in stochastic effects that are not included in the infinite-population differential equations. But this model still involves a homogeneous population in which mutations occurs. The

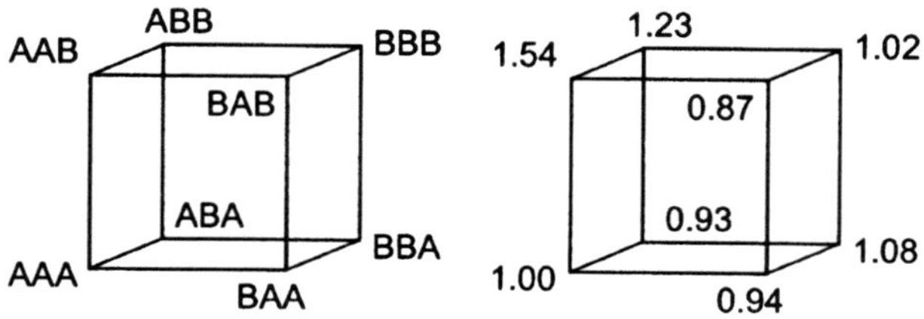

Fig. 5. An example of a fitness landscape of a trimer written in a two-letter alphabet. The left of the diagram shows the available sequences and their connectivities, while the right of the diagram shows the corresponding fitnesses.

polymorphism is a result of the evolutionary dynamics, rather than a critical component. These aspects were to change with the introduction of the theory of quasi-species.

It is easiest to consider the situation by considering a fitness landscape, a term first introduced by Wright [10]. The fitness landscape consists of a sequence space, that is, the space of all possible sequences, combined with a fitness value for each point. Each dimension in the fitness landscape corresponds to one allele or base or amino acid. As a result, the sequence space (as well as the fitness landscape) is extremely high-dimensional. It is a strange space, however, in that only relatively few discrete values in each dimension are allowed. For proteins, for example, each dimension consists of twenty discrete points representing the twenty amino acids. A typical simple fitness landscape for a trimer in a two-letter code (A and B) is shown in figure 5. Again, this diagram does not do justice to the high dimensionality of the space. Another useful but misleading representation is where the discrete nature of the sequence alphabet is ignored, resulting in diagrams such as figure 6. The advantage of this representation is that it provides an intuitive idea of fitness peaks, valleys, and ridges.

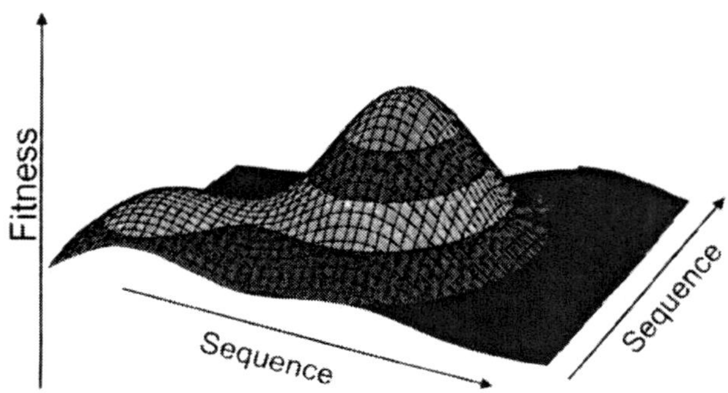

Fig. 6. Fitness landscape with sequences represented as continuous variables.

Eigen modeled evolution by considering a flow reactor containing self-replicating molecules of RNA, with an influx of mononucleotides and an outflow to keep the concentration constant [11–14]. As with biological RNA, the replication rate is not perfect, but mutations naturally arise. The dynamics can be treated with standard approaches from chemical engineering. The result indicated that rather than having a single "wild-type" genotype, instead there would be a cloud of different genotypes centered in the sequence space around a prototypical sequence. The cloud could represent the ultimate steady-state solution – not a transient phenomenon eventually resolved by natural selection. It was then possible to talk about the evolutionary process in terms of changes and competition between these various clouds, which took on the role of species in classical evolutionary theory and thus were named **quasi-species** [13,14]. Eigen's results demonstrated that in order to consider the fitness of any particular quasi-species, it was necessary to consider the fitness of the prototypical sequence as well as the surrounding sequences. A broader, flatter, but lower fitness peak could outcompete a sharper, narrower albeit higher fitness peak, depending upon the overall mutation rate. Finally, there could be qualitative changes in the evolutionary process brought on by quantitative changes in the mutation rate. Specifically, there was a certain critical mutation rate above which the evolutionary dynamics became random and incoherent, with a loss of the genetic information [14].

Again things change when we consider finite populations, due to the resulting stochastic nature of the evolution as well as the discrete nature of the individuals. On a flat fitness landscape an infinite population would expand to fill the available space. In reality, the cloud of members retains its cohesiveness. The edge of the cloud is characterized by a dilute population of members. Such members are highly unstable with respect to evolution, in that any fluctuations in population that take their number down to zero results in an extinction of this subpopulation from which it cannot recover. The center of the cloud is more resistant to these fluctuations. The result is that there is a tendency to eliminate the outlying members of the population cloud, so that the cloud remains centered on the prototypical sequence. The resulting cloud can then wander in a stochastic manner in the fitness landscape [15,16].

One of the more important results of this approach towards evolution is the dependence of the evolutionary process on the fitness landscape. Certain characteristics of the landscape have been especially emphasized by Schuster, Fontana, and their co-workers. (For a good review, see [17].) They investigated the fitness landscape for RNA molecules, taking advantage of rapid algorithms for computing the ground state conformation [18,19]. They found that these molecules had a large degree of neutrality, that is, it was possible to make large changes in the sequence while retaining the same structure. Considering the sequence as genotype and the structure as phenotype, there would be many changes in genotype consistent with a single phenotype. The many-to-few sequence to structure mapping results in large "neutral networks", that is, regions in the sequence space with constant fitness.

The dynamics of the population cloud combined with the large neutral networks can have a large impact on the evolutionary dynamics. The sequence cloud is free to randomly sample the neutral network. The individuals on the tail of the cloud allow the population to sample the fitness landscape at some distance from the prototypical sequence in a large number of different directions. If the tail of this distribution overlaps a region of higher fitness, the whole cloud can adapt in this direction with a small change in sequence.. Note that the resulting dynamics (long periods of neutral evolution, punctuated by rare but rapid adaptive change) is exactly what is described by Kimura's neutral theory.

In this model, the properties of the intersection points between various neutral networks become critical. One important property of the fitness landscape is how close the various neutral networks were to each other. For RNA, it is possible to go from one native structure to almost any reasonably common different structure with only a small change of the sequence, a phenomenon known as **shape-space covering** [17]. As a result RNA can evolve quickly for different structures and possibly different functions. This phenomenon seems not to be true of theoretical models of proteins, in that it is more difficult to go from one structure to another [20–22]. As a result, evolutionarily-related proteins tend to have the same structure, something called "structural inertia". This may be due to the larger number of amino acids compared with the number of RNA bases, or the relatively small number of sequences that will form a viable protein in any structure.

Applications to Proteins

To summarize the previous section:

- We must include the role of stochastic effects resulting from finite population sizes.
- We should be aware of the presence and effect of neutrality in the fitness landscape. One reason for this neutrality is the many-to-few mapping of genotype to phenotype.
- The properties of the fitness landscape such as the size, distribution, and connectiveness of the neutral networks can have a major effect on evolutionary dynamics.
- Techniques drawn from statistical physics are useful in understanding evolution; it is a natural approach to dealing with the general properties of a large number of individuals behaving stochastically.
- Selective pressure does not necessarily result in the organism selecting the highest peak on the fitness landscape. It is not correct to equate evolution with optimization, or even adaptation.
- Even if a feature fulfills some important function, we cannot conclude that feature evolved in an adaptive way to fulfill this function. These features may represent spandrels.

Let us try to apply some of these lessons to understanding protein properties by considering how the sequence translates to phenotype and eventually to fitness

Maximum foldability and the distribution of protein structures. Proteins have three major evolutionary constraints: they must fold to a structure in a reasonable time, the structure they fold to must perform a function, and the folded structure must be stable enough to perform that function reliably while resisting side-reactions such as aggregation and proteolysis. All three of these requirements are complicated. For instance, much experimental and theoretical effort has gone into determining how a protein is able to fold given the vast number of possible conformations, a non-trivial process [23–25]. While a number of simple models have been developed for understanding the qualitative aspects of protein folding, even less is known about the more general properties of protein functionality. Research into protein stability is hampered by our limited understanding of the interactions that determine the folded state. Underlying all of these problems is the fact that proteins are complicated, involving thousands of atoms interacting with the surrounding solvent.

In order to make the system more tractable, we (as others) have looked at simple lattice models of proteins. For instance, we can consider a protein model consisting of a chain of 27 monomers confined to a $3\times3\times3$ three-dimensional cubic lattice as in figure 7. The bonds are all of unit length with adjacent residues existing at adjacent sites. It is possible to enumerate all of the possible self-avoiding walks, not counting reflections and rotations: there are a total of 103,346 walks for the $3\times3\times3$ cubic lattice.

The energy function is of the simple contact form:

$$E = \sum_{i<j} \gamma_{ij}\, u(r_{ij}) \tag{3}$$

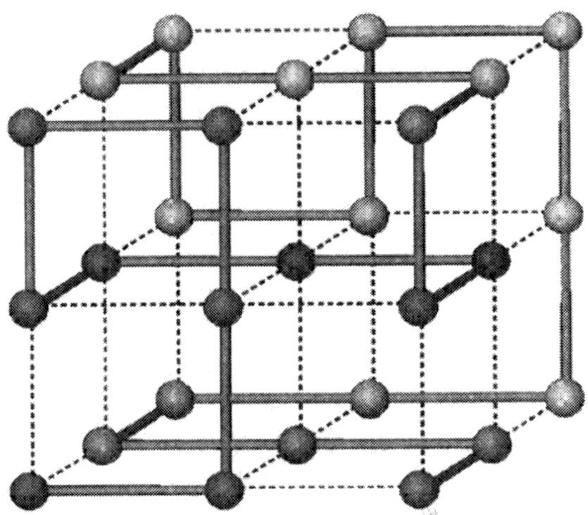

Fig. 7. Lattice model of proteins.

where $u(r_{ij})$ is equal to one if residues i and j are not adjacent in sequence but are on adjacent lattice sites and zero otherwise, and γ_{ij} represents the energy contribution for contact between residues i and j.

As mentioned above, protein structures remain relatively constant in evolutionary time while sequences are surprisingly plastic. This results in an interesting time-scale separation: protein folding takes place in seconds to minutes, too short for much evolution to occur. As a result, on the folding timescale it is the sequence that determines the folded structure. Conversely, on the evolutionary timescale, sequences change but only in such a way that the folded state remains constant. So on the evolutionary timescale, it is the structure that determines (or at least limits) the sequence.

The challenge in folding is to find the native state while avoiding all of the other free-energy minima. Wolynes and co-workers adapted ideas from spin-glass theory to this problem, conjecturing that there were two possible transitions, one to the folded state, the other to a glassy state from which folding was impossible [26–29]. Using the Random Energy Model [30], it is possible to show that folding could be assisted while freezing to the glassy state discouraged by increasing what we called the **foldability** $\mathcal{F}$, that is, the depth of the native state energy minimum compared with the average of the random states relative to the width in the distribution of random state energies [27–29,31]. Shakhnovich and Dill and their respective co-workers used lattice models to demonstrate that a closely-related factor could distinguish between sequences that could and could not fold in numerical simulations [32–38].

As emphasized above, evolution is not equal to optimization, and may even not represent adaptation. Yet for now let us consider how the foldability could be optimized, that is, the limit of infinite selective pressure. In this case it is possible, given a target native-state conformation and an ensemble of random conformations, to solve for the best values of γ_{ij} (that is, the values that maximizes the foldability $\mathcal{F}$) in closed form [27,28,31]. While this solution is exact, it is possible to write down an approximate solution by neglecting correlations between the presence of various contacts. The optimal values of the interactions are given by

$$\gamma_{ij}^{\text{opt}} \approx -\frac{u(r_{ij}^T) - \langle\, u(r_{ij})\rangle}{(1 - \langle\, u(r_{ij})\rangle)\,(\langle\, u(r_{ij})\rangle)} \tag{4}$$

where $u(r_{ij}^T)$ is the value of $u(r_{ij})$ in the native-state conformation (that is, is equal to one if and only if this contact is present in the native state) and the averages are over the random conformations. Note that in the case of contacts present in the native state ($u(r_{ij}^T) = 1$), $\gamma_{ij}^{\text{opt}}$is approximately inversely proportional to the frequency of this contact in non-native states. This results in the prediction that the long-range contacts between very separated parts of the protein chain should be especially strong, and the propensities for local structure should be weak [31,39].

For each structure, we can define a maximum possible foldability, that is, the foldability with the optimal set of interactions. Some structures have very high maximum foldablities; other structures had much lower maximum foldabilities.

In general, the maximum possible foldability is determined by the number of long-range contacts with small values of $\langle u(r_{ij})\rangle$:.

$$\mathcal{F}_{\mathrm{opt}} \approx \sum_{\text{native-state contacts}} \frac{1 - \langle u(r_{ij})\rangle}{\langle u(r_{ij})\rangle} \tag{5}$$

Given the discussion above about the difference between evolution, optimization, and adaptation, why should we care about $\mathcal{F}_{\mathrm{opt}}$? It is because this parameter is an important characteristic of the fitness landscape in the same way that a mountain can be characterized by its height. Let us consider a more neutralist perspective, that the foldability is not optimal but just has to be high enough with foldability $\mathcal{F}$ greater than some critical value $\mathcal{F}_{\mathrm{crit}}$; all sequences with $\mathcal{F} > \mathcal{F}_{\mathrm{crit}}$ would have fitness equal to one while sequences with $\mathcal{F} < \mathcal{F}_{\mathrm{crit}}$ would have fitness equal to zero. In this case, changes in the sequence that maintain a value of $\mathcal{F} > \mathcal{F}_{\mathrm{crit}}$ would be neutral changes. If $\mathcal{F}_{\mathrm{opt}}$ is large, the interactions that determine the folded state ca be far from optimal and still have a foldabililty larger than $\mathcal{F}_{\mathrm{crit}}$. Conversely, you would have to find near-optimal interactions in order to have $\mathcal{F} > \mathcal{F}_{\mathrm{crit}}$ for a structure with a lower $\mathcal{F}_{\mathrm{opt}}$. As a result, there will be many more sequences that will fold into structures with large $\mathcal{F}_{\mathrm{opt}}$ than would fold into structures with lower $\mathcal{F}_{\mathrm{opt}}$; that is, there would be a strong correlation between the optimizability (value of $\mathcal{F}_{\mathrm{opt}}$ for any structure) and the designability (the number of sequences that would successfully fold into that structure) [31,40,41]. This point is made graphically in figure 8. Given the random nature of evolution, we would expect that such highly-designable structures would be more likely to arise through evolutionary dynamics. In addition, we find that there is a strong correlation between the value of $\mathcal{F}_{\mathrm{opt}}$ and the size of the neutral network [21,22]. This would suggest that such highly-optimizable structures would be more resistant to random mutations and more able to evolve to different functions. As $\mathcal{F}_{\mathrm{opt}}$ is correlated with the number of long-range contacts present in the folded state, we would expect proteins that result from evolution to have many such long-range contacts.

Note that this does not imply that structures with many shorter-range contacts fold faster. There is some evidence to the contrary [42] , although this analysis is complicated by the domination of secondary-structure interactions in the calculation, interactions that are omitted from the current theory due to their highly cooperative nature. The theory discussed here would be more appropriate for the assembly of already-formed (or partially-formed) secondary structure elements, where the secondary-structure contacts are already formed.

More recently, designability has been calculated by considering all possible lattice proteins constructed by a two-letter amino acid alphabet [43] . While the results are qualitatively similar, both the overall distribution of designabilities as well as the relative designability of different structures show a strong difference with the results with more complete alphabets [44]. This suggests that it may be dangerous extrapolating from such studies to more realistic protein models.

It is interesting to note that the distribution of designabilities changes as the value of $\mathcal{F}_{\mathrm{crit}}$ is altered [41,44]. For low values of $\mathcal{F}_{\mathrm{crit}}$, the distribution of

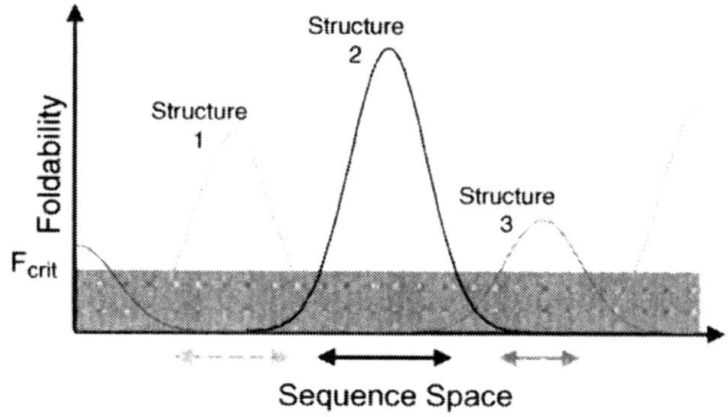

Fig. 8. Representation of the relationship between $\mathcal{F}_{\text{opt}}$ and designability. As the various contact energies are varied away from their optimal values, the foldability of the protein decreases. If $\mathcal{F}_{\text{opt}}$ is larger, there is a larger range of interactions that will have foldabilities larger than $\mathcal{F}_{\text{crit}}$ as shown in the figure – notice the difference in the range of interactions that will fold into structure 2 compared with the other structures. This bias will increase for larger values of $\mathcal{F}_{\text{crit}}$. Given the large dimension of the interaction (as well as sequence) space, small differences in ranges of interaction parameters will result in large differences in the number of sequences folding successfully into one structure or another.

designabilities is rather Gaussian, becoming exponential for larger values of this parameter. Finally, for very large values of $\mathcal{F}_{\text{crit}}$, the distribution of designabilities becomes a highly-stretched exponential of the form

$$\rho\,(\text{designability}) \sim \exp\left(-\alpha \times \ \text{designability}^{\beta}\right),$$

with β less than 1.

Recently, we performed simulations of population evolution [45] . In these simulations, we considered 25 residue proteins confined to a 5×5 two-dimensional maximally-compact square lattice, with each monomer located at one lattice point. This provides us with 1081 possible conformations represented by the 1081 self-avoiding walks on this lattice, neglecting structures related by rotation, reflection, or inversion. We used a simple lattice potential of the form of equation 3 , modified to take into account the possible amino acids at each location

$$E = \sum_{i<j} \gamma(\mathcal{A}_i, \mathcal{A}_j) u(r_{ij}) \tag{6}$$

where $\gamma(\mathcal{A}_i, \mathcal{A}_j)$ is the contact potential between residue type $\mathcal{A}_i$ at position i and residue type $\mathcal{A}_j$ at position j, taken from the statistical study of Miyazawa and Jernigan [46]. Our model of evolution explored how the dynamics changed when a population of proteins was allowed to evolve, similar in spirit to the reactor flow model of Eigen [11]. An initial population of $N = 500$ identical sequences were selected. For all subsequent generations, each residue in every protein sequence was chosen with probability $1/25$ to be mutated to another random residue, so that on average each sequence experienced one mutation per generation. The foldability of the sequences were calculated, and sequences with foldability lower than some critical value $\mathcal{F}_{\text{crit}}$ were considered non-viable and were removed from the population. We then chose 500 members of the next generation randomly (with replacement) from the remaining members of the current generation. The simulations were run with values of $\mathcal{F}_{\text{crit}}$ that still allowed the population to wander over all of the possible structures. After equilibration, we compared the *occupancy* $\mathcal{O}$ (defined as the average fraction of the population folding into a given structure) with the *designability* $\mathcal{V}$ (defined above as the fraction of all viable sequences folding into that structure), finding that the distribution of occupancies is actually more broad than the distribution of designabilities. In our runs with $\mathcal{F}_{\text{crit}}$ =3.5, for example, we found that the occupancy scaled with the designability raised to approximately the 1.45 power. As a result, the highly-designable structures are even more over-represented and the lesser-designable structures less frequent than in cases where population effects are neglected.

Such studies would suggest very few highly-probable structures and many less likely structures [47,31,40,43,20]. This is what is, in fact, observed, with a number of common folds (TIM Barrel, 4-helix bundle, etc.) making up a surprisingly large fraction of all of the observed structures [48–50] . A recent survey of the SCOP database of protein structures shows that this distribution of observed structures can be modeled as a highly-stretched exponential ($\beta = 0.157$) [51]. Conversely, a number of other forms of this distribution (Gaussian, exponential, equilikely) proposed by other investigators [52,49,53,54,50,55,56] can be rejected. The form of this distribution raises difficulties for what is called "structural genomics", the attempt to generate an example of each protein structure, in that we will always be finding new and rarer protein structures. In fact, this study predicts that the total number of existent folds is quite large, on the order of 1500-2000.

Molecular evolution and the thermodynamic hypothesis. Levinthal first pointed out that there are far too many possible protein structures for protein folding to occur through a random search [23]. He concluded that protein folding must occur through a specific folding pathway, and that folding must therefore be under kinetic control. As he stated, "If the final folded state turned out to be the one of lowest configurational energy, it would be a consequence of biological evolution and not of physical chemistry" [57]. Anfinsen concluded, based on his denaturation-renaturation experiments, that proteins in fact fold to the conformation of lowest free energy, a statement that has been termed the *thermodynamic hypothesis* [58]. Much of the work on protein folding has involved a dialog between these two viewpoints, with both experimental [59–64] and theoretical work [65,36,37,26,27,66–72] supporting one view or the other.

These two viewpoints can be reconciled through an evolutionary perspective [73]. For instance, it may be that Levinthal is correct and that biological evolution does result in proteins fulfilling the thermodynamic hypothesis, even if the folding is under kinetic (rather than thermodynamic) control. The basic perspective is that there would be selective pressure on proteins to maintain

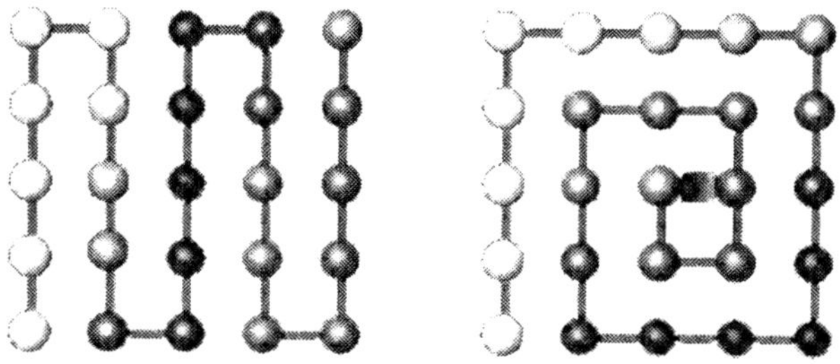

Fig. 9. Protein model used to demonstrate evolution towards fulfillment of the thermodynamic hypothesis. The configuration on the left is the target native state, the conformation to which the protein must fold in order to reproduce. The protein sequence was initialized so that the configuration on the right actually has lower energy. Folding to this latter conformation is prevented by destabilizing the 21-25 bond as shown.

a suitably stabilized native state, at least more stable than other kinetically-accessible states. Other kinetically-inaccessible states might be lower in energy without interfering with the folding process, but there would be no selective advantage in maintaining the stability of these kinetically-irrelevant conformations. As random mutations would be more likely to destabilize irrelevant states than stabilize these states, through evolutionary drift these alternative energy minima would likely disappear leaving the true native state as the global minimum of free energy.

In order to study this phenomena, we performed simulations using simple two-dimensional lattice model proteins, as shown in figure 9 [73].We used a simple lattice potential of the form of equation 6. We first designed a sequence that would repeatedly fold into the structure on the left in figure 9 in lattice simulations, yet had a global energy minimum in the structure on the right, that is, would disobey the thermodynamic hypothesis. Folding into the alternative global energy minimum was prevented by destabilizing the interaction between resides 22 and 25. This demonstrated that the thermodynamic hypothesis was not*required* for protein folding. We then let the sequence evolve, choosing a mutation at random and accepting the mutation if and only if the protein still was able to successfully fold into the native state. Following the ideas described above, the random evolution quickly resulted in a protein that did obey the thermodynamic hypothesis for most of the evolutionary time. This occurred whether or not we allowed residues 22 and 25 to mutate, allowing the original global energy minimum to be kinetically accessible.

For real proteins, as described above, we would expect any conformation would be higher in energy than the native state, even if it is kinetically inaccessible and thus not interfere with the folding process. There may be, however, many more kinetically-inaccessible states than accessible states. We can make statistical models of the number of various states, and by considering typical stabilization energies for biological proteins calculate the probability that any one of the inaccessible states is lower in energy. Such calculations demonstrate that it is incredibly unlikely for an evolutionarily-derived protein to violate the thermodynamic hypothesis if more than a minuscule amount of the conformation space is accessible [73].

This analysis does not consider the situation where there is selective pressure for metastability, as may be the case for serpins [61]. Similarly, this model would obviously not hold for proteins that are not the result of natural evolution [63,64].

Why are proteins marginally stable? Most globular proteins are marginally stable, with a $\Delta G_{\text{folding}}$ of about -10 kcal/mole [74–78] . The general approach to understanding this observation has been to ask how marginal stability fulfills an adaptive role, assuming that the observed property must yield some selective advantage. The most common interpretation of this fact is that marginal stability is correlated with flexibility, and that flexibility is required for functionality [79–81]. As discussed above, it is not necessarily true that observed characteristics represent positive evolutionary adaptation; to assume selective advantage may represent a "Panglossian paradigm" [8]. Does marginal stability represent positive adaptation? Or does it arise through other aspects of the evolutionary process? Is it a spandrel, some property that has arisen for other reasons, that proteins can then adapt to some useful role?

As described above, sequence space is a highly-dimensional space. The majority of possible sequences represent non-viable proteins, with no accessible and stable folded state. We can imagine that there are regions in this space that *do* correspond to viable proteins with a stability sufficient to avoid proteolysis and aggregation. The boundary between this viable region and the surrounding unviable region then represents the boundary between stable and unstable native states. Assuming that the stability varies somewhat smoothly as the sequence is varied, this boundary region must correspond to sequences with *marginal* stability in the ground state. As the dimensionality of this space is quite high, the vast majority of viable protein sequences are in this boundary region. The result of this simple argument is the prediction that the vast majority of stable protein sequences will be marginally stable, at the minimum stability necessary to avoid proteolysis and aggregation.

In a recent paper we investigated these predictions through the use of models of two-dimensional proteins, confined to a 4×4 lattice [82]. We calculated $\Delta G_{\text{folding}} = E_f + kT \ln(Z - \exp(-E_f/kT))$ where E_f is the energy of the folded state (that is, the state of lowest energy [73]) and Z is the partition function, involving the sum over all possible states. As shown in figure 10, the vast majority of protein sequences correspond to unstable proteins with the marginally stable proteins greatly outnumbering the highly stable proteins.

As described above, evolution occurs in populations. We performed simulations with populations of size $N = 3000$. Proteins underwent mutation at a rate of 0.2% of each residues mutating each generation. The protein was considered non-viable if the ground-state structure changed or if a minimum stability ($\Delta G_{\text{folding}} < 0$) was not maintained. As before, the next generation was chosen

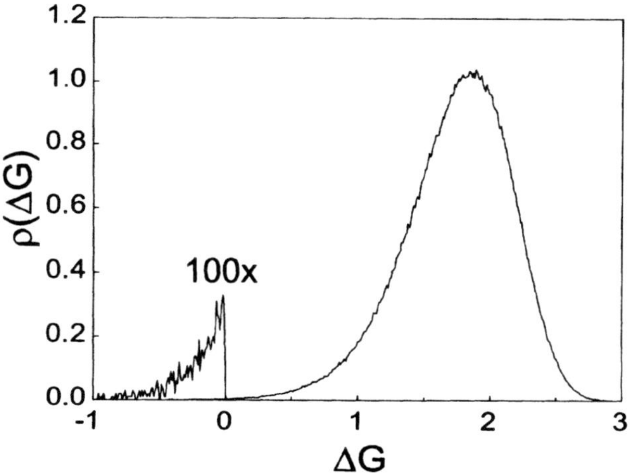

Fig. 10. Distribution of protein stabilities for random sequences.

at random (with replacement) from the surviving members of the current generation. As expected, the population dynamics resulted again in populations of proteins that were marginally stable as shown in figure 11. Again, this was in the absence of any evolutionary pressure favoring marginal stability.

If proteins are naturally marginally stable, as indicated by these simulations, what about including functionality? How would functionality arise in this context? To simulate this, we considered that there might be different ways proteins

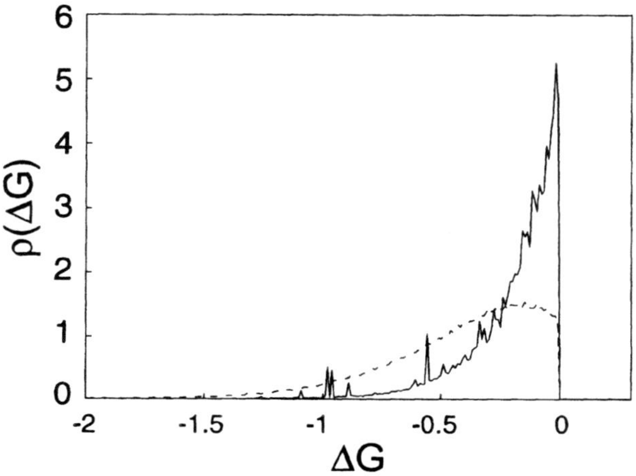

Fig. 11. The distribution of stabilities for random stable ($\Delta G<0$) sequences (solid line) compared with the distribution that results from population evolution (dashed line).

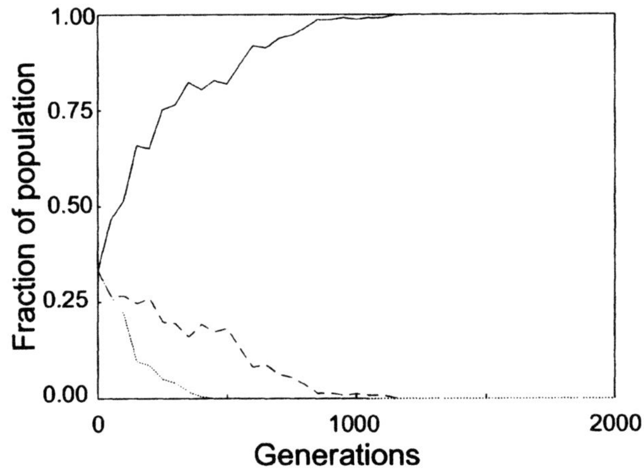

Fig. 12. Typical evolution run, showing how the subpopulation with a mechanism for functionality requiring minimal stability (solid line) has an evolutionary advantage over mechanisms requiring moderate (dashed line) or high stability (dotted line).

could be functional, different mechanisms for protein functionality, corresponding to different requirements for protein stability. Specifically, we modeled three different functional mechanisms, each with both maximum and minimum allowed values of $\Delta G_{\mathrm{folding}}$. We considered three sub-populations (Φ_1, Φ_2, Φ_3) with viability defined as $0 \geq \Delta G_{\mathrm{folding}}(\Phi_1) > -1$, $-1 \geq \Delta G_{\mathrm{folding}}(\Phi_2) > -2$,and $-2 \geq \Delta G_{\mathrm{folding}}(\Phi_3) > -3$, respectively. Each sub-population was created with N =3000 identical sequences, and these three populations were combined into one large population and evolutionary dynamics were simulated. A typical run is shown in figure 12. Of twenty-five separate simulations, in twenty-four the subpopulation Φ_1 with the functionality requiring minimal stability became the sole surviving subpopulation. In only one run did the protein subpopulation requiring moderate stability become dominant, and in every run the subpopulation of proteins with a high stability requirement quickly vanished. Because of the natural propensity for proteins to be marginally stable *independent of functionality,* functional mechanisms compatible with marginal stability are at a great evolutionary advantage.

Marginal stability may represent a "spandrel", a naturally-occurring tendency that biology can use for its own advantage. If so, this tendency has provided biology with a robust system characterized by easy adaptation of new functionalities. In any case, these evolutionary considerations suggest that it is not necessarily true that proteins are marginally stable because it is required for functionality – rather, proteins may require marginal stability to function because they are marginally stable!

Including explicit functionality. We have also neglected the role of other forms of evolutionary pressure such as the need for the folded protein to fulfill its biological function. Being able to fold is crucial but is only one of a long list of requirements a protein must fulfill. The assumption in the work described above is that the requirements of foldability are uncorrelated with these other requirements – two different structures are *a priori* equally likely to be compatible with a given biological activity – and therefore we could neglect functionality when considering protein structures. Is this a valid assumption? How can we bring in functionality into these models? One common aspect of functionality is the ability to recognize and bind some other molecule, including peptides. We developed a model where proteins were evaluated by their ability to bind a specified peptide, as shown in figure 13 [83]. In this example, the peptide was fixed as QIFW and the protein was allowed to evolve. As shown, we considered 16-mer two-dimensional lattice proteins. In order to better represent the thermodynamics, we considered all possible conformations of this protein, including the 69 compact structures that can be fit on a 4×4 square lattice, as well as the 802,016 non-compact conformations. We considered that the protein could be either in an unfolded (non-compact state), in a folded but unbound state, or folded and binding the peptide ligand. We did not consider it possible for an unfolded protein to bind the ligand. There are a total of eight ways of binding the ligand given four sides to the protein and two orientations of the peptide. The energy of all of the various states were calculated using equation 6, with the Miyazawa-Jernigan

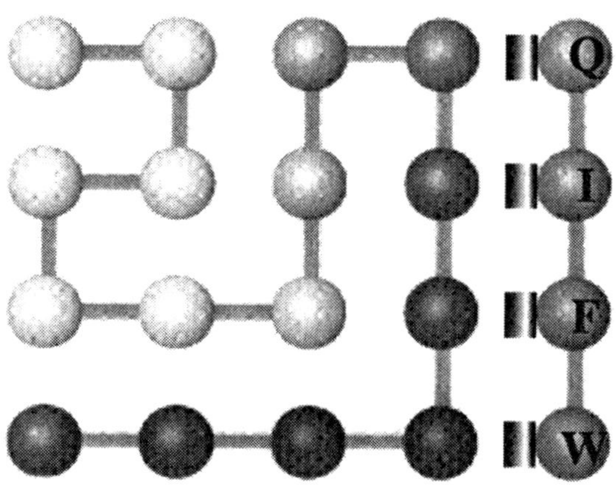

Fig. 13. Model for representing protein functionality, in this case binding of a QIFW tetra-peptide.

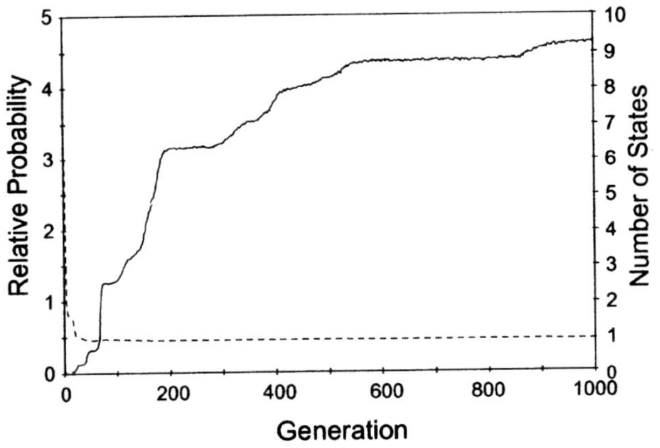

Fig. 14. Evolution of average relative binding probability and effective number of states during a typical simulation, where fitness was represented by the ability to bind the tetra-peptide QIFW.

potentials used for both intraprotein contacts as well as protein-ligand interactions. It is impossible to determine the fraction of proteins that are binding peptide without knowing the relative concentrations of both. We can, however, easily calculate a relative probability of a protein binding a ligand by considering the weak-binding limit. In this limit, the relative probability is given by

$$P(\text{ligand bound}) \propto \frac{\sum_{\substack{\text{compact conformations with}\\ \text{bound peptide}}} \exp(-E/kT)}{\sum_{\substack{\text{all compact and non-compact}\\ \text{conformations}}} \exp(-E/kT)} \tag{7}$$

In previous simulations modeling population evolution, we used neutral evolution with constraints: proteins were either viable or non-viable, and all viable proteins had the same fitness. In this model we considered that the fitness of a protein was proportional to the relative probability of binding the target ligand. A population of 1000 generations was constructed by copying a random sequence. This population was then subjected to random mutations at a rate that averaged 20 total mutations in the population per generation. We then computed the relative probability of binding the target ligand, and chose the next generation randomly with replacement where the probability of choosing any particular individual was proportional to the binding probability. Figure 14 shows how the ability of the protein to bind to the ligand changed during the simulation. We also computed an effective number of compact states by computing the relative probability P_i that a compact state would be in any of the 69 possible compact

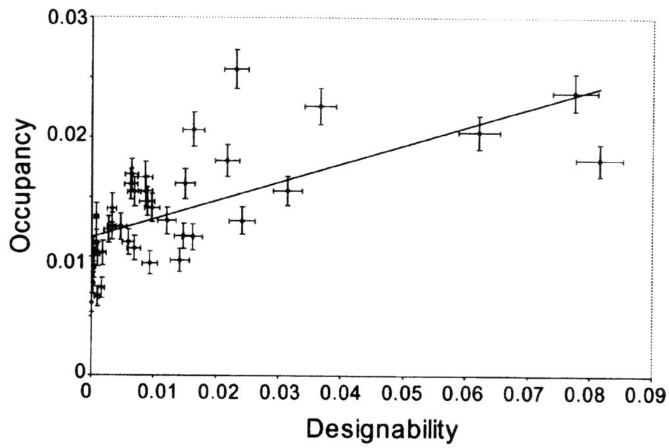

Fig. 15. Correlation between designability $\mathcal{V}$ and occupancy $\mathcal{O}$. The occupancy represents the frequency of sequences forming any given structure at the end of evolutionary simulations where the fitness corresponded to the ability of the protein to bind the tetra-peptide QIFW. The correlation coefficient between these two quantities is 0.66.

states. η, the effective number of compact states, is then given by

$$\eta = \left(\sum_{\text{all compact conformations}} \frac{1}{P_i^2} \right)^{-1} \tag{8}$$

η has the attractive property that it is equal to one if only one state is occupied, and is equal to the total number of states if all are equally occupied. As in the population runs described earlier, we can consider the designability, the fraction of viable sequences that fold into each of the various 69 structures, and compare these values with the occupancy, the fraction of the sequences in the final generation of the evolutionary simulations that folds into that structure. In this case viability for the designability calculations was considered as having a probability of being folded of 50% or more, that is, $\Delta G_{\text{folding}} \leq 0$. This comparison is made in figure 15 for 5000 different simulations of a population of 1000 proteins evolving for 1000 generations to bind QIFW.

There are three observations we can make. The first is that the designability and the occupancy are generally well-correlated. More work is required with different target peptides to better establish this trend. We can also see that there are structures that deviate from the general relationship. Finally, we can tell that the overall distribution of designabilities and the overall distribution of occupancies are quite different. The designabilities follow the trends described earlier, with many rare and a few likely structures. The occupancies are more well balanced, with some proteins resulting from the evolutionary simulation that are effectively never found among random protein sequences. This seems to be at variance with the population simulations described above, which resulted in a more *uneven* distribution of occupancies than designabilities. The difference between these two simulations has to deal with the initial conditions. In the previous study, we allowed the protein population to equilibrate among all of the various possible structures. Sequences that folded into structures that had smaller designabilities would be less resistant to mutations, and would be correspondingly less frequent. In the current simulations, we started with a random set of sequences with low fitness. As the simulation proceeded, the average level of the fitness increased, so that the fitness required for reproducing continually increased. As a result, the volume of the sequence space available to each structure continually decreased. It is interesting to note that, in general, the population very quickly chose a folded native state within tens of generations, as seen in figure 14. This native state was preserved for the remainder of the simulation. As a result, the probability of each of the folded states was decided when the overall fitness required for reproduction was quite low. As described above, low required values of fitness generally results in a more equitable distribution of proteins among the various structures. This indicates that when we try to explain the current observed distribution of structures, we must consider the fitness landscape that was present when the structures were being determined.

Conclusion

Proteins are the result of a long evolutionary process. In order to understand their properties, we must consider the manner in which they arose. There have been many developments in evolutionary theory that can provide interesting and important perspectives. Increasingly, evolutionary theory has focused on random processes, chance events with consequences that, through becoming fixed in a population, increase the "information content" of the species. The randomness of the processes does not mean that we have to give up attempting to understand the underlying theory. Just as we can understand the general patterns of statistical physics, we can comprehend and model the qualitative and quantitative behavior of evolutionary change. And similarly to statistical physics, attention must shift from fitness to "entropy" or other measures of the number of possibilities open to the system. Random processes will preferentially end up in situations corresponding to many of these possibilities, or alternatively, high entropy. That is why there has been recent attention on the designability of different structures as providing an explanation for why certain structures are so over-represented.

We have described in this paper how other phenomena can be explained as well, such as why proteins that fold under kinetic control would still most likely fulfill the thermodynamic hypothesis, why proteins are marginally stable, and why the functionality of proteins may evolve in such a way as to require marginal stability. We have also looked at the effects of explicit evolutionary pressure for simple functionality. It is important to remember during these analyses that non-intuitive aspects may arise, especially as we often have little intuition considering high-dimensional spaces such as the fitness landscape.

Acknowledgments

I would like to acknowledge the co-workers involved in the projects that are described in this paper: Nick Buchler, Sridhar Govindarajan, David Pollock, Ruben Recabarren, Darin Taverna, and Paul Williams. I would also like to thank current and past members of the Goldstein lab for important insights and perspectives, and to Todd Raeker for computational assistance. Financial support was provided by NIH grant LM05770 and NSF equipment grant BIR9512955.

References

1. Li, W. H and Graur, D. *Fundamentals of Molecular Evolution.* (Sinauer, Sunderland) 1991.
2. Volkenstein, M. V. *Physical approaches to biological evolution.* (Springer-Verlag, Berlin) 1994.
3. Page, R. D. M and Holmes, E. C. *Molecular Evolution: A Phylogenetic Approach.* (Blackwell Science, Oxford) 1998.
4. Kimura, M. *Nature (London)* **217** (1968) 624–626.
5. Nei, M and Imaizumi, Y. *Heredity* **21** (1966) 183–190.
6. King, J. L and Jukes, T. H. *Science* **164** (1969) 788–798.
7. Bowie, J. U, Reidhaar-Olson, J. F, Lim, W. A, and Sauer, R. T. *Science* **247** (1990) 1306–1310.
8. Gould, S. J and Lewontin, R. C. *Proceedings of the Royal Society of London, Series B* **205** (1979) 581–598.
9. Fisher, R.A. *J. Animal Ecology* **12** (1943) 54–58.
10. Wright, S. *Int. Proceedings of the Sixth International Congress on Genetics* **1** (1932) 356–366.
11. Eigen, M. *Naturwissenschaften* **58** (1971) 465–523.
12. Eigen, M and Schuster, P. *Naturwissenschaften* **64** (1977) 541–565.
13. Eigen, M and McCaskill, J. *J. Phys. Chem.* **92** (1988) 6881–6891.
14. Eigen, M, McCaskill, J, and Schuster, P. *Adv. Chem. Phys.* **75** (1989) 149–263.
15. Derrida, B and Peliti, L. (1991) *Bull. Math. Biol.* **53**, 355–382.
16. Huynen, M. A, Stadler, P. F, and Fontana, W. *Proc. Nat. Acad. Sci. USA* **93** (1996) 397–401.
17. Schuster, P and FOntana, W. *Physica D* **133** (1999) 427–452.
18. Zuker, M and Stiegler, P. *Nucl. Acids Res.* **9** (1981) 133–148.
19. Zuker, M and Sankoff, D. *Bull. Math. Biol.* **46** (1984) 591–621.
20. Bornberg-Bauer, E. *Biophys J.* **73** (1997) 2393–2403.
21. Govindarajan, S and Goldstein, R. A. *Biopolymers* **42** (1997) 427–438.
22. Govindarajan, S and Goldstein, R. A. *Proteins* **29** (1997) 461–466.
23. Levinthal, C. in *Mossbauer Spectroscopy in Biological Systems,* edited by P. Debrunner, J. C. M. Tsibris, and E. Munck, (University of Illinois Press, Urbana) 1969 pp. 22–24.
24. Unger, R and Moult, J. *Bull. Math. Biol.* **55** (1993) 1183–1198.
25. Hart, W. G and Istrail, S. *J. Comp. Biology* **4** (1997) 1–22.
26. Bryngelson, J. D and Wolynes, P. G. *Biopolymers* **30** (1990) 171–188.
27. Goldstein, R. A, Luthey-Schulten, Z. A, and Wolynes, P. G. *Proc. Nat. Acad. Sci. USA* **89** (1992) 4918–4922.

28. Goldstein, R. A, Luthey-Schulten, Z. A, and Wolynes, P. G. *Proc. Nat. Acad. Sci. USA* **89** (1992) 9029–9033.
29. Goldstein, R. A, Luthey-Schulten, Z. A, and Wolynes, P. G. (1993) in *Proc. 26th Annual Hawaii International Conference on System Sciences*, edited by T. N Mudge, V. Milutinovic, and L. Hunter, (IEEE Computer Society Press, Los Alamitos) Vol. 1, (1993) pp. 699–707.
30. Derrida, B. *Phys. Rev. Lett.* **45** (1980) 79–82.
31. Govindarajan, S and Goldstein, R. A. *Biopolymers* **36** (1995) 43–51.
32. Shakhnovich, E. I and Gutin, A. M. *Prot. Eng.* **6** (1993) 793–800.
33. Shakhnovich, E. I and Gutin, A. M. *Proc. Nat. Acad. Sci. USA* **90** (1993) 7195–7199.
34. Shakhnovich, E. I. *Phys. Rev. Lett.* **72** (1994) 3907–3910.
35. Shakhnovich, E. I. in *Protein Structure by Distance Analysis*, edited by H. Bohr and S. Brunak (IOS Press, Amsterdam) 1994 pp. 201–212.
36. Šali, A, Shakhnovich, E. I, and Karplus, M. J. *J. Mol. Biol.* **235** (1994) 1614–1636.
37. Šali, A, Shakhnovich, E. I, and Karplus, M. J. *Nature (London)* **369** (1994) 248–251.
38. Chan, H. S and Dill, K. A. *J. Chem. Phys.* **100** (1994) 9238–9257.
39. Govindarajan, S and Goldstein, R. A. *Proteins* **22** (1995) 413–418.
40. Govindarajan, S and Goldstein, R. A. *Proc. Nat. Acad. Sci. USA* **93** (1996) 3341–3345.
41. Buchler, N. E. G and Goldstein, R. A. *Journal of Chemical Physics* **112** (2000) 2533–2547.
42. Baker, D. *Nature* **405** (2000) 39–42.
43. Li, H, Helling, R, Tang, C, and Wingreen, N. *Science* **273** (1996) 666–669.
44. Buchler, N. E. G and Goldstein, R. A. *Proteins* **34** (1998).
45. Taverna, D. M and Goldstein, R. A. *Biopolymers* **53** (2000) 1–8.
46. Miyazawa, S and Jernigan, R. L. *Macromol.* **18** (1985) 534–552.
47. Lipman, D. J and Wilbur, W. J. *Proc R Soc Lond [Biol]* **245** (1991) 7–11.
48. Levitt, M and Chothia, C. *Nature (London)* **261** (1976) 552–557.
49. Chothia, C. *Nature (London)* **357** (1992) 543–544.
50. Orengo, C. A, Jones, D. T, and Thornton, J. M. *Nature (London)* **372** (1994) 631–634.
51. Govindarajan, S, Recabarren, R, and Goldstein, R. A. *Proteins* **35**, (1999) 408–414.
52. Wang, Z.-X. *Proteins* **26** (1996) 186–191.
53. Blundell, T and Johnson, M. S. *Protein Science* **2** (1993) 877–883.
54. Alexandrov, N. N and Gō, N. *Protein Science* **3** (1994) 866–875.
55. Zhang, C.-T. *Protein Engineering* **10** (1997) 757–761.
56. Zhang, C.-T. and DeLisi, C. *J. Mol. Biol.* **284** (1998) 1301-1305.
57. Levinthal, C. *J. Chim. Phys* **65** (1968) 44–45.
58. Anfinsen, C. *Science* **181** (1973) 223–230.
59. Kim, P. S and Baldwin, R. L. *Annu. Rev. Biochem.* **59** (1990) 631–660.
60. Dill, K. A. *Biochem.* **29** (1990) 7133–7155.
61. Berkenpas, M. B, Lawrence, D. A, and Ginsburg, D. *EMBO Journal* **14** (1995) 2969–2977.
62. Thomas, P. J, Qu, B, and Pederson, P. L. *TIBS* **20** (1995) 456–459.
63. Mitraki, A, Fane, B, Haase-Pettingell, C, Sturtevant, J, and King, J. *Science* **253** (1991) 54–253.
64. Baker, D, Sohl, J. L, and Agard, D. A. *Science* **356** (1992) 263–265.

65. Honeycutt, J. D and Thirumalai, D. *Proc. Nat. Acad. Sci. USA* **87** (1990) 3526–3529.
66. Gutin, A. M, Abkevich, V. I, and Shakhnovich, E. I. *Proc. Nat. Acad. Sci. USA* **92** (1995) 1282–1286.
67. Leopold, P. E, Montal, M, and Onuchic, J. N. *Proc. Nat. Acad. Sci. USA* **89** (1992) 8721–8725.
68. Bryngelson, J. D, Onuchic, J. N, Socci, N. D, and Wolynes, P. G. *Proteins* **21** (1995) 167–195.
69. Dill, K. A and Chan, H. S. *Nat. Struct. Biol.* **4** (1997) 10–19.
70. Socci, N. D, Onuchic, J. N, and Wolynes, P. G. *J. Chem. Phys.* **104** (1996) 5860–5868.
71. Onuchic, J. N, Luthey-Schulten, Z, and Wolynes, P. G. *Annu. Rev. Phys. Chem.* **48** (1997) 545–600.
72. Plotkin, S. S, Wang, J, and Wolynes, P. G. *J. Chem. Phys.* **106** (1997) 2932–2948.
73. Govindarajan, S and Goldstein, R. A. *Proc. Nat. Acad. Sci. USA* **95** (1998) 5545–5549.
74. Savage, H. J, Elliot, C. J, Freeman, C. M, and Finney, J. L. *Journal of the Chemical Society Faraday Transactions* **89** (1993) 2609–2617.
75. Vogl, T, Jatzke, C, Hinz, H. J, Benz, J, and Huber, R. *Biochemistry* **36** (1997) 1657–1668.
76. Ruvinov, S, Wang, L, Ruan, B, Almog, O, Gilliland, G. L, Eisenstein, E, and Bryan, P. N. *Biochemistry* **36** (1997) 10414–10421.
77. Giver, L, Gershenson, A, Freskgard, P. O, and Arnold, F. H. *Proc. Nat. Acad. Sci. USA* **95** (1998) 12809–12813.
78. Privalov, P. L and Khechinashvili, N. N. *J. Mol. Biol.* **86** (1974) 665–684.
79. Rasmussen, B. F, Stock, A, Ringe, D, and Petsko, G. A. *Nature* **357** (1992) 423–424.
80. Zavodszky, P, Jozsef, K, Svingor, A, and Petsko, G. A. *Proc. Nat. Acad. Sci. USA* **98** (1998) 7406–7411.
81. Tsou, C. L. *Enzyme Engineering XIV* **864** (1998) 1–8.
82. Taverna, D. M and Goldstein, R. A. *Proc. Nat'l Acad. Sci. USA*. submitted.
83. Williams, P. D, Pollock, D. D, and Goldstein, R. A. *Scientific Visualization and Modelling*. in press.

Part II

Phylogeny

The statistical approach to molecular phylogeny: Evidence for a nonhyperthermophilic common ancestor

Nicolas Galtier

UMR 5000 – "Génome, Populations, Interactions", Université Montpellier 2, Place E. Bataillon – CC 63 34095, Montpellier, France

Abstract. The maximum likelihood approach to molecular was proposed as an alternative to the maximum parsimony method after cases of inconsistency of the latter have been described. The maximum likelihood method relies upon statistical modelling of : sites (columns of a set of aligned sequences) are seen as realisations of a random variable whose distribution depends on evolutionary parameters, including the evolutionary tree. The merits and limits of this approach are discussed. The need for appropriate models is stressed, and an example involving a nonstationary process of DNA sequence evolution is presented, allowing discussion about the thermophilic nature of the common ancestor to extant life forms.

1 Introduction

Molecular phylogeny is the field aiming at reconstructing evolutionary trees from the comparison of DNA or protein sequences. Genomic data are widely used for evolutionary purposes. Molecules are especially efficient for the very recent and very ancient time scales, where other data (e.g. morphology) are either not enough variable or irrelevant. The issue of how to reconstruct trees from sequence data has been highly debated over the last 30 years. Three main approaches were proposed: maximum parsimony (MP, e.g. Fitch 1971), distance methods (e.g. Saitou & Nei 1987), and maximum likelihood (ML). This paper aims at discussing the merits of the third approach and stressing the importance of models of sequence evolution in molecular phylogeny.

Building trees more or less involves grouping together sequences similar to each other. Methodological problems arise when some positions (sites) of the analysed molecule used to undergo more than one change Fig. 1. This can result in misleading sites, i.e. sites at which distant sequences show a similar character state by chance. Methods differ in the way they cope with multiple substitutions. The simplest and maybe most natural way to deal with conflicting sites is to assume that misleading sites are less frequent than "good" sites, because they require several convergent changes (i.e. independent changes to the same character state in distinct lineages). This is the principle underlying the MP method.

The MP method, like most methods of molecular phylogeny, involves optimising some criterion over the space of tree topologies. The parsimony criterion is the minimal number of changes required to account for the distribution of the

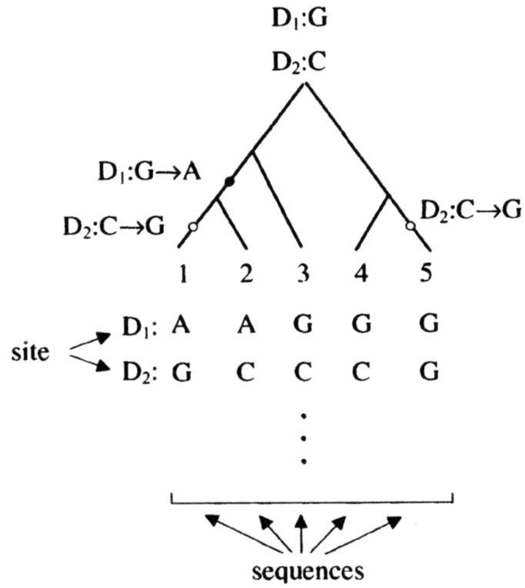

Fig. 1. Evolution of two sites of a set of aligned sequences. Site D_1 correctly supports the (1,2) vs (3,4,5) grouping. Site D_2 wrongly supports (1,5) vs (2,3,4).

variation accross sequences (Fitch 1971). The parsimony score for a tree is calculated by summing the score of every site. The score of a site (for a given tree) is the minimal number of changes required for this site on this tree. In Fig. 1, site D_1 has score 1 and site D_2 score 2 for the tree topology shown. MP seeks the tree that minimises this score. The MP method is quite popular. It can be used on any kind of character and was proposed independently of the emergence of molecular data by the "cladist" school of thought (e.g. see Farris 1982).

2 Maximum parsimony and long branch attraction

A major methodological advance was made when Joe Felsenstein (1978) discovered an important flaw of the MP method, called long branch attraction. Let us illustrate it in the four-species case. With four species three alternative unrooted tree topologies must be considered: (1,2) vs (3,4), (1,3) vs (2,4), and (1,4) vs (2,3) (Fig. 2, top). Only those sites with one character state (say, X) in certain two species and another state (Y) in the other two are informative for MP. Such sites support the tree that groups together the two X species vs the two Y species: a single change is needed for this tree, while alternative topologies require two. Other kinds of sites (e.g. $XXXY$, $XXYZ$) are not informative: they require the same number of changes whatever the assumed tree topology (e.g. one for sites $XXXY$, two for sites $XXYZ$). Recovering the MP tree with four species, therefore, involves numbering three categories of sites, namely (1:X, 2:X, 3:Y, 4:Y), (1:X, 2:Y, 3:X, 4:Y), (1:X, 2:Y, 3:Y, 4:X), henceforth called a-sites, b-sites, and c-sites, respectively, where X and Y are any distinct character states. The MP tree is the tree supported by the most numerous category.

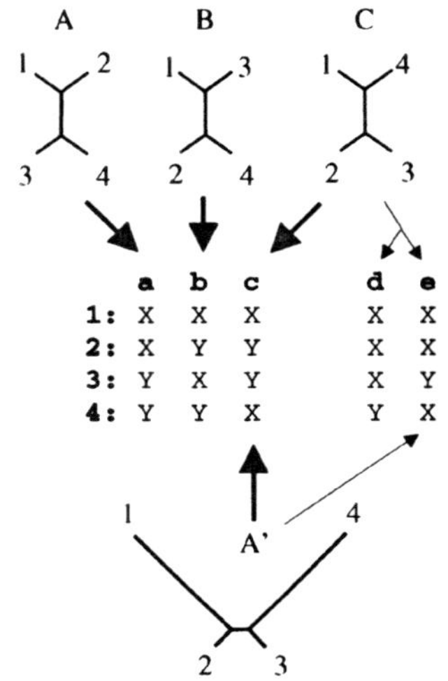

Fig. 2. The long branch attraction problem. Four evolutionary scenarios (tree topology + branch lengths) are considered. *a*-, *b*- and *c*-sites are informative for parsimony. An excess of *a*-sites (respectively *b*-sites, *c*-sites) is expected under scenario A (respectively B, C), consistent with the MP criterion. An excess of *c*-sites, however, is expected under scenario A', confusing the MP criterion. The C and A' scenarios can be distinguished by looking at, say, the number of *d*- and *e*-sites.

Now suppose that sequences evolved according to tree A' of Fig. 2. Here branch lengths represent evolutionary rates, i.e. the expected number of changes for a site evolving from the top to the bottom of the branch. In A', sequences evolve faster in lineages leading to species 1 and 4 than in other. Note that tree A' is an A-tree (1,2 vs 3,4). A good tree-building method should recover an A-topology from the analysis of sequences that evolved according to A'. Let us focus on the behaviour of MP in this case. If the short branches of tree A' are short enough, character states in species 2 and 3 will be identical for nearly all sites: most sites will look like (1:?, 2:Y, 3:Y, 4:?). As far as parsimony-informative sites are concerned, only c-sites (1:X, 2:Y, 3:Y, 4:X) match this motif. It follows that for a high enough ratio of long to short branch lengths in A' MP will support topology C with a high probability, grouping together long vs short branches. This occurs when the probability of two convergent changes in branches leading to 1 and 4 is higher than the probability of one change in the internal branch, a region of the parameter space called the "Felsenstein zone". MP is not consistent (or "positively misleading") in the Felsenstein zone: it will support the wrong tree C with higher and higher probability as the amount of data increases (Felsenstein 1978).

3 The need for models and the maximum likelihood method

Scenarios C and A' both result in an excess of c-sites over a-sites and b-sites. The two scenarios, however, can be easily distinguished if one does not focus only on parsimony informative sites. For example d-sites (1:Y, 2:X, 3:X, 4:X) and e-sites (1:X, 2:Y, 3:X, 4:X) of Fig. 2 have equal expected frequencies if evolution proceeds according to tree C, but very different ones under A'. Examining noninformative sites can yield information about the evolutionary process that used to be followed, allowing a reappraisal of the phylogenetic relevance of informative sites.

This intuitive argument suggests that, as soon as one believes that sequences evolve according to some process more or less common to every site of the analysed molecule, this process should be taken into account for the purpose of building trees: the distribution of the variation accross sequences and sites is determined jointly by the tree and the process. Interestingly, this argument was often made the other way round: it is usually thought that a tree is required to correctly reconstruct character changes (and process). Empirical studies have shown the opposite: as far as molecular data are concerned, inference of the evolutionary process is not dramatically dependent on the assumed topology (as long as "obvious" phylogenetic groupings are kept, e.g. see Yang 1994a), while tree-building heavily relies on the way evolutionary processes are modelled (see below).

Felsenstein (1981) implemented the idea of jointly estimating the tree and process when he designed a maximum likelihood (ML) method for molecular phylogeny. This approach involves (i) assuming that sequences evolve according to some model, (ii) computing the probability of the data as a function of the parameters of the model (topology, branch lengths, parameters of the substitution process) - this function is called the likelihood (Edwards 1972), and (iii) seeking the parameter values that maximise the likelihood.

Time-continuous, four-state Markov models of nucleotide substitution are used. Changes are assumed to occur according to a Poisson process whose rate can vary between branches. The direction of changes (e.g., say, the relative probability of changing to C, G, or T given initial state A) are modelled by a four-by-four matrix including zero to eleven free parameters (Yang 1995). This rate matrix $\mathcal{R}$ must be chosen before the analysis, but distinct models can be tried and compared (Yang 1994a). If $\mathbf{F}$ is the vector $(A(u), C(u), G(u), T(u))^{\intercal}$ of nucleotide frequencies at time u, the dynamics of sequence evolution follows the differential equations:

$$\mathbf{F}(u + du) = \mathcal{R}\,\mathbf{F}(u) \tag{1}$$

where nondiagonal entries r_{xy} of matrix $\mathcal{R}$ are the (positive) rates of change from nucleotide x to nucleotide y and where diagonal entries r_{xx} are such that columns sum to zero. This system is solved by diagonalising $\mathcal{R}$ (e.g. Rodriguez et al 1990):

$$\mathbf{F}(u) = \exp(\mathcal{R}\,u)\,\mathbf{F}(0) \tag{2}$$

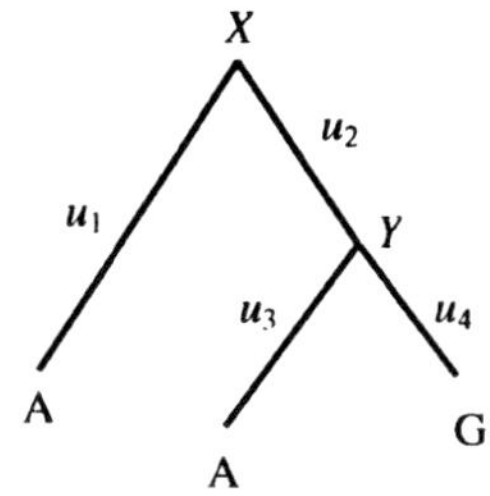

Fig. 3. A model tree for M calculation. Observed nucleotide states at tips are bold-typed. X and Y stand for the unknown ancestral states at internal nodes.

Entry $p_{xy}(u)$ of the $\mathcal{P}(u) = \exp(\mathcal{R}\, u)$ matrix is the probability of reaching final state y after evolving according to $\mathcal{R}$ during time u given initial state x.

Now consider a nucleotide evolving according to a given process $\mathcal{R}$ along a given tree with a given set of branch lengths (u's). We want to calculate the probability that this process result in site D_i, where D_i is the vector of the character states (A, C, G or T) observed in distinct species of the data set at position i. Felsenstein (1981) showed that this is computed by multiplying conditional transition probabilities $p_{xy}(u)$ along connected branches and summing over unknown ancestral states at internal nodes. For instance, the probability of site (AAG) in the three-species tree of Fig. 3 is given by:

$$\text{Prob}(AAG) = \tag{3}$$

$$\sum_x \sum_y \text{Prob}(X = x)\, p_{xA}(u_1)\, p_{xy}(u_2)\, p_{yA}(u_3)\, p_{yG}(u_4)$$

where the summation for x and y is over $\{A, C, G, T\}$. The prior probability of ancestral state x is usually taken from the equilibrium distribution of process $\mathcal{R}$ (i.e. the base composition reached after infinetely long evolution under process $\mathcal{R}$). This involves assuming stationarity of the process (see below for a relaxing of this assumption). More generally, the likelihood calculation can be written as a recursion valid for any tree topology:

$$\text{Prob}(D_i) = \sum_x \text{Prob}(X_R = x)\, \text{Prob}(D_i | X_R = x) \tag{4}$$

$$\text{Prob}(D_i | X = x) = \tag{5}$$

$$\sum_y \sum_z p_{xy}(u1)\, \text{Prob}(D_i | X_1 = y)\, p_{xz}(u_2)\, \text{Prob}(D_i | X_2 = z)$$

where X_R is the nucleotide state at the root node, X the nucleotide state at any internal node, X_1 and X_2 the nucleotide states at the son nodes of node X, and u_1 and u_2 the (X, X_1) and (X, X_2) branch lengths, respectively (summations are again over $\{A, C, G, T\}$). The recursion closes at the tips of the tree by setting $\text{Prob}(D_i|X = x) = 1$ if x is the character observed at tip X in site D_i, 0 otherwise. This calculation can be performed by a single pass on the tree, i.e. in $o(n)$, where n is the number of species. Assuming independent evolution of distinct sites, the probability of a data set D is obtained by multiplying the probabilities of every site D_i. This probability has to be maximised over the parameter space, yielding the ML estimates of the tree, branch lengths, and parameters of the substitution process.

The statistical approach to molecular phylogeny has become more and more popular after Felsenstein's (1981) remarkable paper. Yang (1996) advocated its use and greatly contributed to the incorporation of realistic models of DNA sequence evolution (see below). Several features of genomic sequence data make them suitable for statistical modelling. Molecular characters (sites) are numerous and of similar nature, in contrast with, say, morphological characters. The mutation process and the probability of fixation through random drift are essentially the same for distinct positions of a molecule in a given species. This provides justification for considering sites as realisations of a random variable whose distribution probability is dependent on parameters to be estimated. Natural selection, on the other hand, is essentially unpredictable. It applies differently on distinct sites and in distinct lineages. It can make the evolution of distinct sites nonindependent. The notion that sites evolve according to some common process is less obvious for molecules undergoing complex selective processes (e.g. see Goldstein in this volume, Schuster in this volume). Accounting for selective process is actually the major challenge in our attempt to model molecular evolution. The statistical approach to molecular phylogeny, however, has often proven useful even for genes under strong selective pressure (e.g. see below).

Model-based inference requires good models, i.e. models accounting for the major features of actual evolutionary processes at the cost of as few parameters as possible. A simplistic model can result in biased estimators, while too many parameters tend to increase the sampling variance (i.e. require more data to reach a given level of accuracy). The simplest model is a one-parameter substitution matrix, where all kinds of change are assumed to be equiprobable (Jukes and Cantor 1969, Kimura 1980) added one parameter to account for the fact that transitions (changes within purines [A,G] or within pyrimidines [C,T]) are more frequent than transversions (between purines and pyrimidines). Tamura (1992), after Tajima and Nei (1984), allowed the expected equilibrium base composition to be different from 25%A, 25%C, 25%G, 25%T, greatly increasing the fit to most data sets. Yang (1994a) implemented more general substitution matrices and investigated tools for model selection given a data set.

The literature cited above addresses the problem of seeking the most appropriate substitution matrix $\mathcal{R}$ for a given data set, i.e. improving the assumed evolutionary mode. Models also progressed with respect to how to represent evolutionary rates. A major improvement occurred when Yang (1993, 1994b) relaxed the assumption of rate-constancy accross sites. In these models, the relative rates of distinct sites are assumed to follow some distribution whose shape is optimised from the data. The probability of a given site becomes:

$$\text{Prob}(D_i) = \int_0^\infty f(r)\,\text{Prob}(D_i|r)\,dr \tag{6}$$

where f is the probability density of the assumed distribution of rates, and where $\text{Prob}(D_i|r)$ is obtained from equations 4 after branch lengths have been multiplied by r. Usually, a (discretised) Gamma distribution is assumed, mostly for practical reasons: a Gamma distribution can be either exponential-like or unimodal depending on its shape parameter, providing reasonnable fit to many data sets. Relaxing the hypothesis of constant rates accross sites contributes to accounting for selective effects: natural selection puts contraints on some sites, decreasing their substitution rates, while other sites evolve more or less freely (and therefore rapidly).

4 A nonstationary model of DNA evolution

All of the above-mentionned models share two important assumptions, namely homogeneity and stationarity of the evolutionary process. Homogeneity means that a unique substitution matrix applies in every branch of the tree. Stationarity means that the base composition (frequency of A, C, G and T) has reached the equilibrium frequencies of this process, and therefore is constant in time. The two assumptions imply that the expected base composition of distinct present-day sequences are identical. This requirement of homogeneous, stationary models is not met by every data set: base composition differences among homologous sequences is a common feature of molecular data (e.g. Bernardi 1993, Jermiin et al 1994, Galtier & Lobry 1997). It has been shown that standard tree-building methods are biased when base composition varies among sequences: they tend to group together sequences of similar base compositions whatever their actual phylogenetic relationship (Lockhart et al. 1994, Galtier and Gouy 1995).

To cope with this problem, we implemented a nonhomogeneous, nonstationary model of DNA sequence evolution (Galtier and Gouy 1998). In this model, each branch of the tree has its own substitution matrix $\mathcal{R}$, with its own equilibrium G+C-content. Therefore the expected base composition can vary in time and between lineages. This is achieved at the cost of one additional parameter per branch, namely branch-specific equilibrium G+C-content. Despite the high increase in the number of parameters, this model was found adequate for many data sets in which base composition significantly departed equality among sequences, improving the efficiency of tree-building methods (Galtier and Gouy 1998). Some parameters (e.g., the equilibrium G+C-content of short branches) cannot be accurately estimated from data sets of standard size. They correspond to flat directions of the likelihood surface, making the likelihood maximisation more difficult. Note, however, that these nuisance parameters do not reduce the amount of information available for parameters of interest (e.g., the tree topology), as soon as an efficient enough optimisation algorithm can be sought.

In nonstationary models, the ancestral base composition (Prob(X_R = x) in equation 4) can no longer be deduced from the equilibrium frequencies of the substitution process: it becomes a free parameter that can (has to) be estimated from the data in the ML framework. Somewhat surprisingly, this can be achieved with a high level of accuracy, as shown from a simulation analysis (Galtier and Gouy 1998). Two kinds of data sets were simulated under nonstationary conditions. In the first one, a G+C-rich ancestral sequence was evolved along the branch of a ten-species tree under a process that tends to decrease G+C-content, so that the expected G+C-content was 50% in all present-day sequences. A symmetrical process was used to simulate the second set of data sets: a G+C-poor sequence data was uniformly enriched, again reaching 50% in present-day sequences. When fed with these two categories of data sets, the ML method based on a nonstationary model of evolution was able to distinguish between the two, yielding accurate estimates of the ancestral G+C-content from sequences as short as 500 nucleotides.

The ability of the ML method to distinguish come from the fact that the two data sets, although similar with respect to G+C-content distribution among sequences, are different as far as site frequencies are concerned. The expected proportion of sites for which, say, every sequence shows A is higher in the second case (low to high G+C-content) than in the first. This kind of information is typically used by the ML method to recover the ancestral G+C-content. Applied to a study of the isochore structure in mammals (spatial structure of G+C-content within the genome, Bernardi 1993), this method supported a human-like pattern (i.e. a high level of spatial structure) in the ancestral placental mammal (Galtier and Mouchirou 1998). This was later confirmed by the discovery of a typical isochore structure in crocodile and turtle (Hughes et al. 1999), as well as in birds (e.g. Bernardi 1993), implying an early evolution of well-defined G+C-rich and G+C-poor regions within the genomes of amniotes.

5 The thermophilic nature of the common ancestor

Estimating ancestral G+C-content is not of primary importance for most genes since base composition has usually little functional relevance. The ribosomal RNA (rRNA) of prokaryotes is an exception. The G+C-content of rRNA is positively correlated to optimal growth temperature in prokaryotes (in contrast with genomic G+C-content, Galtier & Lobry 1997). Every hyperthermophilic species (living over 60°) have a G+C-content higher than 60% in the small-subunit rRNA and 57% in the large subunit rRNA, versus an average 54% and 52%, respectively, in mesophilic species. This is because G:C pairs are more stable than A:T pairs thanks to an additional hydrogen bond. Many G's and C's are therefore required within helices to stabilise the secondary structure of rRNA when the temperature is high. This makes the rRNA G+C-content behave like a molecular thermometer: a high rRNA G+C-content is indicative of hyperthermophily.

It has been suggested that the common ancestor to extant life forms was a thermophilic organism (Woese 1987). This hypothesis essentially arose from the examination of the phylogenetic distribution of thermophily among prokaryotes. Thermophily is found in most prokaryotic phyla, and is predominant in deeply-branching lineages. Furthermore, the branches leading to thermophiles in the small-subunit rRNA tree appear somewhat shorter than those of mesophilic lineages, suggesting a higher structural similarity with the ancestral rRNA. The hypothesis of a thermophilic ancestor quickly reached the status of working hypothesis, although a few authors expressed some doubts (Forterre 1996).

To test this hypothesis we estimated the $G+C$-content of small- and large-subunit rRNAs (695 and 1409 unambiguously aligned sites, respectively) in the common ancestor by applying the ML implementation of our nonstationary model to a data sets including 40 species from every known living groups, namely Eukaryotes, Bacteria and Archeae (Galtier et al. 1999). The estimated ancestral $G+C$-contents (56.1% for the small-subunit, 54.0% for the large subunit) were found incompatible with life at very high temperature. Not any present-day species is able to survive over 70° with such low-GC rRNA. Confidence intervals are narrow enough not to overlap the thermophilic zone. Sensitivity to species sampling and tree topology were checked. The most conclusive control, however, came from an analysis of a reduced data set including only $G+C$-rich species. The method consistently recovered a low ancestral $G+C$-content from the analysis of present-day $G+C$-rich species, strongly suggesting that signal rather than noise has been extracted from the data. We therefore argue that the common ancestor to extant life forms was probably not a hyperthermophile. Thermophily has evolved at least twice in prokaryote history. The short branches leading to thermophiles are presumably a consequence of a stronger selection pressure for thermostability on many sites of rRNA in these species.

This study, among other, illustrates the power of the statistical approach to molecular phylogeny and evolution. Inferences of the history is achieved thanks to the use of appropriate models of DNA sequence evolution. The accumulation of many, long genomic sequences advocates further the use of statistical methods for evolutionary purposes.

Adequately modelling the processes of nucleotid (aminoacid) substitution is important for recovering good trees, argued above. It is also relevant to the purpose of approaching the various evolutionary forces that drive genome evolution, and especially natural selection. The effects of natural selection on the substitution process of genomic sequences include the existence of slow versus rapid sites, differences between the rate of accumulation of synonymous versus nonsynonymous substitutions, changes of site-specific evolutionary rates (the so-called covarion process) after the selective constraints applying to a molecule have changed, constraints on the number and nature of states allowed at a specific position. Some, but not all of these effects have been convincingly studied so far. Incorporating them into models of sequence evolution and fitting to actual data would definetely help understanding the way selection proceeds at the genomic level.

References

1. Bernardi G.: The vertebrate genome: isochores and evolution. Mol. Biol. Evol. **10**: 186-204, 1993.
2. Edwards A.W.F.: Likelihood. Cambridge University Press. Cambridge, 1972.
3. Farris J.S.: Simplicity and informativeness in systematics and phylogeny. Syst. Zool **31**: 413-444, 1982.
4. Felsenstein J.: Cases in which parsimony and compatibility methods will be positively misleading. Syst. Zool. **27**: 401-410, 1978.
5. Felsenstein J.: Evolutionary trees from DNA sequences: a maximum likelihood method. J. Mol. Evol. **17**: 368-37.
6. Fitch W.M.: Toward defining the course of evolution: minimum change for a specific tree topology. Syst. Zool. **20**: 406-416, 1971
7. Forterre P: A hot topic: the origin of hyperthermophiles. Cell **85**: 789-792, 1996.
8. Galtier N. & Gouy M.: Inferring pattern and process: maximum likelihood implementation of a non-homogeneous model of DNA sequence evolution for phylogenetic analysis. Mol. Biol. Evol. **15**: 871-879, 1998
9. Galtier N. & Lobry J.: Relationships between genomic G+C content, RNA secondary structures and optimal growth temperature in prokaryotes. J. Mol. Evol. **44**: 632-636, 1997.
10. Galtier N. & Mouchirou D. 1998. Evolution of isochores in mammals: a human-like ancestral pattern. Genetics **150**: 1577-1584, 1998.
11. Galtier N., Tourasse N.J. & Gouy M.: A nonhyperthermophilic common ancestor to extant life forms. Science **283**: 220-221, 1999.
12. Goldstein (this volume)
13. Hughes S., Zelus D. & Mouchiroud D.: Warm-blooded isochore structure in Nile crocodile and turtle. Mol. Biol. Evol. **16**: 1521-1527, 1999.
14. Jermiin L.S., Graur D., Lowe R.M., & Crozier R.H.: Analysis of directional mutation pressure and nucleotide content in mitochondrial cytochrome b genes. J. Mol. Evol. **39**: 160-173.
15. Jukes T.H. & Cantor C.R.: Evolution of protein molecules. pp 121-123 in H.N. Munro, ed. Mammalian protein metabolism. Academic press. New York, 1969.
16. Kimura M. 1980. A simple method for estimating evolutionary rates of base substitutions through comparative studies of nucleotide sequences. J. Mol. Evol. **16**: 111-120, 1980.
17. Lockhart P.J., Steel M.A., Hendy M.D. & Penny D.: Recovering evolutionary trees under a more realistic model of sequence evolution. Mol. Biol. Evol. **11**: 605-612, 1994.
18. Rodriguez F., Oliver J.F., Marin A. & Medina J.R.: The general stochastic model of nucleotide substitution. J. Theor. Biol. **142**: 485-501, 1990.
19. Saitou N. & Nei M.: The neighbor-joining method: a new method for reconstructing phylogenetic trees. Mol. Biol. Evol. **4**: 406-425, 1987.
20. Schuster (this volume)
21. Tajima F. & Nei M.: Estimation of evolutionary distances between nucleotide sequences. Mol. Biol. Evol. **1**: 269-285, 1984.
22. Tamura K.: Estimation of the number of nucleotide substitutions when there are strong transition-transversion and G+C-content biases. Mol. Biol. Evol. **9**:678-687, 1992.
23. Woese C.R. 1987. Bacterial evolution. Microbiol. Rev. **51**: 221-271, 1987.

24. Yang Z.: Maximum likelihood estimation of phylogeny from DNA sequences when substitution rates differ over sites. Mol. Biol. Evol. **10**:1396-1401, 1993.
25. Yang Z.: Estimating the pattern of nucleotide substitution. J. Mol. Evol. **39**:105-111, 1994.
26. Yang Z.: Maximum likelihood phylogenetic estimation from DNA sequences with variable rates over sites: approximate methods. J. Mol. Evol.**39**:306-314, 1994.
27. Yang Z.: On the general reversible Markov process model of nucleotide substitution: a reply to Saccone et al. J. Mol. Evol. **41**: 254-255, 1995.
28. Yang Z.: Phylogenetic analysis using parsimony and likelihood methods. J. Mol. Evol. **42**: 294-307, 1996.

Principles of cophylogenetic maps

Michael A. Charleston

Department of Zoology, University of Oxford,
South Parks Road, Oxford OX1 3PS, United Kingdom

Abstract. Cophylogeny is the study of the relationships between phylogenies of ecologically related groups (taxa, geographical areas, genes etc.), where one, the "host" phylogeny, is independent and the other, the "associate" phylogeny, is hypothesized to be dependent to some degree on the host. Given two such phylogenies our aim is to estimate the past associations between the host and associate taxa. This chapter describes cophylogeny and discusses some of its basic principles. The necessary properties of any cophylogenetic method are described. Charleston [5] created a graph which contains all the potential solutions to a given cophylogenetic problem. The vertices of this graph are associations, either observed or hypothetical, between "host" and associated taxonomic units, and the arcs correspond to the associate phylogeny. A new and more general method of constructing the Jungle is presented, which will correctly account for reticulate host and/or parasite phylogenies.

Keywords: cophylogeny, coevolution, gene tree/species tree, host/parasite coevolution, host switch, horizontal transfer, biogeography.

1 Introduction

Recently, it has come into public awareness that the most likely source of human immuno-deficiency virus (HIV) was a population of chimpanzees in Africa [24]. We frequently hear of influenza virus invading the human population from pigs and chickens [25]. Malaria is also able to move among its host species [6,11].

Understanding the ways in which pathogens can switch between host species is of crucial importance if we are to create the right strategies to combat them. Another topic of recent interest is one of the questions pertaining to genetically modified ("GM") foods – whether genes which have been spliced into food plants could somehow "escape" and enter the food chain to our detriment (quite apart from the grosser ecological effects). Perhaps seemingly unrelated to the above is the problem of reconstructing ancient associations between geographical areas and their endemic species: Where *did* the flightless birds of New Zealand come from? How did amphibious caecilians manage to get to Madagascar from the African continent? What was the pattern of invasion of *Anolis* lizards to the islands in the Caribbean? Such questions are in the area of *biogeography*, and can be answered in terms of the phylogenies of the endemic organisms and the evolutionary histories of their habitats.

The questions of how *phylogenetic events* such as the host switching events mentioned above can occur, and how often they occur, can be answered through studies into the phylogenetic relationships among and between the parasites and their hosts, between the genes and the species which house them, and between organisms and the geographical regions they occupy. We cannot hope to perform the experimentation which would be required to test all cases *in vivo* – rather, we must glean all the information we can from the data we have in the form of known phylogenies and known host ranges of pathogens. We cannot go back in time in reality, yet we do wish to find out what were the ancestral relationships, in order to find where pathogens may have switched hosts in the past, and in order to determine when they might do so in the future.

The problem amounts to reconstructing the best supported pattern of relationships between two phylogenies, one of which is the host and the other is the parasite (or simply "associate") phylogeny. We are usually presented with a host tree H and a parasite tree P and the set of host species each parasite infects. If we observe little or no host specificity, that is, if parasites occur on many different host species, then we cannot talk about coevolution in a meaningful way: the parasites appear to be able to move among hosts easily, or simply have not diverged with the host species even if they do not move amongst host lineages. If on the other hand we observe a perfect match between the two trees and perfect host specificity of the parasites, then we can be reasonably sure that there is little or no successful cross-species transfer of the parasites: they do not appear to survive on other host species. We must be able not only to determine which is the most likely or most parsimonious match between two phylogenetic trees linked by associations at their tips, but also to measure the *degree of agreement* between two such phylogenies. Is 50% cospeciation a lot? A little? Does it depend on the number of host or parasite taxa? These questions remain unanswered at present.

We use *cophylogeny* to mean the study of phylogenetic agreement between the evolutionary histories of two or more groups of organisms. It is only going to be possible to make headway within this area if we can first formalise what exactly we need to know, what we require of a solution to a given cophylogeny problem, and what such solutions will tell us.

Note that this is not as simple as the problem of comparing two phylogenetic trees for the same set of taxa, which might be gained from analyses of different data for example. Such a comparison is *symmetric* because the two trees are treated in much the same way. Our problem here is *asymmetric*: one of the phylogenies – the "host" – is considered to be independent, and the other – the "parasite" – is considered to be dependent to some degree on that host. Thus existing tree comparison metrics (recalling that metrics are by definition symmetric) are not applicable here. The cophylogeny problem requires that we find a kind of "phylogenetic correlation" of parasite phylogeny with host phylogeny. In order to reconstruct the past associations we need to locate all the taxa of the parasite phylogeny on the host phylogeny, that is, we need some kind of *map* from P into H.

Cophylogeny represents another level of difficulty added to the already hard problem of inferring the two phylogenetic trees. Note also that not all phylogenies are trees: in plant systematics for instance there are many instances of hybridization events giving rise to new species. Thus we need methods which can cope with the more general problem of mapping one non-treelike phylogeny into another. Those methods which are based on creating a matrix of pseudo-data from P and optimizing the fit of this matrix on the host tree H, cannot deal correctly with networks. A new kind of mapping is needed.

The statistical issues are also complex and subtle. In order to measure the significance of the degree of fit of P with H the standard approach is to randomize P according to some tree generation model, leaving the observed associations as they are, and create a distribution of degrees of fit from the new P's. However this is computationally intense, and leaves difficult questions open: Since any instance of a cophylogeny problem can yield several potential solutions, any one of which may be optimal depending on how we weight the individual events involved, it is not clear how we should compare statistically *sets* of potentially optimal solutions. Further, there is the issue of uncertainty in H and P: the phylogeny problem is a statistical estimation problem so is prone to uncertainty from sampling error and from systematic bias. The statistics of cophylogeny are tremendously important, but are beyond the scope of this chapter. The interested reader is directed to Huelsenbeck's writing on the subject [10]. For the remainder of this work I concentrate on the mathematical problem of mapping a given P H, leaving aside such difficulties for another time.

2 Terminology

There is little universal terminology in this relatively new problem, but we may draw on some from graph theory and phylogenetics, and with some new terms we can arrive at a sufficiently full toolbox that we can begin work.

2.1 Phylogenetic networks

Let a *phylogenetic network* be a directed, connected graph with a unique root, all but one of whose arcs are directed away from the root, the remaining arc being adjacent to the root, and with no vertices having both in-degree and out-degree equal to one, or having both in-degree and out-degree greater than one. The vertices of out-degree zero usually will correspond to extant taxa; they are called the *tips* (leaves in trees), denoted for a phylogenetic network $\mathcal{N}$ by $L(\mathcal{N})$.

Direction is required to reflect the passage of evolutionary time. The arcs of the phylogenetic network are directed forwards in time, away from the root. The only exception to this is the one arc directed into the root, representing the ancestral lineage before any phylogenetic events took place. We must therefore have the first phylogenetic event being a divergence into at least two descendent lineages – a speciation event. The most "recent" vertices are those which have no arcs directed out of them, and almost always correspond to extant taxa. In rare cases we may obtain information about extinct species (through fossil data or ancient DNA techniques) or ancestral lineages (through epidemiological records

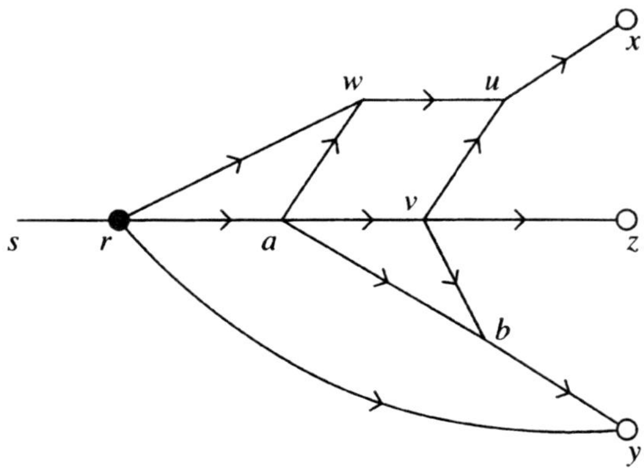

Fig. 1. A phylogenetic network. The above connected, rooted, digraph is a phylogenetic network. Vertex r is the root.

of viruses). The *unique root* represents the common ancestor of the taxa at the tips of the phylogeny.

The requirement that no vertex have both *in-degree and out-degree* equal to one recognises that we cannot generally identify a single species (or some other taxonomic type: population, family etc.) at a particular instance in time. In some cases we may be able to guarantee the existence of a lineage at a particular point in time (through viral records for instance) but if there is no phylogenetic event at that time, such a vertex would give us little more information, and such cases are rare enough that we can safely shelve them for the moment. We represent the phylogenetic events of divergence and recombination (crossing) by vertices since each corresponds to a change in the number of distinct lineages present. Thus in Fig. 1 there are divergences at vertices r, a and v, and recombination events at vertices b, u and w. Vertices x, y and z are tips.

Given two vertices v and u in $V(G)$ we say that v is *ancestral to* u if and only if there exists a directed path through G from v to u. Because we have chosen G to have no directed cycles by dint of having all but one of the arcs directed away from the root, we cannot have any vertex ancestral to itself. We denote that v is ancestral to u in G by $v <_G u$, and if either v is ancestral to u or $u = v$, then we write $v \leq_G u$. We denote by $u \ll_G w$ that u is ancestral to w and not adjacent to it. We also say that edge $e = (a, b)$ is ancestral to vertex u if $a <_G u$ and $b \leq_G u$, and that vertex v is ancestral to e if $v \leq_G a$ and $v <_G b$. The *ancestors* of a vertex u are $\{v : v <_G u\}$ and the *descendants* of u are $\{w : u <_G w\}$. Thus in Fig. 1, $a < v$ and $a < w$, the ancestors of w include a, r and s, and the descendants of a are $\{b, u, v, w, x, y, z\}$. The *parents* $\pi(v)$ of a vertex v are those vertices which are both adjacent and ancestral to v; the *children* $\kappa(v)$ are those vertices which are both descendent from and adjacent to v. Thus from Fig. 1, $\pi(b) = \{a, v\}$, and $\kappa(a) = \{b, v\}$. A *common ancestor* of a set S of vertices is a vertex v such that $v <_G s$ for all $s \in S$. A *most recent common ancestor* (mrca) of a set S of vertices in G is a common ancestor of S such that no descendent of v is also a common ancestor of S. In Fig. 1, a is a common ancestor of b and u, but is not a most recent common ancestor since v, a descendent of a, is also a common ancestor of b and u.

Note that since we allow non-tree-like phylogenies, a most recent common ancestor of a group of vertices need not be unique.

We say that two vertices u and v in $V(G)$ are *not comparable* if neither $u \leq_G v$, nor $v <_G u$. If there exists a vertex z which is a parent of both u and v, then u and v are said to be *siblings*. A sibling set is a set $S = \{s_1, \ldots, s_k\}$ of vertices which have at least one parent in common.

G may be *dated*: in that case some or all of the vertices of G have a real-valued date t associated with them. These dates must be compatible with the ancestor/descendent relations implied by the arcs $A(G)$ of G. That is, the partial ordering of $V(G)$ implied by $A(G)$ must not contradict the partial ordering of $V(G)$ implied by the dates: if $v <_G u$ then either $t(v) < t(u)$ or t is not defined for both u and v, and if $t(v) < t(u)$ then either $v <_G u$ or u and v are not comparable.

If $t(v) < t(u)$ or $v <_T u$ then v *precedes* u in T; we write this as $v \prec_T u$. The precedence relation relates information both from ancestry (the arcs) and date (the times) on the vertices of G. Note that $v \prec_T u \Rightarrow t(v) < t(u)$, though these times may not be defined. Note also that while if v is ancestral to u it must precede it, the converse is not true: just because v precedes (is older than) u, that doesn't mean v is an ancestor of u.

Let a *phylogenetic tree* be a rooted tree with no vertices other than the root with degree two. Define a *planted phylogenetic tree* to be a phylogenetic tree whose edges have been directed away from the root, and with the subsequent addition of a single arc directed into the root (see Fig. 2).

There is an obvious bijection between phylogenetic trees and planted phylogenetic trees, and all planted phylogenetic trees are phylogenetic networks: if T is a planted phylogenetic tree, then it has a unique root and satisfies the degree requirements since each vertex has in-degree 1 and out-degree $\neq 1$.

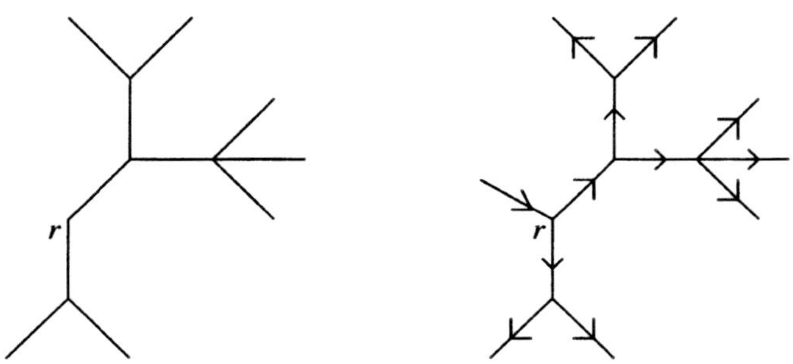

Fig. 2. Relationship between rooted trees and Phylogenetic Networks. The above shows the bijection between rooted trees and phylogenetic networks, described in the text.

3 Cophylogeny

We consider a pair of phylogenetic networks, one which is designated the *host*, labelled H, and the other is designated the *parasite* P, in accordance with the common interpretation of this problem in host/parasite coevolution (there are many other interpretations but we shall not go into them here – the interested reader is directed to [13,18]). We also have a function φ which maps a subset ν of the vertices (usually the leaves) of P into the space $\xi = \nu \times V(H) \times \{1,2\}$.

The elements of ξ can be of *type 1* or of *type 2* according to the value of the last component in the ordered triple above. These types correspond to locations at the host vertex (type 1) and on the edge ancestral to the host vertex (type 2). We call these elements *j-vertices*.

In my earlier paper on jungles [5], I denoted j-vertices by $(p : h)^*$ and $(p : h)$, whereas now they are written $(p, h, 1)$ and $(p, h, 2)$ respectively. The transition should pose no problems to the reader, and makes the manipulations in this article much more manageable.

We shall impose a partial ordering on the elements of ξ, which is consistent with the partial ordering of the vertices of P and H: given two j-vertices $u = (p_u, h_u, t_u)$ and $v = (p_v, h_v, t_v)$,

$$u \prec v \Leftarrow \begin{cases} p_u \prec_P p_v, \text{ or} \\ p_u, p_v \text{ are not comparable and } h_u \prec_H h_v. \end{cases} \tag{1}$$

In addition we define $u \prec v$ if $u = (p_i, h, 2)$ and $v = (p_j, h, 1)$, since in this case p_i is located on the edge ancestral to h, whereas p_j is located at h itself. However without further knowledge we would be unable to compare u with, say, $w = (p_k, h, 2)$ since either could occur earlier on the edge.

We define a *subtree match* as a case where two subtrees H' and P' are isomorphic, and the isomorphism is preserved by $\varphi|_\nu$, as shown in Fig. 3. We define a *subphylogeny match* in the obvious way, extending the definition to include isomorphic networks.

3.1 Cophylogenetic events

For reasons which should become evident, we shall define four *cophylogenetic events*, that is, biological events relating to the macroscopic coevolution of two groups of species. These events are *codivergence*, *duplication*, *lineage sorting* or *loss*, and *host switching*.

A *codivergence* occurs when a divergence (speciation) in the host phylogeny is reflected by a contemporaneous divergence (speciation) in the associate phylogeny. A *duplication* event occurs when the associate diverges independently of the host. A *loss* occurs when the associate is missing from one of the descendent lineages of a host lineage it occupied. A *host switch* occurs after a duplication and means that one descendent associate lineage changes its host. Figure 4 shows these events.

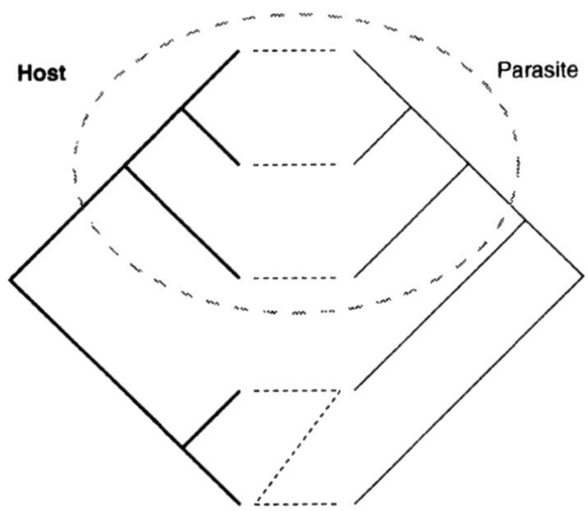

Fig. 3. Tanglegram with a subtree match. In the above tanglegram the top three parasite and host taxa (unlabelled here, for clarity) form a subtree match, since φ preserves the isomorphism between those parts of H and P which are encircled by the light dashed line.

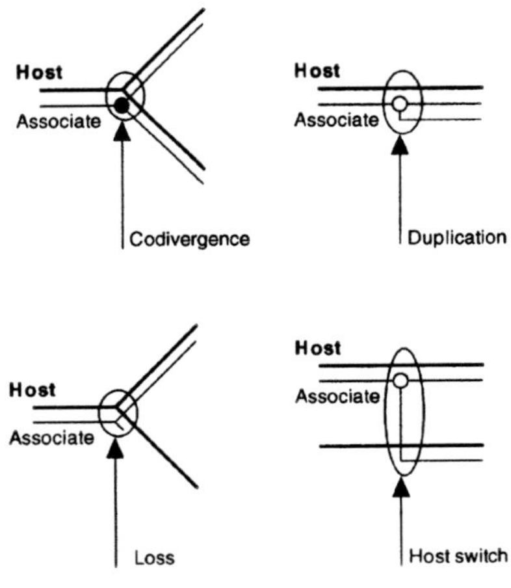

Fig. 4. Four kinds of cophylogenetic events. (Time is from left to right.)

Our aim in constructing a cophylogeny mapping is to estimate such events as above as best we can. We shall see shortly how these are the only kinds of events we can hope to recover with this approach.

3.2 Recovery of events

It is not possible to recover all the possible events in a cophylogeny. There are several reasons for this, but rather than go through the reasons in turn, we shall explore the events themselves and then explain how we must either miss them, or not be able to distinguish them from other events.

Bear in mind that for the most part we are not usually dealing with dated trees – many phylogenies, particularly the older ones and those based on non-molecular data, had no way of estimating the times of divergence of the lineages in the tree. In fact even molecular estimates of dates are known to have enormous margins of error, and should generally be treated with a pinch of salt.

3.3 "Missing the boat" and extinction

It would be nice to be able to distinguish the case in which a parasite lineage simply fails to be in the right place at the right time to continue to infect the new lineages of a host species after it diverges – "missing the boat" *sensu* Paterson [21] from the case in which the parasite does indeed "catch the boat", but then fails to survive the trip, dying out at some point on that lineage. Observing these events as they are represented in Fig. 5, it is clear that we cannot tell them apart without at least some very precise information about the dates of appearance and disappearance of lineages, such precision not being a usual feature of the fossil record. We would also require that a reasonable amount of evolutionary time must have passed from the divergence of the host and the subsequent extinction of one of the new parasite lineages, such that the short-lived lineage will even show up. The odds are stacked against us mathematically as well, since the cophylogeny problem is presented as a pair of phylogenies (almost always as trees), whose tips are generally contemporaneous with ourselves – we simply do not have the extinction events in our parasite (or host) trees to begin with. We must therefore lump both kinds of events together, and call them a *loss*.

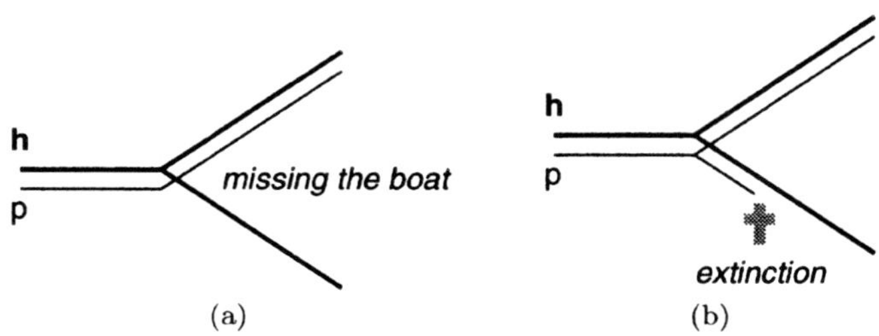

Fig. 5. Loss: Missing the boat and extinction. We cannot distinguish between the two phylogenetic events above, of missing the boat or of simply becoming extinct on one lineage. The former lies at the end of a scale of possible histories, when, effectively, it *does* get to the new host lineage but then survives for an arbitrarily short time.

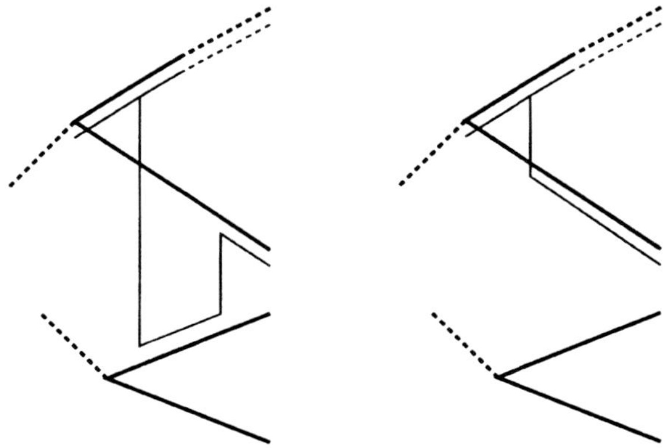

Fig. 6. Untraceable "short stay" parasite. On the left the parasite switches to a distant host lineage, then switches back to the nearer host, leaving no trace on the more distant one. This is indistinguishable from simply switching to the nearer host lineage.

Short stays: A *short stay* is a case where a parasite lineage switches to one host, possibly after a duplication, and then switches away again leaving no descendants behind. It is clear to see why we cannot hope to reconstruct this from the triple of H, P and φ, in Fig. 6.

3.4 Pandora's box: Sampling, ghosts, sources and sinks

Another spanner in the works is caused by the inevitable difficulties of *sampling* all taxa: we cannot guarantee to find all cases of parasitological infestation among all lineages of host, nor can we guarantee to find all host species at all. Some parts of the parasite life cycle may be on other hosts entirely, in different ecological niches, and we may simply miss their presence by chance.

If we are missing host taxa (either through extinction or lack of sampling) then our reconstruction of the past associations between host and parasite are likely to contain deflated numbers of lineage sorting events: we shall have missed them simply because we are unaware of the ancient host species, now missing from our phylogeny.

Given that we do not know the relationship between numbers of taxa and the probability distribution of number of cophylogenetic events, we cannot then ignore the possibility that missing out some host taxa will adversely affect the level of significance which we ascribe to the match between host and parasite phylogenies. We expect (but we do not yet know) that 50% (say) codivergence will be more significant in larger trees than in smaller ones, so clearly the issue of sampling effects must be borne in mind in all our cophylogenetic studies.

If we are missing associate taxa then there may be lineages in the host phylogeny which have no parasites on them, in which case, why should we include them in a study of coevolution at all? These host lineages are "empty" – but note that they may *in the past* have hosted parasites currently in our study.

Host lineages may even have appeared, however briefly, acted as a refuge for a parasite lineage, and after the parasite lineage has exited, then gone extinct. These "*ghost*" lineages may never be found – the only chance of doing so would be in the fossil record – and so dealing with them statistically would have to effectively integrate over all possible ghost lineages with a probabilistic model of their creation and extinction, and the probabilities of host switching, codivergence, duplication and loss.

It is but a tiny step to realise also that we need not restrict ourselves in the above lineages to ghosts: completely different species may act as refugia for parasites, a perpetual *source* and *sink* of host switching parasites. These sources and sinks further complicate the problem.

Which events can *we recover?* Fortunately, not all is lost. We are able to recover the four events listed above, even though we are unable to distinguish among the variations of events (lineage sorting, missing the boat, extinction, sampling) leading to a loss. The way we do this is by reconstructing a history of associations between parasite and host, the relationships among which determine the cophylogenetic events undergone by each parasite lineage. This reconstruction is a *cophylogeny mapping.*

4 The cophylogeny mapping

We define a cophylogeny mapping and the properties which it must satisfy to be soluble; to be well-defined and consistent, and the properties it must also satisfy given the underlying hypotheses which are made in cophylogeny, to whit, that there are certain processes involved in evolution which give rise to interactions between the phylogenetic histories. These interactions mean that we expect phylogenies to match by default, and the events which give rise to a mis-match are in some sense "special".

A *cophylogeny mapping* is a function Φ mapping the vertices and edges of P as $v \in V(P) \mapsto \Phi(v) \in \xi$ and $e = (u, v) \in E(P) \mapsto (\Phi(u), \Phi(v)) \in \xi \times \xi$, and which has the properties listed below. We denote the image set of some $B \subseteq V(P)$ under Φ as $\Phi(B)$, that is, $\Phi(B) = \{\Phi(b) : b \in B\}$. We also write $\Phi(P)$ to mean the graph whose vertices are $\Phi(V(P))$ and whose arcs are $\{(\Phi(x), \Phi(y)) : (x, y) \in E(P)$.

I have divided the properties into those which are largely "mathematical" and those which are largely "biological". The "Mathematical" conditions follow:

Condition M_0: The "lifting" condition: $\Phi|_\nu = \varphi$. The lifting condition requires that Φ, when restricted to the domain of the original φ, is the same as φ – this is an obvious requirement for any mapping which extends φ;

Condition M_1: The "consistency" condition: if $P \cong H$ and φ preserves the isomorphism, then Φ preserves the isomorphism also;

Condition M_2: The "order preserving" condition: for any $p, q \in V(P)$ with $p \prec_P q$, $\Phi(p)$ must precede $\Phi(q)$ in H.

And now the "Biological" conditions:

Condition B_0: The "isomorphism" condition: $\mathcal{G} = \Phi(P)$ is isomorphic to P. In that way we may consider Φ to be simply a function assigning to each vertex v in $V(P)$ a point (h,t) in $V(H) \times \{1,2\}$, which amounts to identifying either a vertex $h \in V(H)$ when $t = 1$, or the edge immediately ancestral to h when $t = 2$. Thus Φ is precisely the sort of map which we need to identify ancestral locations of nodes (vertices) of the parasite phylogeny on the host phylogeny;

Condition B_1: If $S = \{p_1, \ldots, p_k\}$ are siblings, mapped by Φ to (p_i, h_i, t_i) for $i = 1, \ldots, k$, and if p' is a parent of all the p_i with $\Phi(p') = (p', h, t)$, then $h <_H h_i$ for at least one $i \in \{1, \ldots, k\}$. Putting it another way, if p maps to (p, h, t) then at least one of the descendants of p must map to a descendent of h. If we do not make this condition then it will be impossible to reconstruct the possible Φ from the original P, H and φ (see Fig. 6). This is also known as the condition of being "traceable".

Condition B_2: If $u = (p, h, 1) \in \Phi(V(P))$ then for all children p_i of p, $\Phi(p_i) = (p_i, h_i, t_i)$ satisfies $h <_H h_i$. Informally, all of the children of a type 1 j-vertex must be mapped to descendants of that j-vertex. Quite apart from the difficulty in interpreting an "event" such as represented in Fig. 7, we must conclude that for a parasite lineage to both diverge with the host lineage and for one of the new parasite lineages to switch a different host at precisely the same time, must be beyond the scope of likelihood. The figure shows a situation which might be interpreted as

1. codivergence of host and parasite at u with subsequent host switch of one of the new parasite lineages, leaving no trace (thus failing condition B_1);
2. host switch leaving a parasite lineage p behind, followed immediately by a lineage sorting event at u;
3. lineage sorting event at u, followed immediately by a duplication of s and host switch of q.

Diagrams for each of these interpretations are shown in Fig. 8.

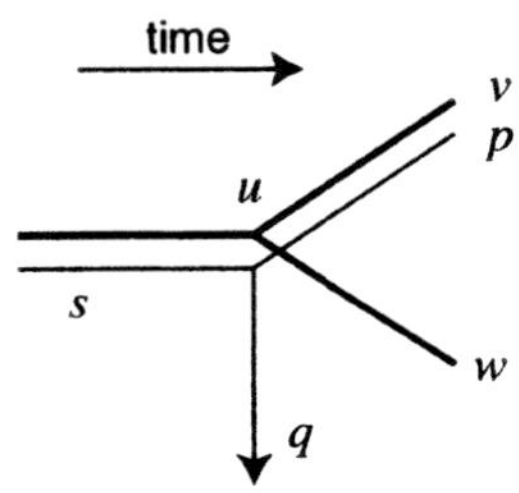

Fig. 7. Concurrent divergence and switches are precluded. The above shows a situation which we preclude for a number of reasons, not least being that there are several different interpretations of which events actually took place, and which are covered by other cases.

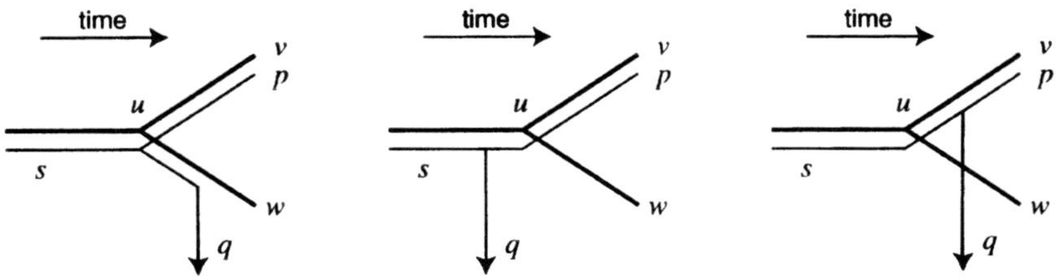

Fig. 8. Three interpretations of "concurrent" divergence and switch. The three figures above show the possible interpretations of the situation represented in Fig. 7, above.

5 Problems

With all the subtleties of terminology demonstrated above it is easy to lose track of what we hope to obtain from our cophylogeny maps. The following problems may seem obvious, and you may be surprised that their solutions are not, for the most part, known at present. This reflects the mathematical difficulty and newness of this area. Of the problems below I give a solution to the first; the others still demand attention.

For a given pair of phylogenetic trees H and P and mapping $\varphi : \nu \mapsto \nu \times L(H) \times \{2\}$,

1. is it always possible to find a cophylogeny mapping $\Phi : V(P) \mapsto \xi$?
2. how many possible maps are there? how does this number grow with (i) number of host taxa? (ii) number of parasite taxa? (iii) decreasing host specificity? (iv) increase in permitted host switches?
3. what is the frequency distribution of the number of cospeciations?
4. given numbers of host and parasite taxa, how many codivergences could we expect to occur by chance?

5.1 Existence

We asked above, is it always possible to find a cophylogeny mapping under the conditions of H, P and φ which were prescribed. This can be answered in the affirmative without too much difficulty.

Suppose we have H, P and φ as above. We need to find a mapping $\Phi : P \mapsto \mathcal{G}$ such that $\mathcal{G}$ satisfies the mathematical and biological conditions.

Case I: If $P \cong H$ by isomorphism η, then we set $\Phi(p) = (p, \eta(p), 1)$.

Case II: If $P \ncong H$, then let an internal vertex of P be p, and let it have children $\kappa(p) = \{c_1, \ldots, c_k\}$. We let $\Phi|_\nu = \varphi$ (thus satisfying condition M_0). Now we construct Φ by considering the internal (non-tip) vertices of P in the canonical reverse partial order[1], going from the tips to the root of P. Let $\Phi(c_i)$ be (c_i, h_i, t_i)

[1] That is, if $p <_P q$ then we consider q first, and if p and q are not comparable then it doesn't matter which order we choose.

for all the children in $\kappa(p)$. Let $K = \{h_i\}$, the union of the h_i. Then let $\Phi(p)$ be (p, h, t) where $h \leq_H K$. That is, h is equal or ancestral to all the hosts h_i. If $h \in K$ then we put $t = 2$ since h must precede all the h_i ($h \prec h_i$). If $h \notin K$ then we put $t = 1$.

We now show that Φ as defined above satisfies all the conditions of a cophylogeny mapping.

Case I: If $P \cong H$ with isomorphism η, then we defined Φ to be the mapping $\Phi(p) = (p, \eta(p), \delta(p))$, where $\delta(p)$ is 2 if p is a tip and 1 otherwise. This automatically satisfies the lifting condition, since if η is an isomorphism between P and H, then φ must be $\varphi(p) = (p, \eta(p), \delta(p))$ for all $p \in \nu$. M_1 applies trivially, and M_2 holds by Eq. (1). B_0 is true by the definition of Φ, and both B_1 and B_2 hold since by construction all children of p are mapped by Φ to descendants of $\Phi(p)$.

Case II: Clearly M_0 is satisfied since we set $\Phi|_\nu = \varphi$. Condition M_1 cannot apply here since $P \not\cong H$. Now consider the "order preserving" condition, M_2: this states that for any $p \prec_P q$, we must have $\Phi(p) \prec \Phi(q)$. By our construction of $\Phi(p) = (p, h, t)$ we have the host h being equal or ancestral to all the h_i with which the children of p have been associated. If h is equal to one of these h_i then we set $t = 2$. Since $p \prec_P \kappa(p)$ we have $\Phi(p) \prec \Phi(\kappa(p))$ also.

The biological conditions hold for the same reasons as in Case I.

Thus we can safely say that there is indeed always a cophylogeny map for a given P, H and φ.

5.2 The number of cophylogenetic maps

In order to answer these questions it is necessary either (i) to determine aspects of the computational complexity of the cophylogeny problem, or (ii), somewhat less satisfactorily, to perform simulation tests and assess the number of possible solutions directly. Ronquist [23] has shown that the number of possible maps from parasite tree into host tree rises more than exponentially with the number of taxa in each.

In general though we are not interested in the number of cophylogenetics maps we could expect for *random* trees or networks: rather, we shall in general be investigating cases of apparent cophylogenetic evolution where we have good cause to do so. That is, we are not likely to test for significant apparent codivergence between two completely unrelated phylogenies, so in most cases the "expected" complexity is far worse than we *really* expect. It will be valuable to test a number of artificial cases in which there are just a few host switches, or only a slight mis-match between the phylogenies, in order that we may estimate the number of solutions available for more "realistic" cases.

The number of possible cophylogeny maps is at least as many as the number of solutions to the jungle constructed for a given H, P, φ, as in [5]. That construction limits the maps in order to eliminate from consideration as many as possible of the maps which are "guaranteeably non-optimal". Thus we may use jungles to assess the relative difficulty of all instances of the cophylogeny problem. At present although the theory is adequate, the software is unable to deal with reticulate (non-tree-like) phylogenies, so this section is restricted to tree-like phylogenies.

5.3 What are the statistical properties of cophylogeny maps?

It is of interest to the combinatorist how many maps there are satisfying various properties, but of little practical interest to a biologist who needs to determine the level of significance of cophylogenetic match between two phylogenies of interest. It is an open problem exactly how many cophylogenetic maps there are corresponding to a given number of codivergence events, and the author is currently investigating this in an empirical framework.

6 Optimality

So far I have discussed only the properties of the cophylogeny map which we need it to satisfy, and have not yet touched on the way in which we may determine which map provides the best explanation of the historical relationships between the phylogenies.

It is very easy to find an optimality criterion which provides a perfect score for a perfect fit, but much harder to find one which provides a meaningful measure when the fit is *not* perfect. The method developed by Brooks (known as Brooks' Parsimony Analysis or BPA) does indeed have a minimal cost when the parasite and host trees match (the method is not defined for networks), but the interpretation for imperfect matches is problematic in the extreme.

In the current approach, optimality is determined by a total cost, that is, to each cophylogenetic event is assigned a cost, and as the complete map Φ induces certain numbers of each of these events on the parasite phylogeny, so the total cost is simply the sum of the event costs multiplied by their counts. It is therefore conceivable that under different cost schemes, different maps may be optimal. What would be very handy is a method by which we could retrieve *all* such potentially optimal solutions from the set of all cophylogenetic maps. The first implementation of the jungle was able to locate all those reconstructions which were optimal for a given cost scheme, but this was not really adequate, since it is generally very hard to estimate the best values for the event costs.

(In the original jungle construction it was a requirement that the codivergence cost c_c should be strictly less than the other costs, and for convenience at the time, $c_c < 0$. More recently in order to implement a branch-and-bound in the TreeMap program [16], this has been altered such that $c_c = 0$ and the other costs are all strictly positive. This has no effect on the jungle construction if no bound is given, but permits a faster search if a bound is used: the jungle can be abandoned if no Φ can be recovered from it which has either (i) a sufficiently low total score, or (ii) few enough cophylogenetic events, or (iii) both. A detailed description of this new construction is in preparation by the author.)

A better solution would be to find all the reconstructions which could be optimal under *any* cost scheme, and consider them as a set of alternate hypotheses which we may then test. This is implemented in the latest version of TreeMap [16], by filtering out all the solutions with a total cost which is definitely higher than the rest. For instance, if a solution is recovered with 4 codivergence events, 2 host switches and 3 losses, then it is definitely more costly then a solution with 4 codivergences, 2 host switches, 4 losses and a duplication, but not necessarily any more costly than one with 2 codivergences, 2 host switches and 2 losses.

The costs can be likened to the negative log of their probability, so that extremely unlike events are given a high cost. In that sense the minimal cost solution is the most likely one, for that cost scheme.

Once we have a set of all the possible cophylogenetic maps which can conceivably be optimal under some scheme of event costs, we can answer questions about what costs are required for a given solution to be optimal. For instance if we have two solutions, one with a single host switch and one with no host switches, we can determine how high we must make the cost of a host switch in order for them to be precluded altogether. In a sense this is like a sensitivity analysis of the given cophylogeny problem, since it takes into account a range of models.

Without a proper probabilistic model of the event costs, that is, a prior distribution on the kinds of event costs (negative log probabilities) we should expect to observe in real life, it would seem that we cannot do much more than to recover the "best possible" cophylogeny maps and interpret them in light of biological information relevant to each case.

7 Existing methods of finding cophylogeny maps

There are, to my knowledge, no cophylogeny maps already in existence which are defined on non-treelike phylogenies. In principle Brooks' Parsimony Analysis [2,3] could be extended to networks, but it does not even work well on trees, so it seems pointless to pursue that extension. There are some other maps though:

7.1 Page's "reconciled trees"

Page adopted the term "reconciled tree" to indicate a particular kind of cophylogeny mapping, defined on trees, and which excluded all host switching events [17]. The scheme was devised as a solution to the problem of reconstructing gene/organism relationships from one or more gene tree and one organismal tree. I contended in 1998 [5] that the term be extended to include the more general case allowing host switches, but shall avoid potential confusion in this discussion at least, and talk about cophylogeny maps, not tree reconciliation.

7.2 Brooks' parsimony analysis

Also occasionally known as Hennig's Parasitological Method [2], this method is not a true cophylogeny mapping. It is a somewhat knee-jerk reaction to the problem of estimating the most parsimonious explanation as to the cause of differences between host and parasite phylogeny. It treats the cophylogeny problem as one of minimising "homoplasy", which, while it has a place in the realm of phylogenetic reconstruction, has no meaning in cophylogeny. I think it for this reason that BPA is indeed a "valiant failure" [12] – while it works perfectly well when P and H are isomorphic and φ preserves that isomorphism, it becomes uninterpretable when attempting to deal with host switches. The interested reader is directed to the debate between Page and Brooks in various papers [3,12,13].

7.3 Ronquist's solution

Ronquist has done a great deal of the analytical work on cophylogeny, and has calculated the computational complexity of many different approaches. Some of that is summarised in [23].

Much has been made of the good match between the phylogenies of pocket gophers and their chewing lice [1,9,14,15,19,20].

Ronquist has also given it some attention, but he differs from Page in his interpretation of the event costs. Effectively Ronquist assigns a *zero* cost to a duplication event, whereas Page assigns a *positive* cost. His approach is a true cophylogenetic map, though it does not take advantage of the implicit assumption of many methods, that the parasite lineages, once diverged, are independent of each other. For that reason his method requires a list of all the possible maps and configurations of host switches, which makes it slower.

In the paper of 1998 [5] are shown two solutions, each of which can be optimal under some cost scheme. Page's solution [14] is there, embedded in the jungle and shown separately, and Ronquist's solution [22] is in the jungle too, though technically this need not always be true (recall that in our construction of event costs we restricted ourselves to strictly positive duplication event costs).

8 Jungles and scary creatures

We now consider the set of *all* possible cophylogeny mappings for a given P, H and φ. This comprises a collection of sets of j-vertices, each set carrying with it an image of P.

The *jungle* [5] for a given P, H and φ is a graph $\mathcal{J}$ which contains as subgraphs a subset of all the possible Φ. The paradigm behind its construction is that, since many solutions will have the same hypothesised host/parasite associations (the j-vertices), and since the arcs in the graph induced by Φ are deemed to be independent of each other, it makes little sense to repeat them for each Φ. Instead the complete set of all Φ which are *not guaranteeably non-optimal* is constructed by finding the *feasible parent set* of each sibling set in the parasite tree. We can avoid some mappings as being definitely non-optimal by some obvious short-cuts such as were outlined previously [5], and by implicit use of the conditions which I described above.

Once the jungle is constructed, it can be traversed from the images of the tips of P to images of its root, that is, from the j-vertices in $\varphi(\nu)$ to all the possible $\Phi_i(r)$ where r is the root of P.

The original construction is applicable to the case in which P is a tree, since each non-root vertex in a tree has but a single parent, and each set of siblings trivially has a unique mrca. The construction proceeds from tips to root in P and is complete when all the possible feasible locations have been found for the root of P. This is adequate for multifurcating H and P, but not for cases in which either of the phylogenies are reticulate, that is, non-treelike.

8.1 A new jungle construction

Suppose we are presented with two phylogenetic networks, P and H, and we are told that P is the parasite. We are also presented with a mapping $\varphi : \nu \mapsto \xi$. Since we are attempting to deal with the most general possible cophylogeny formulation, and P is not necessarily constrained to be a tree, but only a phylogenetic network, we must determine Φ somewhat differently from the methods used previously. Consider an arc $a = (p, q)$ in P, such that p is a parent of q. p need not be the only parent of q, and q need not be the only child of p.

In the same spirit of nomenclature as existed at the time of the invention of *jungles*, we introduce the *liana* as the set of all possible j-vertices (p, h, t) given that q is associated with host h_q in j-vertex $\Phi(q) = (q, h_q, t_q) = u$.

Definition 1. The *liana* $\mathcal{L}(p, u)$ is

$$\mathcal{L}(p, u) = \{(p, h, t) : h \leq_H h_q, (p, h, t) \prec u\}.$$

Appropriately, we shall construct the entire jungle from lianas.

In other words, given an arc $e = (p, q)$ of the parasite tree in which the ancestral parasite is p, and given a j-vertex $u = (q, h, t)$, the liana $\mathcal{L}(p, u)$ is the set of all possible j-vertices placing p on any of the locations on the path(s) from h to the root of H.

We shall first see how this is consistent with the previous implementation of jungle construction, and then see how it extends to cope properly with reticulate (non-treelike) host and parasite phylogenies.

Consider the small problem shown in Fig. 9. In this example, we have parasites a and b infecting host species s and t respectively. The parent of a and b, in this case unique, but not necessarily so, is p, and the most recent common ancestor of s and t (also unique) is x. To illustrate the formation of lianas from different children, we denote the vertex at the ancestral end of edge (s, a) by s_a, and the vertex at the ancestral end of (s, b) by s_b. In this case $s_a = s_b$. We construct the liana $\mathcal{L}(p, (a, s, 2))$, which is listed as $\{(p_a, s, 2), (p_a, u, 1), (p_a, u, 2), (p_a, v, 1) \ldots (p_a, z, 2)\}$.

We must now in some way acknowledge the fact that a and b do share a common parent, in this case unique. Recall condition B_2, which states that for a type 1 j-vertex $v = (q, m, 1)$, all of its child j-vertices must be of the form

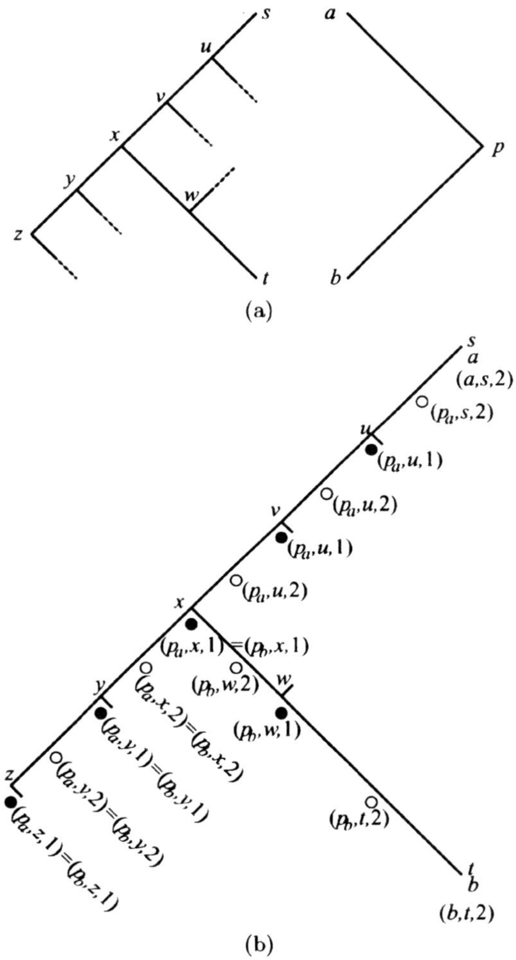

Fig. 9. The construction of two lianas. In the above figure we begin with a simple tanglegram on the pair of host and parasite taxa, $\{s, t\}$ and $\{a, b\}$ respectively (a). In (b) we construct the lianas $\mathcal{L}(p, (a, s, 2)$ and $\mathcal{L}(p, (b, t, 2))$. These are all the possible j-vertices denoting the positions of the parent p of a (resp. b) with respect to *any* other siblings sharing that parent.

(p, h, t) where p is a child of q and h is a descendent of m. Thus we may remove all the type 1 j-vertices from our construction which are not ancestral to both s and t. Also, we know from earlier work [5] that it is guaranteeably non-optimal to have j-vertices which are ancestral to the common ancestor of s and t, so we may remove them also from consideration.

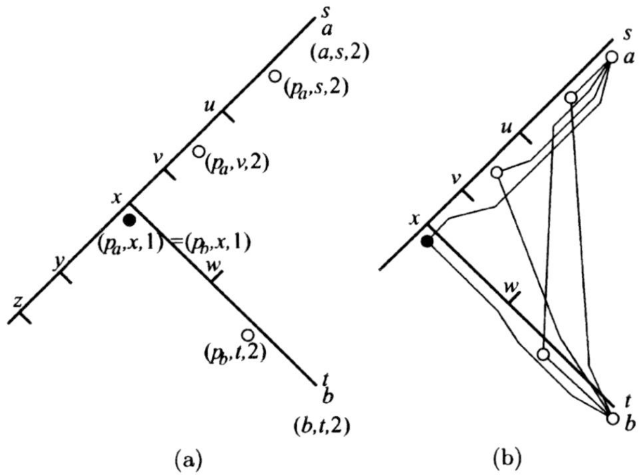

Fig. 10. Twisting two lianas together.In part (a) we see the remaining j-vertices after the identical ones have been coalesced and those which are guaranteeably non-optimal have been removed. Part (b) shows the cophylogeny mappings which correspond to this remainder

We thus arrive at a reduced set of j-vertices, in a sense "twisting" the lianas together. For a set X of j-vertices $v_i = (q_i, h_i, t_i), i = 1, \ldots, k$, such that $\{q_i\}$ is a sibling set with parent p, let the *meet* $\wedge$ of two or more lianas $\{l_i\}$ be that set of j-vertices $\wedge(l_1, \ldots, l_k) = \{u = (p, m_j, t_j) : u \in \mathcal{L}(p, v_i)$ for some $i, m_j \not\ll_H \{h_i\}\}$. This is clearer in an example. In Fig. 10 the lianas are "twisted" together by identifying the same parasite vertices ($p_a = p_b$) and removing all those j-vertices which we can guarantee do not occur in any optimal cophylogeny mapping.

To construct the entire jungle in this way we perform the above procedure from the more recent vertices of P to its root, for each arc (p, q) of P and each j-vertex v of the form (q, h_q, t_q) already in the graph J, calculating the liana $\mathcal{L}(p, v)$. For each $u \in \mathcal{L}(p, v)$ we then find each of its sets of child j-vertices (recall that $v = (q, h_q, t_q)$ is a child of $u = (p, h_p, t_p)$ precisely when q is a child of p, by construction), and create arcs from u to them. We must also take into account the partial ordering which host switches impose on the host phylogeny, since some combinations are simply impossible (they are *strongly incompatible*) while others must incur extra loss events on H (these are *weakly incompatible*) [5].

The procedure for networks is very lengthy, and unfortunately is beyond the aim of this current article. The author is currently preparing a manuscript detailing the construction of cophylogeny maps specifically in the case of non tree-like networks.

9 Examples

I now provide two examples of cophylogeny studies, using jungles to find the optimal maps from P into H. The reader interested in more of the the finer details of jungle construction is directed to [5].

9.1 A simple three taxon case

The tanglegram shown in Figure 11 shows P and H with three leaves each and the associations $\varphi : L(P) \mapsto L(H)$. To construct all the possible cophylogeny maps from P into H we must follow the conditions laid out earlier. We proceed in the same canonical order which was used in the argument above for the existence of a cophylogeny map (that map will also be recovered by this process), that is, from the tips of P to its root.

We first consider the sibling j-vertices $(q, b, 2)$ and $(r, c, 2)$, and construct the set of allowable j-vertices for the parent parasite s, with respect to $(q, b, 2)$ and $(r, c, 2)$. This *feasible parent set* (FPS) is

$$\{(s, b, 2), (s, c, 2), (s, d, 2), (s, e, 1)\}.$$

These are all the j-vertices whose hosts are ancestral to b or c, excluding all type 1 j-vertices which are not ancestral to both [5]. It turns out that the FPS of a parasite p with respect to j-vertices $\Phi(\kappa(p)) = \{v_1, v_2, \ldots\}$ is $\wedge\{\mathcal{L}(p, v_1), \mathcal{L}(p, v_2), \ldots\}$ when H and P are trees, which illustrates how lianas help us in reconstructing cophylogenetic maps for non tree-like phylogenies.

Next we find the FPS for each pair of sibling j-vertices. First we begin with $FPS(\{(p, a, 2), (s, b, 2)\}) = \{(t, d, 1)\}$. By a simple argument [5] this is the the only allowable j-vertex for these two siblings.

Now

$$FPS(\{(p, a, 2), (s, c, 2)\}) = \{(t, a, 2), (t, c, 2), (t, d, 2), (t, e, 1)\}.$$

Next,

$$FPS(\{(p, a, 2), (s, d, 2)\}) = \{(t, d, 2)\}.$$

And finally

$$FPS(\{(p, a, 2), (s, e, 1)\}) = \{(t, e, 2)\}.$$

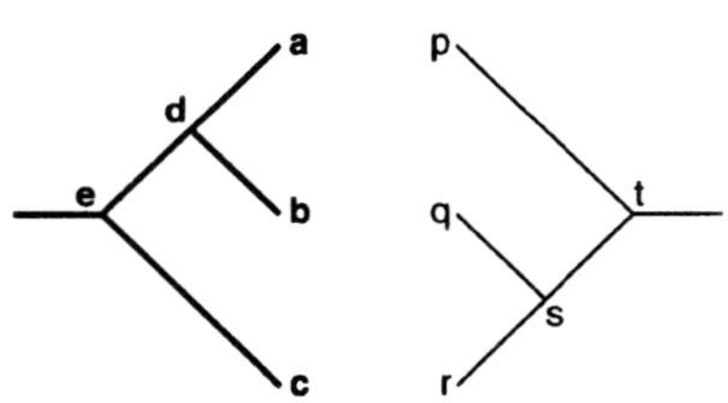

Fig. 11. Tanglegram of two three-taxon trees

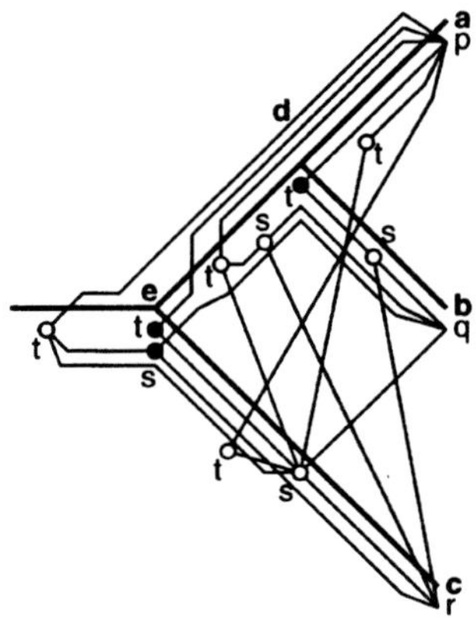

Fig. 12. Jungle for three taxon tanglegram. This jungle contains all the cophylogenetic maps which are allowed under our conditions. Several of these are guaranteeably non-optimal (see text).

Table 1. Cophylogeny maps for three taxon case

Solution	j-**vertices**	codivergences	duplications	losses	host switches
Φ_1	$\{(s,b,2),(t,a,2)\}$	1	0	0	1
Φ_2	$\{(s,b,2),(t,d,1)\}$	1	0	0	1
Φ_3	$\{(s,c,2),(t,a,2)\}$	0	0	0	2
Φ_4	$\{(s,c,2),(t,c,2)\}$	0	0	0	2
Φ_5	$\{(s,c,2),(t,d,2)\}$	0	0	1	2
Φ_6	$\{(s,c,2),(t,e,1)\}$	1	0	1	1
Φ_7	$\{(s,c,2),(t,e,2)\}$	0	1	3	1
Φ_8	$\{(s,d,2),(t,d,2)\}$	0	1	2	1
Φ_9	$\{(s,e,1),(t,e,2)\}$	1	1	3	0

The cophylogeny maps are embedded in the jungle J in Figure 12. By definition and our conditions they must be all the subtrees of J which are isomorphic to P under an isomorphism Φ which extends φ. They are listed in Table (1), excluding the j-vertices which were given by φ and which are in every cophylogeny mapping.

As we can see from the table, some solutions cannot be optimal. Φ_6 must cost more than Φ_1 and Φ_2; so we may discard it. Similarly we may lose Φ_5, Φ_7 and Φ_8; we are left with five solutions, each of which may be optimal under some scheme of event costs.

9.2 Lizards and malaria in the caribbean

Now we move on to a real example. This is taken from a recent study in which two varieties of *Plasmodium azuruphilum* were discovered to show significant

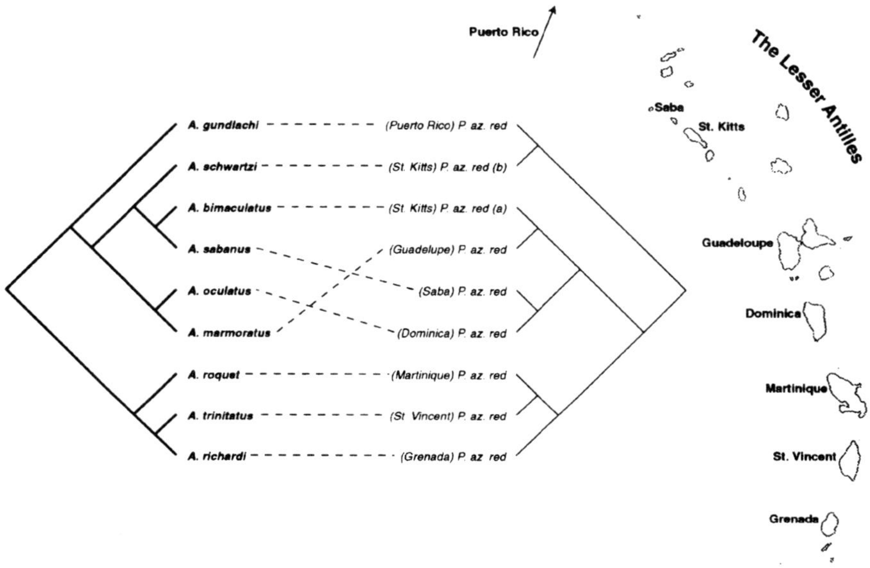

Fig. 13. A tanglegram for *Anolis* and *P. azurophilum.* The above shows the tanglegram depicting the phylogenetic relationships estimated by Charleston and Perkins [6], and in a map of the Lesser Antilles where they are found.

amounts of codivergence with their *Anolis* lizard hosts, through a group of Eastern Caribbean islands [6]. The next figures are adapted from that study.

The host taxa are several species of *Anolis* lizard, which show a high degree of local specialisation in their island territories. The malaria which afflicts them, *Plasmodium azurophilum*, appears to occur in two varieties which form distinct phylogenetic groups from each other, these being the "red" type and the "white" type. I show only the results of the comparison of *Anolis* with the former variety, *P. azurophilum* (red).

There were only three possibly optimal solutions to this problem, and they are represented in Figure 14.

We are met with the difficulty of how to choose among these solutions – which is the best? Which, if any, is the real answer? We do not have a *statistical* method to choose among these solutions, but we can say something about the mode of host switching of these varieties of malaria, since all of these potentially optimal solutions share the characteristic of a host switch from south to north. This is in the direction of the prevailing current in that region, which leads us to suspect that the malaria is able to "raft" from one island to the other (c.f. Censky *et al.* [4]). Also since the second solutions involve host switches *against* the current, we may consider them to be less likely to represent the true history of these organisms.

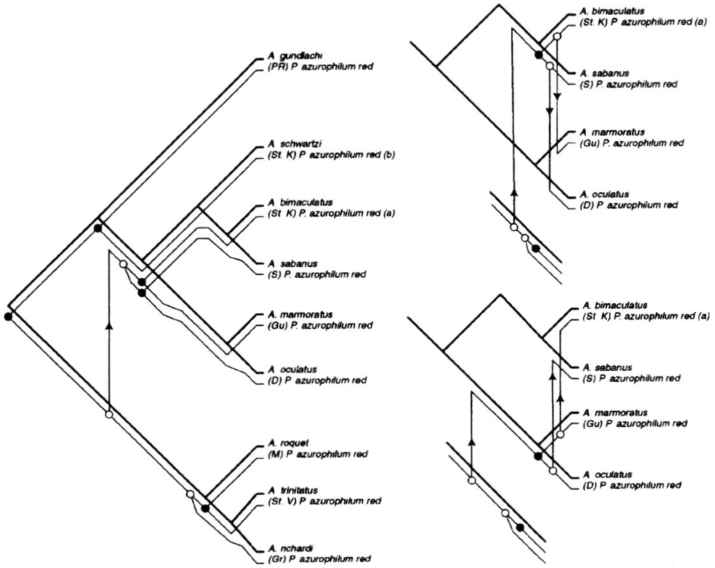

Fig. 14. Three potentially optimal cophylogeny maps from *P. azurophilum* (red) into *Anolis*

10 Discussion

There is a great deal of work to be done in this relatively new area. It requires new statistical methods, new mathematics, and new understanding in biology. It also presents some wonderful challenges both in terms of the intricacies of the theory and the subtle biological issues involved. There are so many questions, it is hard to know where to start!

There remain some major hurdles to jump, only some of which have been touched on here: The computational complexity of jungles and of the number of potentially optimal cophylogeny maps remain to be established, in particular as they relate to the number of taxa involved and the degree of mis-match between P and H; assignment of event costs themselves, and in fact the need to incorporate more phylogenetic information in those costs, is problematic; dealing with unseen host and parasite lineages; significance testing of cophylogenetic match; incorporation of more than two phylogenies into a cophylogenetic study. These are all very difficult, but it is my belief these hurdles are not impassable.

Further issues not mentioned in this brief introduction include the problem of dealing with parasites' *failure* to diverge with the host; detection of causes of apparent cophylogenetic match other than codivergence [7]; empirically estimating the most appropriate event costs; and dealing with limited "space" on the host, such that there can be no more than a certain number of parasite lineages per host lineage (as in Goddard and Burt's yeast problem [8]). In fact we have been guilty of the same implicit assumption of nearly all the current cophylogenetic methods, that the parasite lineages do not interact once they have diverged from each other. If we were to lose that assumption then the construction of the jungle would suffer greatly, since we could no longer have exactly the same components of a cophylogeny map (the j-vertices and arcs) in different solutions, nor would the arc costs remain constant across those solutions. Not being able to make those savings would hugely increase the computational complexity of constructing a jungle.

Further, we have restricted ourselves herein to the consideration of *pairs* of phylogenies, the parasite and the host, and mapped the former into the latter according to our well-defined ideas. However, things in nature do not, of course, fall into such neat cases. We may be unsure as to the relationship between the phylogenies – we may not know which is behaving like the "parasite" and which like the "host", and we may have more than just two phylogenies. In such cases it is clear that the construction and search outlined in this discussion will not be adequate. The computational complexity of trying all possible combinations of maps from T_1 into T_2, and T_2 into T_3 (say) is prohibitive. Instead we must pay closer attention to the rules which have been derived in order to minimise the construction of a "simple" jungle (when there are just two phylogenies) and extend them to cope with more complex cases. This hypothetical structure has been termed a *hyperjungle*, but remains to be developed...

On the other hand, a great deal of progress has been made already by several workers mentioned above. Alternative approaches using probabilistic models show some promise, though I believe there are just too many unknowns and too many complications in individual cases for these methods to gain us much understanding. I would, however, love to be convinced otherwise!

As I have mentioned, the assignment of event costs is difficult. The host switching cost should be related to phylogenetic or physical distance between hosts – there is much anecdotal evidence that host switching is much more likely to be successful between hosts which are more closely related than between more distantly related hosts. In a sense we can treat the filtering out of all guaranteeably non-optimal solutions (maps) and retention of all potentially optimal maps as a *sensitivity analysis*: without a prior distribution on the event costs we cannot compare them and say one is better than the other, but we can make conclusions about the specific linear combinations of event costs for which a given solution is optimal. This will help us towards better understanding of the biological processes involved.

Significance testing is possibly the thorniest issue at present. Ideally we might consider modelling the behaviour of the parasite phylogeny P given H, in a sense "growing" P along it, allowing for certain rates of duplication and extinction, and (conditional) probabilities of codivergence, lineage sorting and host switching. We could then see how many times according to our model we observed the same degree of phylogenetic congruence between H and our randomly grown P's. However there is an enormous variation in the behaviour of P's under such models, which suggests that the ability of such tests to recover with confidence precise estimates of the parameters of the model would be greatly limited. We would therefore have to conduct vast numbers of replicates in order to get good estimates of the probability of observing the same or better degree of phylogenetic match which we see in our data. Each biological system is different: there are many kinds of ecological interaction among groups of organisms and they cannot all be encompassed by one simplistic model. Should we therefore conduct our Monte Carlo tests for each empirical case? That would seem both impractical and useless, at this time.

Perhaps the empirical complexity of a given jungle, and/or the number of potentially optimal solutions, could form a useful guide to the true significance of fit: since in the case of a perfect match there is only one optimal solution (that of complete codivergence, for a total cost of zero), and for increasing degrees of mismatch between P and H there are more and more potentially optimal solutions, this might form a useful guide to the total

All in all the cophylogeny problem is complex, subtle, widely applicable in many different areas of biology, and becoming much more relevant now that more and more phylogenies are becoming available. The global melt-down of ecological diversity is leading to greater chances of unrelated organisms interacting, leading in turn to greater potential of new pathogens crossing the species barrier into the human population. Understanding the way in which such cross species transmissions occur is of fundamental importance and it is through phylogenetic tools such as cophylogenetic maps which will shed the light we need.

Acknowledgements

This work was presented at the International Workshop on Biological Evolution and Statistical Physics, May 10-14, 2000, in the Max Planck Institut Für Physik Komplexer Systeme, Dresden, Germany. It greatly benefited from conversation with Mike Steel at that meeting. It was subsequently helped on its way by absolutely no-one, but in the comfort of my office in Oxford and the patience of my fiancée – now my wife. The author is a Royal Society University Research Fellow.

References

1. S. C. Barker. Lice, cospeciation and parasitism. *International Journal for Parasitology*, 26:219–222, 1996. Reply to Page et al. 1996.
2. D. R. Brooks. Hennig's parasitological method: A proposed solution. *Systematic Zoology*, 30:229–49, 1981.
3. D. R. Brooks. Parsimony analysis in historical biogeography and coevolution: methodological and theoretical update. *Systematic Zoology*, 39:14–30, 1990.
4. E. J. Censky, K. Hodge, and J. Dudley. Over-water dispersal of lizards due to hurricanes. *Nature*, 395:556, 1998.
5. M. A. Charleston. Jungles: A new solution to the host/parasite phylogeny reconciliation problem. *Mathematical Biosciences*, 149:191–223, 1998.
6. M. A. Charleston and S. L. Perkins. Lizards, malaria, and jungles in the caribbean. In *Tangled trees: phylogeny, cospeciation, and coevolution*. University of Chicago Press, 2000.

7. M. A. Charleston and D. Robertson. An alternative hypothesis to explain the apparent codivergent history of primate lentiviruses. *x*, x(x):x, x. (in prep.).
8. M. R. Goddard and A. Burt. Recurrent invasion and extinction of a selfish gene. *Proceedings of the National Academy of Sciences of the U.S.A.*, 96(24):13880–13885, 1999.
9. M. S. Hafner and S. A. Nadler. Phylogenetic trees support the coevolution of parasites and their hosts. *Nature*, 332:258–259, 1988.
10. J. P. Huelsenbeck, B. Rannala, and Z. Yang. Statistical tests of host-parasite cospeciation. *Evolution*, 51(2):410–419, 1997.
11. J. J. Nino-Vasquez, D. Vogel, R. Rodriguez, A. Moreno, M. E. Patarroyo, G. Pluschke, and C. A. Daubenberger. Sequence and diversity of DRB genes of *Aotus nancymaae*, a primate model for human malaria parasites. *Immunogenetics*, 51(3):219–230, 2000.
12. R. D. M. Page. Component analysis: A valiant failure? *Cladistics*, 6:119–36, 1990.
13. R. D. M. Page. Maps between trees and cladistic analysis of historical associations among genes, organisms, and areas. *Systematic Biology*, 43:58–77, 1994.
14. R. D. M. Page. Parallel phylogenies: Reconstructing the history of host-parasite assemblages. *Cladistics*, 10:155–73, 1994.
15. R. D. M. Page. Temporal congruence revisited: comparison of mitochondrial DNA sequence divergence in cospeciating pocket gophers and their chewing lice. *Systematic Biology*, 45:151–67, 1996.
16. R. D. M. Page and M. A. Charleston. TREEMAP program. Version 2.0d. Platforms: Macintosh®. In development.
17. R. D. M. Page and M. A. Charleston. From gene to organismal phylogeny: Reconciled trees and the gene tree/species tree problem. *Molecular Phylogenetics and Evolution*, 7(2):231–240, 1997.
18. R. D. M. Page and M. A. Charleston. Trees within trees: phylogeny and historical associations. *Trends in Ecology and Evolution*, 13(9):356–359, 1998.
19. R. D. M. Page, D. H. Clayton, and A. M. Paterson. Lice and cospeciation: A response to barker. *International Journal for Parasitology*, 26:213–8, 1996. Response to Barker 1994, he replied 1996 Response to Barker 1994, he replied 1996.
20. R. D. M. Page and M. S. Hafner. Molecular phylogenies and host-parasite cospeciation.: gophers and lice as a model system. In J. M. S. P. H. Harvey, A. J. Leigh Brown and S. Nee, editors, *New uses for new phylogenies*, pages 255–70. Oxford University Press, Oxford, 1996.
21. A. M. Paterson, R. D. Gray, and G. P. Wallis. Parasites, petrels and penguins: Does louse presence reflect seabird phylogeny? *International Journal for Parasitology*, 23(4):515–526, 1993.
22. F. Ronquist. Reconstructing the history of host-parasite associations using generalised parsimony. *Cladistics*, 11:73–89, 1995.
23. F. Ronquist. Parsimony analysis of coevolving species associations. In *Tangled trees: phylogeny, cospeciation, and coevolution*. Chicago University Press, 2000. (in press).
24. P. Sharp, E. Bailes, D. L. Robertson, F. Goa, and B. H. Hahn. Origin and evolution of AIDS viruses. *Biol. Bull.*, 196:338–342, 1999.
25. K. F. Shortridge, N. N. Zhou, Y. Guan, P. Gao, T. Ito, Y. Kawaoka, S. Kodihalli, S. Krauss, D. Markwell, K. G. Murti, M. Norwood, D. Senne, L. Sims, A. Takada, and R. G. Webster. Characterization of avian H5N1 influenza viruses from poultry in hong kong. *Virology*, 252(12):331–342, 1998.

Accounting for phylogenetic uncertainty in comparative studies of evolution and adaptation

Mark Pagel[1] and François Lutzoni[2]

[1] School of Animal and Microbial Sciences, University of Reading, Reading RG6 6AJ UK
[2] Department of Biology, Duke University, Box 90338 Durham, NC 27708 USA

Abstract. We describe the application of Markov Chain Monte Carlo (MCMC) methods to two fundamental problems in evolutionary biology. Evolutionary biologists frequently wish to investigate the evolution of traits across a range of species. This is known as a comparative study. Comparative studies require constructing a phylogeny of the species and then investigating the evolutionary transitions in the trait on that phylogeny. A difficulty with this approach is that phylogenies themselves are seldom known with certainty and different phylogenies can give different answers to the comparative hypotheses. MCMC methods make it possible to avoid both of these problems by constructing a random sample of phylogenies from the universe of possible phylogenetic trees for a given data set. Once this sample is obtained the comparative hypotheses can be investigated separately in each tree in the MCMC sample. Given the statistical properties of the sample of trees – trees are sampled in proportion to the probability under a model of evolution – the combined results across trees can be interpreted as being independent of the underlying phylogeny. Thus, investigators can test comparative hypotheses without the real concern that results are valid only for the particular tree used in the investigation. We illustrate these ideas with an example from the evolution of lichen formation in fungi.

1 Introduction

Phylogenetic trees describe the pattern of descent amongst a group of species. With the rapidly accumulating quantities of DNA sequence data, more and more phylogenies are being constructed based upon sequence comparisons (Fig. 1). The combination of phylogenies and statistical models for the analysis of trait evolution provides investigators with a means to reconstruct the probable ancestral states and trajectories of traits as they evolved in the past, and to test hypotheses about correlations among pairs of traits (Pagel, 1997, 1999a). Comparative methods comprise one of biology's most enduring set of techniques for investigating evolution and adaptation (Harvey and Pagel, 1991). They are widely used in evolutionary biology, molecular evolution, animal behaviour, ecology and conservation.

Recent applications of statistical comparative models of trait evolution to phylogenies include reconstructing the nucleotide content of the common ancestor to life on Earth (Galtier, Tourasse, and Gouy, 1999), predicting ancestral

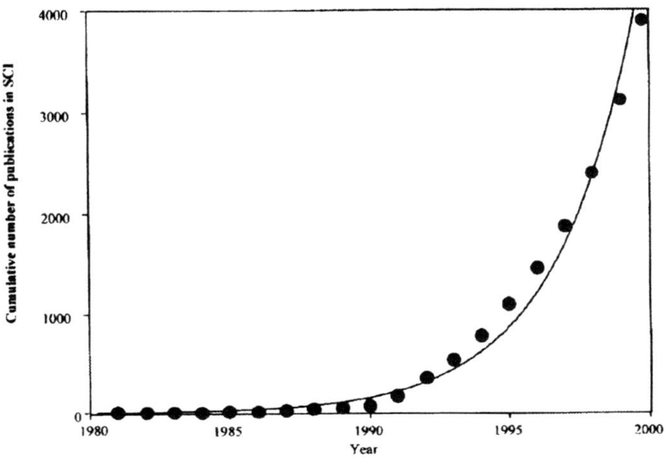

Fig. 1. Plot of the total number of articles reporting the key words "molecular phylogeny" in their title, keywords, or abstract in the years 1981 to 2000. Data from the Science Citation Index. The data are well approximated by exponential curve

ribonuclease enzyme sequences in artiodactyls (Jermann et al. 1995; Schluter, 1995), investigations of parallel molecular evolution in the opsin genes of the visual system (Chang and Donoghue, 2000), detection and reconstruction of ancestral 'signature sequences' that identify common ancestry amongst a group of organisms (van Dijk, et al., 2001), estimation of timings of key events in the history of evolution (Bromham and Rambaut, 1998; Cooper and Penny, 1997; Hedges, et al., 1996), detection of correlated evolution among different sites of the same gene (Krakauer, et al., 1996; Pollock, Taylor, and Goldman, 1999), and of shifts in rates of nucleotide substitution as a consequence of transitions to mutualism in fungi (Lutzoni and Pagel 1997).

Comparative analyses of this kind presume that the phylogeny is known without error. This assumption has long plagued the field because phylogenies are inferences from data, they are subject to error and uncertainty, and different estimates of the phylogenetic tree can return different answers to the comparative question. As a result, all conclusions derived from comparative analyses performed on a single phylogenetic tree are conditional upon that phylogeny.

Attempts to resolve the problem of phylogenetic uncertainty frequently involve trying out the statistical inference on a sample of 'best' or 'favoured' trees, or on all equally most parsimonious trees (e.g., Hibbett, Gilbert, and Donoghue, 2000). Another approach is to construct a consensus tree from some subset of suitably chosen trees. These approaches are all limited by lacking any clear probabilistic basis. The consensus tree is by definition not the most probable tree (there may be some exceptions), and there is no particular reason to believe that the best or most probable tree according to some goodness of fit criterion (such as maximum likelihood or parsimony) is necessarily the true tree. The vast number of possible phylogenetic trees even for small numbers of species or lineages (over 34 million for a tree with only ten tips, over 1076 for a tree of 50 lineages) makes it impractical, save for the smallest cases, to enumerate and analyse all possible trees.

2 Markov-chain Monte Carlo methods

Markov-Chain Monte Carlo Methods Markov-chain Monte Carlo (MCMC) methods offer a formal statistical procedure for taking phylogenetic uncertainty into account in comparative studies. The MCMC approach is to construct a Markov-chain which, if allowed to run long enough, produces states in direct proportion to the equilibrium distribution of states in the model (Gilks, et al. 1996). Applied to phylogenetic trees the goal is to construct a Markov-chain whose states are the possible phylogenetic trees in the universe of all possible phylogenetic trees. If this chain is allowed to reach equilibrium, then successive states of the chain will sample trees in proportion to their 'equilibrium' probabilities, that is, in proportion to their probabilities under some model of evolution.

Given a random sample of phylogenetic trees, the comparative parameters of interest (rates of evolution, ancestral states, trends, correlations, and so on) can then be estimated over the sample, yielding their density distributions. Inferences about trait evolution based upon these distributions will be independent of any particular phylogenetic hypothesis. Owing to characteristics of the MCMC sampling, the statistical distribution of the comparative parameters, if weighted by an appropriate prior probability, can be interpreted as the Bayesian posterior probability distribution.

3 Bayesian statistics and the MCMC approach to phylogenies

MCMC methods implemented in a phylogenetic context, attempt to produce a Markov-chain that if allowed to run long enough will randomly visit sites in the universe of phylogenetic trees in proportion to their probabilities under the model of evolution. If all of these possible sites (trees) are thought *a priori* to be equally likely then the distribution of tree probabilities in the MCMC sample directly estimates the true distribution of tree-probabilities, that is, the probability density distribution of trees.

However, some trees may be thought more or less likely on *a priori* grounds. The MCMC sample then provides a way to update those prior beliefs. Bayesian statistical logic provides the formal framework within which one uses a realised set of outcomes to update a set of prior beliefs, the result being a posterior set of beliefs.

Let S represent a set of aligned gene-sequence data on a set of species, and let ω be a phylogenetic tree, specified by the parameters of a model of gene-sequence evolution and its branch lengths. We may wish to say something about the distribution of ω (that is, about the probability distribution of phylogenetic trees) as a function of S. Formally, Bayes' Rule states that,

$$P(\omega|S) = \frac{P(\omega)P(S|\omega)}{P(S)}, \tag{1}$$

where $P(\omega|S)$ is the posterior probability of ω given S, $P(\omega)$ is the prior probability of ω (in the absence of any knowledge about S), $P(S|\omega)$ is the probability of the sequence data S given ω and $P(S)$ is the probability of the sequence data. $P(S)$ is calculated over all possible trees. In the absence of any prior beliefs is typically set to $1/N$ for all trees where N is the total number of trees.

To accomplish the sampling needed to estimate the frequency distribution or probability density of $p(\omega|S)$, a Markov-chain is constructed whose states are the possible phylogenetic trees. At each step in the chain a new tree is proposed by a tree-proposal algorithm that alters characteristics of the current tree. New trees are accepted or rejected with probabilities determined by the Metropolis-Hastings (M-H) algorithm (Metropolis et al., 1953; Hastings, 1970):

$$\frac{p(S|\omega^*)p(\omega^*)}{p(S|\omega)p(\omega)}\frac{q(\omega^*,\omega)}{q(\omega,\omega^*)}, \tag{2}$$

where asterisks denote 'new' trees (sets of model parameters and branch lengths) and $q(\omega,\omega^*)$ is the probability of moving in the parameter-space from ω to ω^*.

If the M-H ratio is greater than 1, the new tree is accepted with probability 1. If the ratio is less than 1, the new tree is accepted with probability equal to the ratio. If the new tree is not accepted the chain remains in the 'old' state. The M-H algorithm ensures that the Markov-chain, if allowed to run long enough, visits successive states in the universe in proportion to their likelihoods. The chain is run until a large number (say 500,000) of trees is generated, from which a smaller number (say 5000) is sampled to ensure independence among successive trees in the chain. Then the distribution of ω is estimated from the smaller sample.

The emphasis on estimating the distribution of ω rather than on finding the single best (e.g., highest likelihood, most parsimonious, shortest distance) tree distinguishes MCMC approaches from conventional 'single-tree' studies. MCMC-phylogenetic methods do not seek the best tree, rather they seek to sample in an unbiased way from the probability distribution of trees. The logic underlying this approach is that there is no particular reason to believe that the best tree under some model of evolution corresponds to the true phylogenetic tree. For example, the 'best' tree under a model of parsimony is the one that yields the fewest evolutionary transitions. However, if events of parallel or convergent evolution occur in a number of independent lineages, then seeking the shortest tree may not return the best estimate of the true phylogeny.

The MCMC approach can be contrasted with existing approaches for sampling the universe of trees. The best known of these procedures is the non-parametric bootstrap (Felsenstein, 1985). The bootstrap procedure derives a sample of phylogenetic trees by resampling repeatedly from S. If the data matrix is of length n sites, a single bootstrap sample is created by sampling n

sites at random with replacement from S. If this is repeated, say, 100 times, and only one most optimal tree was recovered from each resampled data set, then a bootstrap sample of 100 trees can be constructed. There has been much debate about precisely what the formal statistical properties of the bootstrap are (Felsenstein and Kishino, 1993; Newton, 1996). Some authors suggest that it approximates a MCMC sample with a uniform prior (i.e., all trees equally likely *a priori*). Whatever its correct interpretation, the computational effort to create a bootstrap sample of trees using maximum likelihood procedures is prohibitive (Larget and Simon, 1999), and the MCMC procedure is not restricted to uniform priors.

MCMC methods have begun to be used to infer aspects of phylogenies (e.g., Yang and Rannala, 1997; Larget and Simon, 1999; see Lewis, 2001 for a recent review) and to estimate population genetic parameters on genealogies (e.g., Wilson and Balding, 1998). How to combine MCMC methods for phylogenetic inference with comparative methods for investigating evolutionary processes has received very little attention. Our own work on the evolution of lichen-formation and loss of lichenization in fungi (Lutzoni, Pagel, and Reeb 2001) and a demonstration of estimating gains and losses of horned soldiers in aphids (Huelsenbeck et al., 2000), are to our knowledge the only attempts to combine comparative methods and MCMC sampling of phylogenies.

4 MCMC and phylogenetic-comparative methods

The MCMC approach to comparative methods must somehow include the comparative data and associated evolutionary model parameters in the Markov-chain along with the sequence data and model of evolution used to construct the phylogenetic trees. Let D be the set of comparative data and μ be the parameters of the model of evolution used to analyse the comparative data (μ might for example contain variances and covariances, the parameters of correlations and regression coefficients). The comparative data might be a set of quantitative traits such as body size and life history variables, or traits that adopt a finite number of states, such as mating system or diet. The goal is to estimate some feature of the comparative data - for example, the ancestral state of some trait at a specified interior node of the phylogeny - simultaneously accounting for the uncertainty in the phylogeny.

The most obvious way to combine the comparative data analysis with estimation of the phylogeny is to build a Markov-chain that simultaneously samples the comparative parameters and the phylogenies (Huelsenbeck et al., 2000). That is, the parameters in μ and ω are estimated simultaneously on S and D. The Markov-chain is made to traverse simultaneously the space of phylogenetic trees and comparative outcomes (e.g., ancestral states, rates of evolution, etc). This is accomplished by adding a comparative-parameter-proposal mechanism to the usual tree-proposal mechanism. The Metropolis-Hastings algorithm is then applied to accept or reject new combinations of trees and comparative relationships. If left to run long enough, this approach will visit trees and their comparative results in proportion to their joint probabilities in the universe. The posterior distribution of the comparative parameters, such as a correlation or regression, or an ancestral state, can then be calculated directly from the sampled Markov-chain.

Although this is the formal MCMC approach, it may have drawbacks. There is no particular reason to believe that the comparative data will in general provide good information about the phylogeny. Traits are often selected for investigation in comparative studies because they evolve independently a number of times on the tree, a phenomenon known as homoplasy. The more homoplasy a character shows the less information it has about phylogenetic history. Most phylogenetic tree reconstruction algorithms will try to minimise the amount of homoplasy. This will influence the way the Markov-chain traverses the tree-space because combinations of trees and comparative results that return the highest likelihood (least homoplasy) will be preferred.

Another way to describe this problem is that the MCMC algorithm will traverse a different sample of phylogenetic trees depending upon the comparative data used in combination with the gene-sequence data. Different hypothesis tests may be based upon different kinds of samples. The argument in favour of including the comparative data is that any information that is available should be included when searching the tree space.

A second approach to comparative-phylogenetic MCMC samples the phylogenetic trees independently from the comparative outcomes. In this approach a sample of phylogenetic trees is produced by MCMC from a set of aligned gene-sequence data (S). Then the comparative data D are analysed on each tree to derive the posterior distribution of the parameters in μ. The values of mu cannot influence the sampling of phylogenies, but the approach retains the desirable feature that the comparative outcomes are automatically weighted by the probability of a given tree type in the posterior distribution of trees. If the parameters of the comparative model are independent of the phylogenetic tree topology, this method will approximate very closely the Bayesian posterior, and return results similar to those of procedure (i). We have employed this approach in the Lutzoni et al. (2001) investigations and is the technique we shall report below.

5 Application to the evolution of lichen formation

The lichen symbiosis consists of an obligate mutualistic association of a fungus species with an alga, a cyanobacterium, or with species of both photobiont types. We wished to investigate the evolution of lichen symbioses in the Ascomycota fungi. This phylum contains approximately 98% of all known lichens. One of us (FL) obtained sequence data on the small and large subunit nuclear ribosomal DNA (SSU and LSU nrDNA) of 54 fungi species. Fifty two of these species are in the Ascomycota phylum and two Basidiomycota were used as outgroups. In addition, we recorded for each species whether it was lichen-forming or not.

5.1 MCMC phylogenetic tree sampling

We used Markov-chain Monte Carlo (MCMC) methods (Larget and Simon, 1999) to approximate the posterior probability density of phylogenetic trees. Given a set of aligned gene sequences S, and following Bayes' theorem, the posterior probability of the ith tree sampled ω_i is

$$P(\omega_i|S) = \frac{L(S|\omega_i)\,P(\omega_i)}{\sum_{j=1}^{N} L(S|\omega_j)\,P(\omega_j)}, \tag{3}$$

where $L(S|\omega_i)$ is the likelihood of the sequence data given the tree, and $P(\omega_i)$ is the prior probability of the tree (here assumed to be uniform at $1/N$). The summation in the denominator is over the N possible trees for the set of species. The likelihood of the sequence data given the tree is integrated over all possible combinations of branch lengths and parameters in the model of sequence evolution. The posterior probability, $P(\omega_i|S)$, is the probability that tree i would arise given the model of sequence evolution.

For a tree of 54 taxa the summation in the denominator of $P(\omega_i|S)$ is vast. The MCMC procedure is used to approximate $P(\omega_i|S)$ by drawing a random sample of trees. We used the general time reversible model of gene-sequence evolution combined with gamma rate heterogeneity to estimate the likelihood of each tree (Hillis et al. 1996). Following convergence of the Markov-chain we generated 200,000 trees. We excluded information on the state (lichen-forming/non-lichen-forming) of each species from the MCMC sampling procedure to ensure that the distribution of trees was not influenced by this trait. A series of runs using the BAMBE (Larget and Simon, 1999) 'global' and 'local' options was conducted to ensure that the Markov-chain converged to the same region in the universe of trees.

5.2 Reconstruction of gains and losses, and ancestral states

We employed a continuous time Markov-model of trait evolution, as implemented in the computer program Discrete (Pagel, 1994), to investigate the evolutionary rate of gains and losses of lichenization. The Markov- model approach in Discrete permits independent gains (q_{01}) and losses (q_{10}) in each branch of the phylogenetic tree. We calculated these separately for each tree sampled in the MCMC procedure. Because trees are represented in the sample in proportion to their probability, investigating the rates over all trees automatically weights our results by the probability of a particular tree.

The gain and loss parameters q_{01} and q_{10} contain the information required to reconstruct ancestral states. The calculation of the posterior probability of the ancestral state at a node follows Pagel (1999b) using the 'local' method and was found from

$$P(n_i = s_j|D,t) = \frac{L(D|n_i = s_j,t)\,P(n_i = s_j)}{\sum_{j=0}^{1} L(D|n_i = s_j,t)\,P(n_i = s_j)}, \tag{4}$$

where $P(n_i = s_j|D, t)$ is the probability that node i takes state j, given the data (lichen-forming/non-lichen-forming) and phylogenetic tree, t, $L(D|n_i = s_j, t)$ is the likelihood of the data given that node i takes state j on tree t, and $P(n_i = s_j)$ is the prior probability that node i takes state j, here assumed to be 1/2. The summation in the denominator is over the two possible states.

5.3 Convergence of the Markov-chain

Fig. 2 shows a typical run of the Markov-chain of trees, evaluated by their log-likelihoods. The chain quickly ascends from a region of trees that produce very poor fits of the data to the model of evolution, into a region in which the chain reaches an asymptote. The enlarged region of this part of the chain (inset) shows how the converged chain then 'wanders' around tree space, sampling trees as directed by the Metropolis-Hastings algorithm. This serves to underscore that the MCMC procedure is not designed to find the best tree, but rather visits trees in proportion to their probabilities under the model of evolution. We sampled 20,000 trees from 200,000 trees generated from the converged chain. We then removed the first 100 to ensure that no trees were included prior to convergence of the chain. Fig. 3 shows the frequency distribution of log-likelihoods for the remaining 19,900 trees.

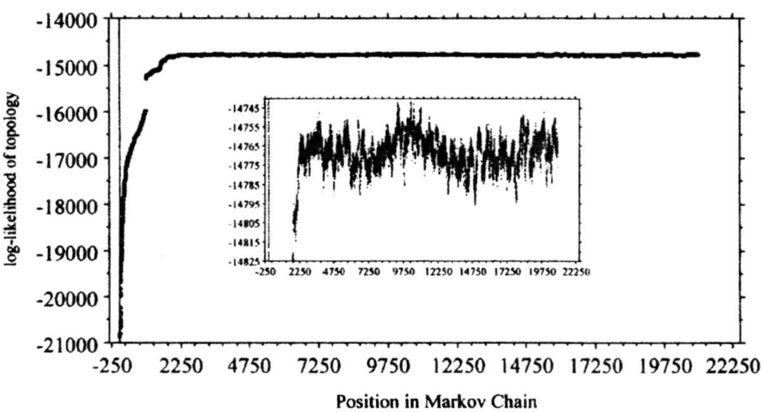

Fig. 2. The convergence of a Markov-chain. The y-axis plots the log-likelihood of successive phylogenetic trees of 52 species of Ascomycota fungi (plus two basidiomycete outgroup species) in the Markov-chain. Likelihoods were calculated from a model of gene-sequence evolution allowing unequal rates of transitions and transversions and allowing for unequal rates of evolution at different sites (Hillis, et al., 1996 [14]). Data were small and large subunit nuclear ribosomal DNA. The inset shows how the converged Markov-chain 'wanders' the tree-space in the converged region.

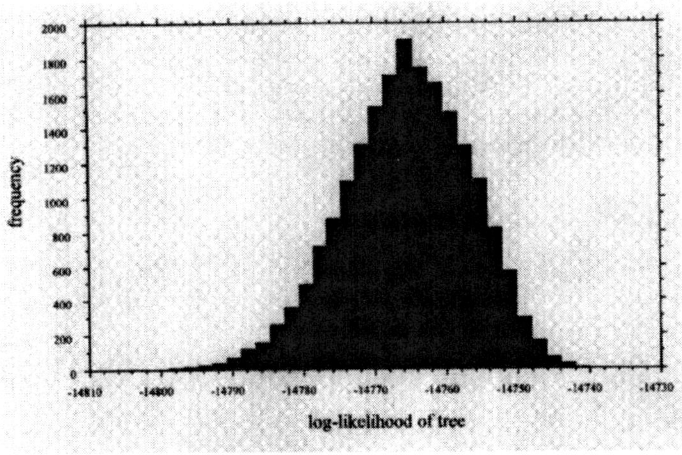

Fig. 3. The frequency histogram of 19,900 trees sampled from the converged Markov-chain.

5.4 Rates of gains and losses of lichenization

Fig. 4 shows that in 18,029 of 19,900 (90.6 %) trees sampled from the Markov-chain the estimated rate of loss exceeds the rate of gain (i.e., the loss/gain ratio is greater than one, and therefore is above the diagonal line in Fig. 4. The average ratio of the rate of loss to the rate of gain is 1.56 ± 0.53 and is positively skewed towards higher ratios (range $= 0.56 - 3.24$). The highest rates of loss are associated with the lowest rates of gain (Fig. 4); $r = 0.40$) when examined across trees. The ratio of losses to gains is, however, independent of the phylogenetic tree topology (correlation between ratio of rate of losses to gains and log-likelihood of trees $= -0.024$).

These results indicate that approximately 1.5 times as many losses of the lichen symbiotic state as gains are expected to have occurred during the evolution of the Ascomycota. This conclusion can be drawn without reference to any given phylogenetic tree. Previous work on the evolution of the lichen symbiosis based upon single phylogenetic trees and non-stochastic models of trait evolution based upon parsimony (Gargas, et al., 1995) suggested that lichens evolved independently on many separate occasions, with few losses.

For purposes of comparison with the conventional 'single-tree' approach to comparative studies, we estimated the same rates of gain and loss of lichenization on the consensus tree of the 19,900 trees. The Table shows the estimates and 95% confidence intervals for the single consensus tree and for the data reported in Fig. 2. The rate of loss of lichenization exceeds that of gains on the consensus tree, but the 95% confidence intervals are wide and overlap. Estimates derived from the MCMC procedure are more similar to each other but confidence intervals are narrower and do not overlap the value of the opposite parameter. Thus, the single-tree approach would not allow one to reject the hypothesis of equal rates

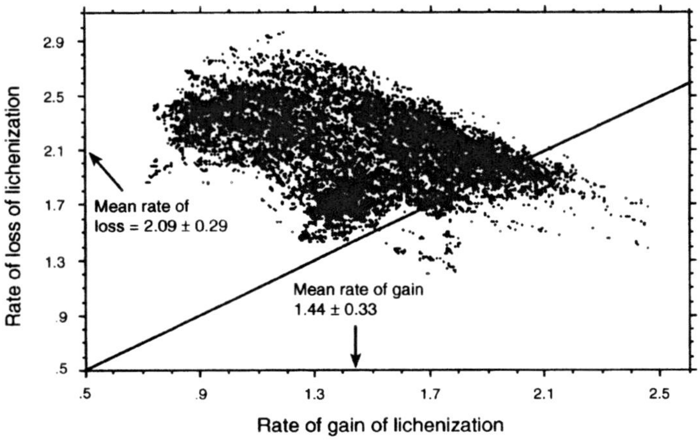

Fig. 4. The rate of loss of lichenization exceeds the rate of gains of lichenization, independently of tree topology. Data are for 19,900 MCMC trees. Solid line is the 1:1 relationship.

Table 1. Rates of gain and loss of lichenization estimated on consensus tree and MCMC sample

Transition Rate	Consensus tree of 19,000 MCMC trees	Separate estimates on each MCMC tree
gain	1.04 (0.05-4.5)	1.44 (0.85-2.05)
loss	2.41 (0.7-5.6)	2.09 (1.55-2.64)

of gains and losses, whereas strong evidence against this hypothesis emerges from Fig. 4.

5.5 Probable phylogenetic position and number of gains and losses

Fig. 1 (right panel) shows the majority-rule consensus phylogenetic tree as derived from our MCMC sample. We show this tree not to propose a particular phylogeny, but to provide a vehicle for identifying events of probable gains and losses of lichenization. The numbers above each internal branch correspond to the node to which a branch points. These numbers represent the proportion of trees in our MCMC sample in which that node was observed. Nodes with values of 100 define a collection of species all of which and only those of which appeared in every one of the trees sampled from the Markov-chain. Other nodes were less certain.

We reconstructed the most probable ancestral states (Pagel 1999b) of fourteen nodes. These nodes identify groups of lichen-forming and non-lichen-forming species in such a way as to make it possible to put reasonable upper and lower limits on the number of independent gains and losses of lichenization. Based upon these reconstructions, we can infer that green areas of the tree are regions of lichen symbiosis, red areas are regions in which the ancestral state is uncertain, and the remaining (uncoloured) branches correspond to non-lichenized regions of the tree.

The left panel of Fig. 5 plots the ancestral state for each of these labeled nodes separately across the 19,900 sampled trees (average probability across all trees labeled on right y-axis of each plot). The reconstructed ancestral state is independent of the phylogeny for some nodes, but for others, the particular tree topology can exert a substantial effect.

The pattern of reconstructed ancestral states implies that a minimum of one and a maximum of three gains of lichenization occurred during the evolution of the Ascomycota. If lichen formation evolved immediately after node 65 (labeled # on phylogeny), then one gain of lichenization is implied for the Ascomycota. Two origins of lichen symbiosis are implied (labeled * on phylogeny) if lichenization evolved independently at node 89 and again at the base of the clade that includes the Lichinales (LI), Arthoniales (AR) and Pyrenomycetes-Dothideales (PD). Three independent gains are implied if the closely related AR and LI groups independently evolved lichenization (labelled † on phylogeny).

By comparison, a minimum of three and possibly four losses of the lichen symbiosis have occurred in the Ascomycota. Nodes 93, 97, and 101 high posterior probabilities of being lichenized (left panel of Fig. 5), and each is followed by an unambiguous loss of the lichen symbiosis. For these three nodes the MCMC procedure leaves little doubt that a loss of lichenization quickly followed in the descendant species. A fourth loss of lichenization is implied at the base of the PD group if lichen-formation indeed originated at the points labeled '#' or '*' on the tree.

6 Conclusions and discussion

We have shown how to combine estimation of the phylogenetic tree with a statistical model of trait evolution to account for phylogenetic uncertainty when investigating historical events of evolution. MCMC sampling makes it possible to derive the posterior probability distribution of parameters that are relevant to testing hypotheses of evolution and adaptation. This can add statistical power to inferences and allows one to distinguish between relatively certain and less certain results about the evolution of a given trait.

Our results for the evolution of the lichen symbiosis overturn the conventional wisdom that lichens evolved independently on many separate occasions. Rather, our results suggest that lichens evolved earlier than previously believed and that some of the major fungal lineages that are strictly composed of non-lichenized extant species are derived from lichen ancestors. The minimum of three losses of the lichen symbiosis that we have identified indicate that entire orders of non-lichen-forming fungi are in fact derived from lichen-forming ancestors. This result serves to emphasize that an important distinction must be drawn between ancestrally non-lichen forming and secondarily derived non-lichen-forming fungi. Intriguingly, many of the non-lichen forming Ascomycota fungi that have impor-

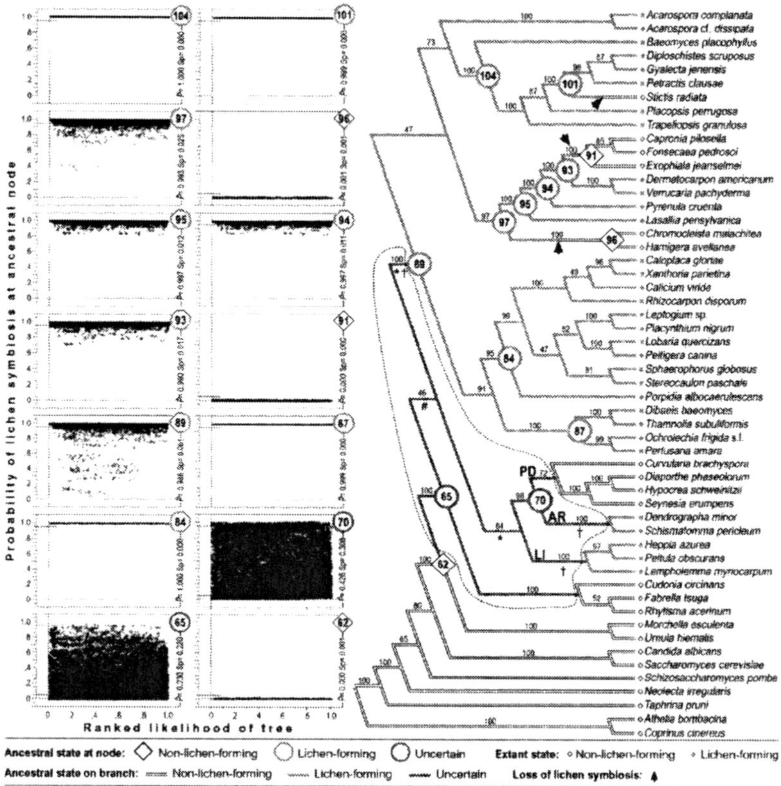

Fig. 5. Bayesian posterior probabilities for reconstructed evolution of the lichen symbiosis and for each node of the Ascomycota phylogeny. Numbers within each of the node symbols (e.g., 62, 65, ... 104) refer to specific nodes and connect nodes in the tree with their respective graphs. Left panel (set of 14 graphs), reconstructed probability that ancestral state was lichen-forming at specified node calculated on each of 19,900 trees generated by MCMC sampling. The average probability and standard deviation are provided on the y-axis to the right of each graph. Right panel (phylogeny), Ascomycota majority-rule consensus of 19,900 MCMC sampled trees based on SSU and LSU nrDNA sequences. Numbers above each internal branch correspond to the posterior probability (%) of the node to which it points. The region of the tree for which the ancestral states of branches could not be extrapolated, because of uncertainty associated with specific nodes, is delimited by a dotted line. The pattern of ancestral states indicates that lichenization has been gained and then lost in the same tree. See text for description of symbols (#, *, +) associated with various evolutionary scenarios for gains of the lichen symbiosis. Pagel and Lutzoni Phylogenetic Uncertainty in Comparative Studies page 15.

tant medical or health benefits to humans are from this group of secondarily derived non-lichen forming species (Lutzoni, Pagel and Reeb, 2001).

MCMC methods are relatively new to biology and in particular to phylogenetics and comparative methods. Owing to the vast number of possible phylogenetic trees for samples of even moderate numbers of species, MCMC methods cannot always be counted on to converge to the optimal region of the universe (e.g., Larget and Simon, 1999). New developments in MCMC sampling, notably Metropolis-coupled MCMC (or MCMCMC) may improve convergence especially in large samples (Gilks and Roberts, 1996).

Acknowledgements

This work was supported by a grant from the Leverhulme Trust, the Natural Environment Research Council and the Biology and Biotechnology Research Council of the UK (MP). F.L. is supported by the US National Science Foundation (DEB-9615542). The molecular work was performed at The Field Museum's Pritzker Laboratory of Molecular Systematics and Evolution, operated with support from an endowment from the Pritzker Foundation.

References

1. Bromham, L., Rambaut, A., Fortey, R., Cooper, A. & Penny, D. Testing the Cambrian explosion hypothesis by using a molecular dating technique. Proc. Natl. Acad. Sci. USA **95**, 12386- 12389, 1998.
2. Chang BSW and Donoghue MJ. Recreating ancestral proteins. Trends in Ecology and Evolution **15**, 109-114, 2000
3. Cooper, A. & Penny, D. Mass survival of birds across the Cretaceous-Tertiary boundary: molecular evidence. Science **275**, 1109-1113, 1997.
4. Felsenstein, J. Confidence limits on phylogenies - an approach using the bootstrap. Evolution **39**, 783-791, 1985.
5. Felsenstein, J. and Kishino, H. Is there something wrong with the bootstrap? A reply to Hillis and Bull. Syst. Biol. **42**, 193-200, 1993.
6. Galtier, N., Tourasse, N., & Gouy, M. A nonhyperthermophilic common ancestor to extant life forms. Science **283**, 220-221, 1999.
7. Gargas, A., DePriest, P.T., Grube, M. &Tehler, A. Multiple origins of lichen symbioses in fungi suggested by SSU rDNA phylogeny. Science **268**, 1492-1495, 1995.
8. Gilks, W.R. and Roberts, G.O. Strategies for improving MCMC. In, Markov Chain Monte Carlo in Practice (Gilks, W.R., Richardson, S., Spiegelhalter, D.J. eds). Chapman and Hall, 1996.
9. Gilks, W.R., Richardson, S., Spiegelhalter, D.J. Introducing Markov chain Monte Carlo. In, Markov Chain Monte Carlo in Practice (Gilks, W.R., Richardson, S., Spiegelhalter, D.J. eds). Chapman and Hall, 1996.
10. Harvey, P.H. & Pagel, M. The comparative method in evolutionary biology. Oxford: Oxford University Press, 1991.
11. Hastings, W. Monte Carlo sampling methods using Markov chains and their applications. Biometrika **57**, 97-109, 1970.
12. Hedges, S.B., Parker,P.H., Sibley, C.G. & Kumar, S. Continental breakup and the ordinal diversification of birds and mammals. Nature **381**, 226-229, 1996.

13. Hibbett DS, Gilbert LB, and Donoghue MJ. Evolutionary instability of ectomycorrhizal symbioses in basidiomycetes. Nature **407**, 506-508, 2000
14. Hillis, D.M., Moritz, C. & Mable, B.K. Molecular Systematics, 2nd edition. Sinauer: Sunderland, Ma. , 1996.
15. Huelsenbeck, J., Rannala, B., and Masly, J.P. Accmodating phylogenetic uncertainty in evolutionary studies. Science **288**, 2349-2350, 2000.
16. Jermann, T.M., Opitz, J.G., Stackhouse, J. & Benner, S.A. Reconstructing the evolutionary history of the artiodactyl ribonulcease superfamily. Nature **374**, 57-59, 1995.
17. Krakauer D.C., Pagel M., Southwood T.R.E., et al. Phylogenesis of prion protein. Nature **380**, 675-675, 1996
18. Larget, B. & Simon, D.L. Markov chain monte carlo algorithms for the Bayesian analysis of phylogenetic trees. Mol. Biol. Evol. **16**, 750-759 , 1999
19. Lewis, P. O. Phylogenetic systematics turns over a new leaf. Trends in Ecology and Evolution **16**, 30-37, 2001.
20. Lutzoni, F. and Pagel, M. Accelerated molecular evolution as a consequence of transitions to mutualism. Proc. Natl. Acad. Sci. USA **94**, 11422-11427, 1997.
21. Lutzoni, F. ,Pagel, M., and Reeb, V. 2001. Major fungal lineages derived from lichen-symbiotic ancestors. Nature, **411**, 937-940.
22. van Dijk, M. A.M., Madsen, O., Catzeflis, F., Stanhope, M.J., de Jong, W.W. and Pagel, M. Protein sequence signatures support the 'African clade' of mammals. Proceedings of the National Academy of Sciences **98**, 188-193, 2001.
23. Metropolis, N., et al. Equation of state calculations by fast computing machines. J. Chem. Phys. **21**, 1087-1092, 1953.
24. Newton, M. Bootstrapping phylogenies: large deviations and dispersion effects. Biometrika **83**, 315-328, 1996.
25. Pagel M. Inferring the historical patterns of biological evolution. Nature **401**, 877-884, 1999a.
26. Pagel, M. Detectinq correlated evolution on phylogenies: a general method for the comparative analysis of discrete characters. Proceedings of the Royal Society (B) **255**, 37-45, 1994.
27. Pagel, M. Inferring evolutionary processes from phylogenies. Zoologica Scripta **26**, 331-348, 1997.
28. Pagel, M. The maximum likelihood approach to reconstructing ancestral character states of discrete characters on phylogenies. Syst. Biol. **48**, 612-622, 1999b.
29. Pollock DD, Taylor WR, and Goldman N. Coevolving protein residues: Maximum likelihood identification and relationship to structure. J. Mol. Biol. **287**, 187-198, 1999.
30. Yang, Z. and Rannala, B. Bayesian phylogenetic inference using DNA sequences: a Markov chain Monte Carlo method. Mol. Biol. Evol. **14**, 717-724, 1997.
31. Schluter, D. Uncertainty in ancient phylogenies. Nature **377**, 108-109,1995.
32. Wilson, I, and Balding, D. Genealogical inference from microsatellite data. Genetics **150**, 499- 510, 1998.

The 'shape' of phylogenies under simple random speciation models

Mike Steel and Andy McKenzie

Biomathematics Research Centre, University of Canterbury,
Christchurch, New Zealand

Abstract. We describe some discrete structural properties of evolutionary trees generated under simple null models of speciation, such as the Yule model. These models have been used as priors in Bayesian approaches to phylogenetic analysis, and also to test hypotheses concerning the speciation process.

Here we describe new results for four properties of trees generated under such models. Firstly, for a rooted tree generated by the Yule model we describe the probability distribution on the depth (number of edges from the root) of the most recent common ancestor of a random subset of k species. Secondly, for trees generated under the Yule and uniform models, we describe the induced distribution they generate on the number C_n of *cherries* in the tree, where a cherry is a pair of leaves each of which is adjacent to a common ancestor. Next we show that, for trees generated under the Yule model, the approximate position of the root can be estimated from the associated unrooted tree, even for trees with a large number of leaves. Finally, we analyse a biologically-motivated extension of the Yule model and describe its distribution on tree shapes when speciation occurs in rapid bursts.

1 Introduction

Phylogenetic trees are widely used in biology to represent evolutionary relationships between species. In these trees the leaves represent extant species, and the internal vertices represent hypothesised speciation events. There is much interest in the process of speciation, and the extent and manner in which the distribution of phylogenetic tree shapes can be modelled by a random process. Several simple stochastic models of speciation have been proposed and several investigators have aimed to test or refine such models by comparing their predictions with published phylogenetic trees [1–8]. These models make predictions about the shape of the phylogenetic tree connecting the extant species, and they are also used as a basis for calculating the probability of certain configurations under random speciation [9]. These probabilities may then be useful in testing hypotheses concerning the speciation process. Speciation models can also provide prior probabilities for phylogenetic trees in Bayesian approaches to tree reconstruction [10–12].

In this paper we will consider just the model's predictions regarding the discrete underlying tree structure, without regard to the lengths of the edges. While such an approach may neglect some informative characteristics of the tree, the approach has two motivations - firstly, the predictions regarding the discrete tree remain valid under a much wider class of models (they are insensitive to

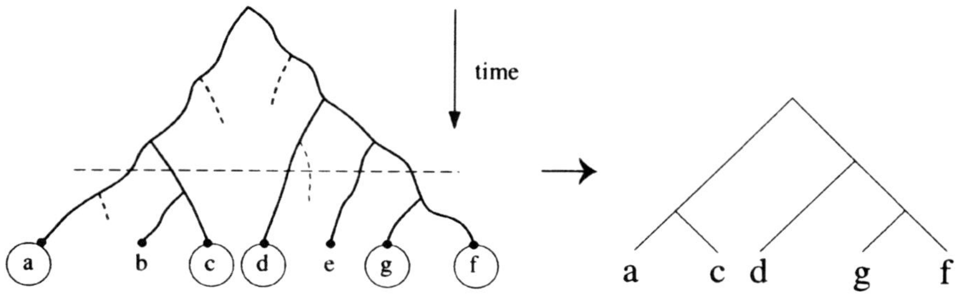

Fig. 1. A hypothetical sequence of speciation and extinction events is illustrated on the left. The descending (dashed) lineages become extinct, while seven species survive to the present. At the time corresponding to the horizontal dashed line there are six extant species. The circled present-day species (a, c, d, g, f) are those that are selected by the investigator for study. The induced phylogenetic tree on these five selected species is shown at right.

underlying parameters) and, secondly, we are interested in isolating out the information that is conveyed solely by the discrete tree shape. One may also make a random selection of species from the set of extant species, to obtain the (induced) phylogenetic tree (as illustrated in Fig. 1) and for certain models this does not alter the induced distribution on (discrete) tree shapes.

We will initially concentrate on some properties of the Yule model, which is perhaps the simplest stochastic model for speciation. In this model whenever a speciation event occurs, each of the species existing at that moment is equally likely to give rise to the new species. However the rate of speciation (and extinction) may vary arbitrarily with time. For example, in Fig. 1 each of the six species that exists at the time indicated by the horizontal dashed line has an equal rate of speciation at that moment.

Despite the generality of such a model it leads to a well-defined probability distribution on the (discrete) shapes of phylogenetic trees. In this paper we investigate certain predictions that this (and other) models make regarding some of the properties of phylogenetic trees that can be determined simply by considering their shape.

One of the problems we will consider is how to recover the position of the root of such a tree, when one is only given the associated unrooted tree. This is particularly relevant in phylogenetic analysis, since most tree reconstruction methods return an unrooted tree. Somewhat surprisingly, it is possible to estimate the approximate position of the root fairly accurately even for very large trees generated by the Yule model (a similar result for a related, but different, process was established by [13]).

It is useful to contrast the Yule distribution with (i) the distribution on tree shapes obtained by selecting a phylogenetic tree uniformly at random (the "uniform distribution") and (ii) the shapes of published phylogenetic trees that have been reconstructed from biological data. Several studies (see for example

[14]), have suggested that published trees tend, on average, to be less balanced than the Yule model would predict, yet more balanced than the uniform model would allow. Of course a published tree is only an estimate of the true species tree. Thus, it may also be important to determine biases in tree shape that arise due to particular tree reconstruction methods, and other factors, such as the non-random selection of species by the investigator [8].

Nevertheless, with a view to obtaining less balanced trees than those generated by the Yule model, we will consider a simple modification to this model. In this modified model a species that has recently speciated is more likely to speciate again than one that has existed without undergoing speciation for a long time. Such a model appears to lead to less balanced trees, and we prove that in a sufficiently extreme case it leads precisely to the uniform model (which, incidentally, provides a natural way in which the uniform model may be regarded as a speciation model). It would be interesting, for future work, to evaluate precisely how well these types of models can account for the shapes of published trees, and also to explore the effect of extinctions in such models.

The structure of the paper is as follows. We begin by introducing some basic terminology for phylogenetic trees (Sect. 2). The Yule model is then introduced, and some of its properties are described (Sect. 3). We then consider the probability distribution on the number of edges separating the root of a tree from the most recent common ancestor of a randomly selected subset of size k (Sect. 4). Following this we investigate the induced probability distribution for the number of cherries under the Yule and uniform models (Sect. 5). Exact formulae are given for small trees, and the asymptotic distribution is found to be normal. Next, a maximum likelihood approach to edge-rooting an unrooted tree is presented, and it is shown that even for large unrooted trees the approximate location of the root can be identified with high probability (Sect. 6). Following this a modification of the Yule model is considered in which the rate of speciation of a lineage is dependent on the time back to the last speciation event on that lineage (Sect. 7). We show that this modified model reduces to the uniform model under the condition of "explosive radiation". The results we describe here are mostly based on [7] for Sect. 5 and [15] for Sects. 4,6,7 where full proofs and details may be found.

2 Terminology

Evolutionary relationships are often represented by rooted or unrooted *binary (phylogenetic) trees* [16]. Such trees consist of labeled *nodes* of degree 1 called *leaves* and unlabelled *internal* nodes of degree 3 (also, in case the tree is rooted, it contains an additional *root node* of degree 2, so that every node can be regarded as having exactly two descendants). A pair of leaves adjacent to a common node is called a *cherry.* Edges adjacent to a leaf are called *pendant edges*, while all other edges are *internal.* A (tree) *shape* is the unlabeled tree obtained by dropping the labeling of the leaves of a binary phylogenetic tree. For further clarification of these terms see Fig. 2. The number of *rooted* leaf-labeled binary trees on n leaves

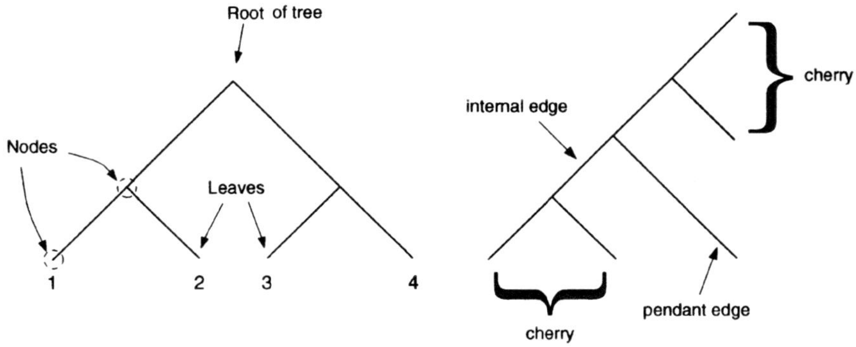

(a) **(b)**

Fig. 2. **(a)** A labeled rooted tree with 4 leaves. **(b)** An unrooted tree shape with 5 leaves

is given by $(2n-3)!! = 1 \times 3 \times 5 \cdots \times (2n-3)$, and the number of *unrooted* leaf-labeled binary trees on n leaves is given by $(2n-5)!! = 1 \times 3 \times 5 \cdots \times (2n-5)$ [17,18].

Throughout we will use T to denote a phylogenetic tree, and τ to denote a tree shape. We will frequently use the asymptotic expression $f(n) \sim g(n)$ to denote $\lim_{n\to\infty} \frac{f(n)}{g(n)} = 1$. As usual, $\mathbb{P}[A]$ (resp. $\mathbb{P}[A|B]$) denotes the probability of event A (resp. the conditional probability of event A given B), and $\mathbb{E}[X]$ (resp. $\mathbb{E}[X|Y]$) denotes the expectation of random variable X (resp. the conditional expectation of X given Y).

3 The Yule model

A simple model of speciation is to assume the exchangeability condition that, at any given time, each of the then-extant species are equally likely to give rise to one new species. The 'rate' of speciation may vary with time, or with the present and past number of species in the tree. Also we may allow extinctions (or random sampling of extant taxa) provided that a similar exchangeability criterion applies - that is, whenever an extinction event occurs each of the then-extant species is equally likely to go extinct. Depending on how the various parameters are set in such a model, we obtain various probability densities over all edge-weighted trees that connect a group of extant species. However, if we simply regard these trees as unlabeled discrete graphs without edge length (tree shapes) then the underlying parameters and details do not affect the resulting discrete probability distribution, provided the exchangeability criteria still apply (see [1]). This distribution on tree shapes is often called the *Yule model* and it has been widely studied [3,19–21].

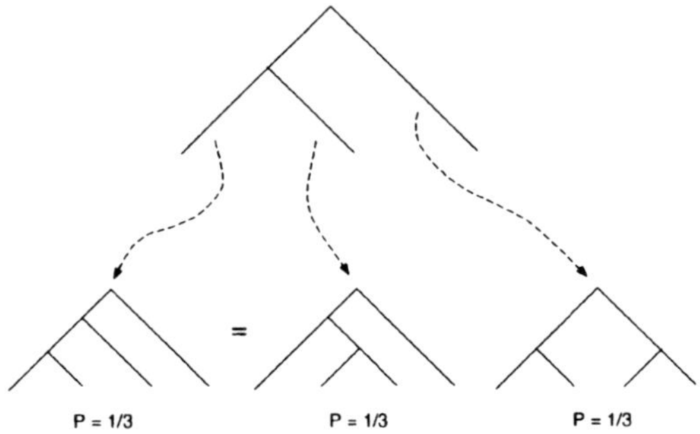

Fig. 3. The Yule model probabilities for shapes with 4 leaves. A shape on 4 leaves is formed by the splitting of one of the pendant edges of the shape on 3 leaves. Each pendant edge has the same probability of splitting, so for the shape on 3 leaves each pendant edge has a probability of 1/3 of splitting. The resulting symmetric shape on 4 leaves has a probability of 1/3. The other two shapes on 4 leaves are the same (up to rotation about internal vertices), and so the probability of this shape is 2/3

We can reformulate this model in the discrete setting, by evolving a (discrete) tree shape under the following rule. We start with the rooted tree on two leaves and repeat the following procedure until the tree has n leaves:

For the tree shape so far constructed, select a leaf randomly and uniformly, and make it the direct ancestor of two new descendent leaves.

Alternatively, we may attach an edge added uniformly and randomly to a pendant edge at each step. This process is illustrated in Fig. 3.

This process provides a probability distribution on rooted tree shapes and also on unrooted tree shapes (by suppressing the root). Also if species are assigned to the leaves in random order we also obtain probability distributions on rooted and unrooted phylogenetic trees ([3]).

The Yule model arises in a number of seemingly different ways. For example, in the context of population genetics, one has the *coalescent* model [1,22,23]. In this model one starts with n objects, then picks two at random to coalesce, giving $n-1$ objects. This process is repeated until there is only a single object left. If this process is reversed, starting with one object to give n objects, then it is equivalent to the Yule model. Note that in the coalescent model there is commonly a probability distribution for the times of coalescences, but in the Yule model we ignore this element.

Another closely-related realisation of the Yule model is obtained as follows. Given a rooted binary phylogenetic tree T, let $\mathring{V}$ denote the set of internal vertices of T. A *ranking* of T is a function r that associates to each vertex $v \in \mathring{V}$ of T a unique element from the set $\{1, 2, \ldots, |\mathring{V}|\}$ in such a way that $r(v_1), r(v_2), \ldots$ is strictly increasing along any sequence $v_1, v_2, \ldots$ of vertices directed away from the root. Thus we might regard r as describing the order of the speciation events that are represented by the internal vertices of T. Observe that a phylogenetic tree having the shape of the right-most tree in Fig. 2 has exactly two possible rankings, while for the left-most tree there is just one possible ranking. The pair (T, r) is sometimes called a *labeled history*. If we now select a labeled history (T, r) on n species uniformly and consider just T, then this once again leads to the Yule distribution on rooted binary phylogenetic trees. Furthermore, if we consider just the shape of T we obtain the Yule model on rooted tree shapes.

This connection with labeled histories provides a convenient tool for describing the probability distribution of a tree shape τ, since it is possible to count the number of labeled histories, and rankings on a given tree. For a vertex v of a rooted binary phylogenetic tree, let $\delta(v)$ denote the number of internal vertices (including v) that are descendants of v (v' is a descendant of v if the path from v' to the root includes v). Note that $\delta(v)$ is equal to one less than the number of leaves of the tree that are descendants of v.

Then, for a rooted binary phylogenetic tree with n leaves, the number of associated labeled histories is precisely

$$\frac{(n-1)!}{\prod_{v \in \mathring{V}} \delta(v)} \tag{1}$$

where $\mathring{V}$ denotes the set of internal vertices of the tree. This result is from [2,24]. It is also possible to give an exact expression for the total number of labeled histories on a set of species. From [25] the number of labeled histories on n species is

$$\frac{n!(n-1)!}{2^{n-1}}. \tag{2}$$

A further important property of the Yule model is that it satisfies the following *hereditary* property. Let us generate a rooted binary tree T according to the Yule model, and let t_1, t_2 denote the two subtrees of T incident with the root. Let S denote a fixed subset of species. Then, conditional on the event that S is the set of species labeling the leaves of t_1 the probability distribution on t_1 is also the Yule distribution. This property follows from a particular case of the *group elimination property*, described by [1].

4 Depth of a most recent common ancestor (MRCA)

Suppose we evolve a rooted phylogenetic tree T on n extant species under the Yule model, and we select a random subset S of k extant species. Let $X_{n,k}$ denote the number of edges separating the root of T from the vertex in T that corresponds to the most recent common ancestor of S. In this section we investigate the probability distribution of $X_{n,k}$ for various values of k, particularly in the limit as n becomes large. Some of the reasons why a biologist might be interested in such questions are discussed by Sanderson [26]; the cases $k = 1$ and $k = 2$ are also of some independent interest as we will see.

Note that although we will regard S as a random subset of the n species, our results would apply even if we regard S as a fixed set of species, since we are investigating properties of S in a tree that is generated by a model that assigns equal probability to all possible labelings of the leaves by the n species. Also, whenever we talk about the *distance* between two vertices in a tree, we are referring to the number of edges separating the two vertices (also called the *graph distance*). The following summarises results from [15].

4.1 Distance of MRCA from root

The case $k = 1$ corresponds to the distance of a randomly selected leaf from the root, and has been analysed before [27,28]. In the following theorem $c(n, q)$ denotes the unsigned Stirling number of the first kind, which is the number of permutations on n elements that have exactly q cycles [29].

Theorem 1. *[27,28] Let P^q_{n+1} be the probability that a randomly chosen leaf from a tree on $n+1$ leaves has distance q from the root. Under the Yule model we have*

$$P^q_{n+1} = \frac{2^q c(n, q)}{(n+1)!} \, .$$

Furthermore, the mean (μ_n) and variance (σ^2_n) of this distribution are given by

$$\mu_n = 2\sum_{j=2}^{n} \frac{1}{j}; \qquad \sigma^2_n = 2\sum_{j=2}^{n} \frac{1}{j} - 4\sum_{j=2}^{n} \frac{1}{j^2}$$

where $\mu_1 = 0$ and $\sigma^2_1 = 0$.

For $k > 1$, the asymptotic probability $\mathbb{P}[X_{n,k} = 0]$ was determined by Sanderson [26] who showed that

$$\lim_{n\to\infty} \mathbb{P}[X_{n,k} = 0] = 1 - \frac{2}{k+1} \, .$$

Here we provide an exact, closed-form expression for $\mathbb{P}[X_{n,k} = 0]$. In addition, as n becomes large, $X_{n,k}$ has a geometric distribution with parameter $2/(k+1)$.

Theorem 2. *[15]*

1.

$$\mathbb{P}[X_{n,k} = 0] = 1 - \frac{2(n-k)}{(k+1)(n-1)}$$

2. For $k > 1, r \geq 0$,

$$\lim_{n\to\infty} \mathbb{P}[X_{n,k} \geq r] = \left(\frac{2}{k+1}\right)^r$$

The case $k = 2$ allows us to obtain an expression for the expected distance between two randomly selected leaves in a rooted binary tree with n leaves generated by the Yule model. Recall that the distance $d(v_1, v_2)$ between a pair of vertices v_1, v_2 of T is being measured by the number of edges separating them.

Let $v_{ij} \in \mathring{V}$ denote the most recent common ancestor of i and j in the tree, and let ρ denote the root of the tree. Then for leaves i, j of T,

$$d(i,j) = d(i,\rho) + d(j,\rho) - 2d(v_{ij},\rho)$$

and so $\mathbb{E}[d(i,j)] = \mathbb{E}[d(i,\rho)] + \mathbb{E}[(j,\rho)] - 2\mathbb{E}[d(v_{ij},\rho)]$. Now, $\mathbb{E}[d(i,\rho)] = \mathbb{E}[d(j,\rho)] = \mu_n$ (see Theorem 1), while $\mathbb{E}[d(v_{ij},\rho)] = \mathbb{E}[X_{n,2}]$, so we have $\mathbb{E}[d(i,j)] = 2\mu_n - 2\mathbb{E}[X_{n,2}]$. By Theorem 2, $\lim_{n\to\infty} \mathbb{E}[X_{n,2}] = 2$ and so if μ_n^* denotes the expected distance between two randomly selected leaves, then

$$\lim_{n\to\infty} 2\mu_n - \mu_n^* \sim 4 \, .$$

Actually, it is possible to derive an exact expression for $\mathbb{E}[X_{n,2}]$, namely

$$\mathbb{E}[X_{n,2}] = 2\left(1 - \frac{\mu_n}{n-1}\right)$$

which provides an exact expression for $\mathbb{E}[d(i,j)]$.

5 Probability distribution for cherries

A *cherry* is a pair of leaves, each of which is adjacent to a shared vertex. In this section we describe the distribution of the number of cherries for a tree that evolves under the Yule model. We also compare this to the distribution of the number of cherries for a tree selected uniformly at random from the set of all (unrooted) leaf-labelled trees. Note that when we suppress the root of a rooted tree (and thereby convert an rooted tree into an unrooted one) the number of cherries is either unchanged or it increases by at most 1, and this difference has a negligible effect on the asymptotic results described.

Under either model we will let the random variable C_n denote the number of cherries in the randomly generated tree. By realizing the process of cherry formation in these two models by extended Polya urn models it can be shown that C_n is asymptotically normal. We also give exact formulas for the mean and standard deviation of C_n in these two models. This section is based on [7], where proofs of the main results may be found.

5.1 Yule Model

Theorem 3. *[7,30] Let μ_n be the mean number of cherries for a rooted binary tree on n leaves, and σ_n^2 be the variance for the number of cherries. Under the Yule distribution we have the recursions, for $n \geq 2$:*

$$\mu_{n+1} = 1 + \mu_n(1 - \frac{2}{n}); \quad \sigma_{n+1}^2 = \sigma_n^2(1 - \frac{4}{n}) + \frac{2}{n}\mu_n(1 - \frac{2}{n}\mu_n)$$

which may be solved exactly to give

$$\mu_n = \frac{n}{3} \;\; (n \geq 3); \quad \sigma_n^2 = \frac{2n}{45} \;\; (n \geq 5) \, .$$

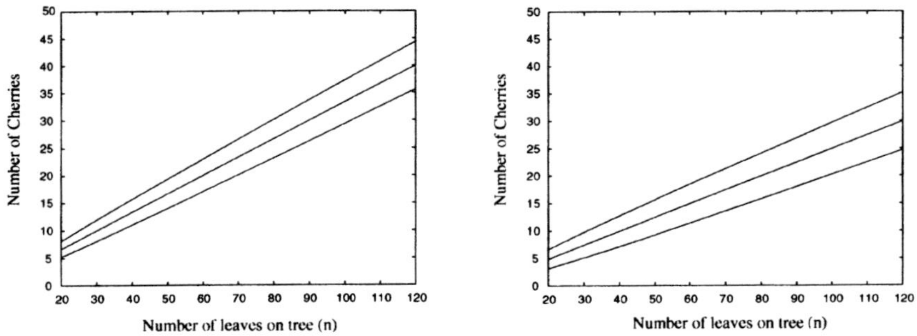

(a) Yule model (b) Uniform model

Fig. 4. Rejection limits for large n of the Yule and uniform null hypotheses at the $\alpha = 0.05$ level. The solid line represents the mean number of cherries, while the dashed lines are the lower and upper limits for rejection of the null hypotheses. The rejection limits are based upon a normal approximation which is valid for n $\gtrsim 20$. **(a)** Yule model **(b)** Uniform model

Asymptotically we have

$$\frac{C_n - n/3}{\sqrt{2n/45}} \to \mathcal{N}(0,1) .$$

The Yule model can be used as a simple null hypothesis to explore patterns in phylogenetic trees. A simple two-tailed test of the Yule null hypothesis, for a given tree, can be made based on the number of cherries in the tree. If the number of cherries is below some lower critical value, or above some upper critical value, then the Yule null hypothesis is rejected.

For small n, the rejection limits may be calculated exactly using a recursive formula for the probabilities. For larger values of n ($n \gtrsim 20$) a normal approximation is valid. In this case, based on the asymptotic result in Theorem 3, the rejection region for a two-sided test at the α level is given by

$$C_n < \frac{n}{3} - Z_{\frac{\alpha}{2}}\sqrt{\frac{2n}{45}} \qquad \text{and} \qquad C_n > \frac{n}{3} + Z_{\frac{\alpha}{2}}\sqrt{\frac{2n}{45}} .$$

The lower and upper critical values for rejection at an $\alpha = 0.05$ level are shown in Fig. 4. If the Yule model is rejected then this implies that one or more of the assumptions upon which it is based is invalid. Often it is assumed that the assumption of equal probability of speciation is the invalid assumption, but this need not be the case [21].

5.2 Uniform model

In the uniform model equal probability is assigned to each possible leaf-labeled binary tree on n leaves. Thus the uniform model distribution may be used to model the frequency of outcomes that would occur if the process of tree reconstruction did no better than random selection from the set of possible binary trees on n leaves.

Theorem 4. *[7,30–32] Let μ_n be the mean value of C_n for an unrooted binary tree on n leaves, and σ_n^2 be the variance for C_n. Under the uniform model, for $n \geq 4$,*

(a)

$$\mathbb{P}[C_n = k] = \frac{n!(n-2)!(n-4)!2^{n-2k}}{(n-2k)!(2n-4)!k!(k-2)!}, \quad k \geq 2$$

(b)

$$\mu_n = \frac{n(n-1)}{2(2n-5)} \sim \frac{n}{4}; \quad \sigma_n^2 = \frac{n(n-1)(n-4)(n-5)}{2(2n-5)^2(2n-7)} \sim \frac{n}{16}$$

Asymptotically we have

$$\frac{C_n - n/4}{\sqrt{n/16}} \rightarrow \mathcal{N}(0,1)\,.$$

A test of the uniform model null hypothesis may be constructed based on the number of cherries in a tree. For small n the probability distribution given in Theorem 4 may be used to calculate the rejection limits. For larger n ($n \gtrsim 20$) a analysis similar to that for the Yule model, but based on the asymptotic result in Theorem 4, gives as the rejection region:

$$C_n < \frac{n}{4} - Z_{\frac{\alpha}{2}}\sqrt{\frac{n}{16}} \qquad \text{and} \qquad C_n > \frac{n}{4} + Z_{\frac{\alpha}{2}}\sqrt{\frac{n}{16}}\,.$$

The lower and upper critical values for rejection at an $\alpha = 0.05$ level are shown in Fig. 4.

5.3 An example

Figure 1 in [33] is a rooted phylogenetic tree for 34 species of *eureptantic nemerteans* (ribbon worms). This tree has 7 cherries (rooted or unrooted). For the Yule model null hypothesis test at the $\alpha = 0.05$ level the lower rejection limit is 8 cherries or less, and the upper rejection limit is 15 cherries or more. So for the ribbon worm tree the Yule model null hypothesis is rejected. For the uniform model null hypothesis test at the $\alpha = 0.05$ level the lower rejection limit is 5 cherries or less, and the upper rejection limit is 13 cherries or more, and so the test does not reject the uniform model null hypothesis. In any hypothesis test, however, it is important to note that a reconstructed tree is only an estimate of the underlying species tree. Consequently a more refined analysis would take into account the uncertainty and possible biases in phylogeny reconstruction [8,34].

6 Rooting an unrooted tree

Typically, construction of an evolutionary tree for a set of species is a two stage process. In the first stage, using biological data of some sort, an unrooted tree is constructed. In the next stage, the unrooted tree is rooted at some point. Commonly this is done by outgroup comparison, or using some auxiliary data (for example embryological or fossil data) [35].

However, in some circumstances an outgroup is not available, or the auxiliary data is unclear. Furthermore, the choice of outgroup can strongly influence the accuracy of tree reconstruction [36]. In these circumstances heuristic methods provide an alternative way to root the tree. For example, in the *midpoint method*, the root is located at the point halfway between the two leaves that are the furtherest distance apart [37,38]. In another approach the root is located at a point where the mean distance to the species on either side is the same (for example, the program described in [39] uses this method). Here we explore a third alternative, based on the structure of the trees under the Yule model.

Before proceeding further we introduce some terminology. For a rooted binary tree T' the associated unrooted binary tree T is obtained from T' by suppressing the root and identifying the two edges incident with the root to form a single edge e – we call this edge the *root edge* of T. Given T and its root edge, one can easily recover the rooted tree by *subdividing* e – that is, by placing a new (root) vertex at the midpoint of the root edge.

In applications, one is typically given just the unrooted tree T and one would like to estimate which edge is the root edge, or at least find a small subset of edges that contains the root edge with high probability. For detailed proofs of the results that follow see [15].

6.1 Maximum likelihood estimation of the root edge

Suppose we have a stochastic model (such as the Yule model) for the generation of rooted binary phylogenetic trees. Given an unrooted binary tree, T and an edge e, let $\mathbb{P}[T, e]$ denote the probability of generating the rooted binary tree obtained by subdividing edge e of T. Let $\mathbb{P}[T] = \sum_e \mathbb{P}[T, e]$ which is the probability of generating a rooted binary tree which produces T when the root is suppressed (this provides a probability distribution on unrooted binary phylogenetic trees). Finally, let

$$\mathbb{P}[e \mid T] = \frac{\mathbb{P}[T, e]}{\mathbb{P}[T]} \ . \tag{3}$$

Note that $\mathbb{P}[e \mid T]$ is the probability that edge e is the root edge of T, given that T is the unrooted tree obtained by suppressing the root.

For example, consider a labeled unrooted tree on 4 leaves (Fig. 5). The probability of this tree ($\mathbb{P}[T]$) is 1/3. For the internal edge, the probability of the corresponding labeled rooted tree is 1/9, thus the conditional probability ($\mathbb{P}[e \mid T]$) for the internal edge is 1/3. For the pendant edges the probability of the corresponding labeled rooted tree is 1/18, thus the conditional probability ($\mathbb{P}[e \mid T]$) for each pendant edge is 1/6.

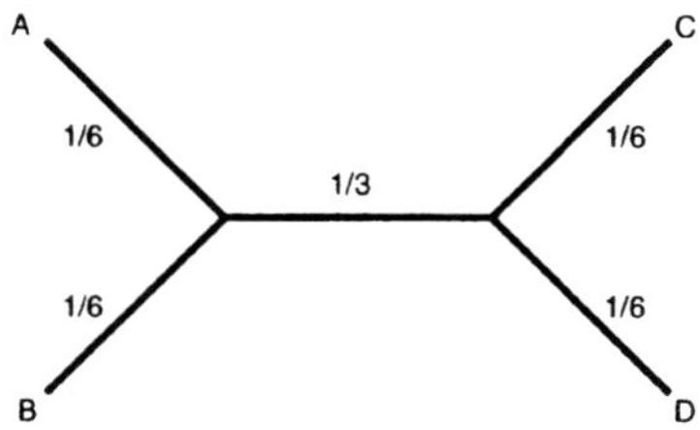

Fig. 5. Conditional probabilities ($\mathbb{P}[e \mid T]$) for the edges of a labeled unrooted tree on 4 leaves

Given an unrooted binary tree T, the *method of maximum likelihood* selects as its estimate of the root edge any edge e that maximizes $\mathbb{P}[e \mid T]$. We let $E_{\max}(T)$ denote the set of edges of T that maximize $\mathbb{P}[e \mid T]$, and we let $e_{\max}(T)$ denote any edge in $E_{\max}(T)$. It is possible, for example when symmetry is present, that $|E_{\max}(T)| > 1$. However, we will describe below that, for the Yule model, $|E_{\max}(T)| \leq 3$.

6.2 Probability of locating the root edge

Suppose we generate a rooted binary tree T' on n leaves according to the Yule distribution, and we let $u(T')$ denote the unrooted binary tree obtained from T' by suppressing the root. Let $\varepsilon(n)$ denote the probability that a particular maximum likelihood edge (e_{max}) of $u(T')$ is the root edge of $u(T')$. By the law of total probability,

$$\varepsilon(n) = \sum_{T} \mathbb{P}[e_{\max} \text{ is the root edge of } T|u(T') = T]\mathbb{P}[u(T') = T],$$

where the summation is over all unrooted binary trees on the set of n species, and hence,

$$\varepsilon(n) = \sum_{T} \mathbb{P}[e_{\max}(T) \mid T] \; \mathbb{P}[T] \; . \tag{4}$$

One might expect that $\varepsilon(n)$ would converge to 0 as n tends to infinity, since the number of edges (and so possible root edges) grows without bound. Indeed we will see that $\mathbb{P}[e_{\max}(T) \mid T]$ can converge to 0 for certain ("caterpillar") trees as the number of leaves grows.

However we will see that $\varepsilon(n)$ has a non-zero limit. This parallels similar non-zero asymptotic behaviour for an analogous model, the Yule-Furry model, in which edges are added at random to vertices [13]. Furthermore, although the limit $\varepsilon(n)$ is small (about 0.15) the fact that it is non-zero suggests that one should be able to locate the root edge to within a small (edge) distance of $e_{\max}(T)$ with high probability, and this is confirmed by simulations.

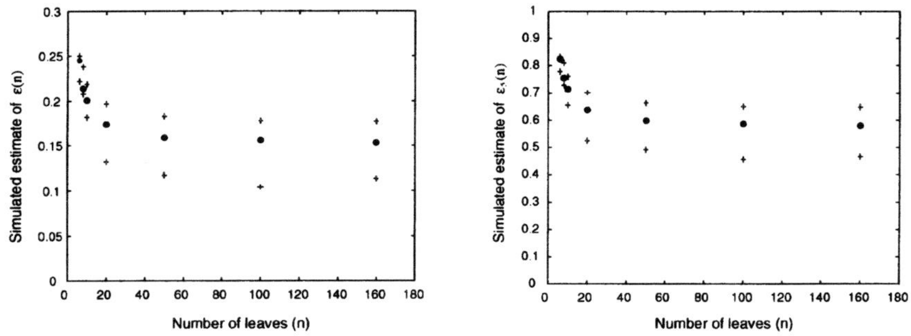

Fig. 6. Simulation results for the conditional probability of edges. Two hundred unrooted trees were randomly generated for different values of n. The trees were produced by unrooting the rooted tree generated by a Yule process. The minimum and maximum probabilities for each simulation are represented by crosses (+) **(a)** Estimate of the mean probability that $e_{\max}(T)$ contains the true root. **(b)** Estimate of the mean value for the sum of the five largest conditional probabilities for a tree

For n small, $\varepsilon(n)$ can be explicitly calculated, but for larger values $\varepsilon(n)$ was approximated by simulation. Simulated values were calculated by the formula

$$\varepsilon(n) \approx \frac{1}{N}\sum_{i=1}^{N} \mathbb{P}[e_{\max}(T_i) \mid T_i] = \frac{1}{N}\sum_{i=1}^{N} \frac{\mathbb{P}[T_i, e_{\max}(T_i)]}{\mathbb{P}[T_i]}, \tag{5}$$

where T_i is a labeled unrooted binary tree on n leaves obtained by generating a rooted tree according to the Yule process, then unrooting it, and where N is the number of trees generated.

The simulation results suggest that $\lim_{n\to\infty} \varepsilon(n) \approx 0.15$ (Fig. 6a). The five edges with the largest conditional probabilities for a tree were always an internal edge and the four edges adjacent to it. Let $\varepsilon_5(n)$ denote the mean value for the sum of the five largest conditional probabilities for a tree. The simulations suggest that $\lim_{n\to\infty} \varepsilon_5(n) \approx 0.58$ (Fig. 6b). Thus, even for a large unrooted tree, the location of the root may be narrowed down to a small cluster of five edges, of which one is more likely than not to be the true root. Progressively extending the radius further it appears from simulations that the limiting expected probability that the root edge is within a given (edge) distance d from $e_{\max}(T)$ continues to increase towards 1. For example, when $d = 3$ the limiting probability appears to be close to 0.9.

6.3 Exact asymptotic value of $\varepsilon(n)$

Given an edge e of an unrooted phylogenetic tree, let $H(e)$ denote the number of labeled histories associated with the rooted tree that arises from T by subdividing edge e. Let edge e be an internal edge of an unrooted binary phylogenetic tree

T. Denote the four subtrees of T adjacent to e by A, B, C, D, and let a, b, c, d respectively denote the number of leaves in these trees. Then $H(e) \geq H(e')$ for each of the four edges e' incident with e precisely if both the following two inequalities hold:

$$a + b \geq max\{c, d\}; \quad c + d \geq max\{a, b\} . \tag{6}$$

Furthermore, $H(e) > H(e')$ for all e' precisely if these two inequalities hold as strict inequalities. It follows that any two edges in $E_{\max}(T)$ are adjacent, and consequently, $|E_{\max}(T)| \leq 3$. Furthermore, if both the inequalities in (6) are strict, then $|E_{\max}(T)| = 1$.

We now describe the exact asymptotic value of $\varepsilon(n)$. This value was calculated by embedding the discrete process of rooting a tree into a continuous analogue involving 'stick breaking'.

Theorem 5. *[15] Generate a rooted binary tree with n leaves randomly under the Yule model, and let T denote the tree obtained by suppressing the root. The probability that edge $e_{\max}(T)$ is unique and equal to the root edge of T convergences to the value $4\ln(4/3) - 1$ (≈ 0.15) as $n \to \infty$.*

We end this section by noting that $\mathbb{P}[e_{\max}(T) \mid T]$ may be arbitrarily close to zero, for a tree T with a sufficiently large number of leaves. For example, consider a *caterpillar tree*, which is any unrooted binary phylogenetic tree that reduces to a path (a tree having vertices of degree 1 or 2) once the pendant edges and leaves are deleted. The simulation results suggest that caterpillar trees are the trees for which $\mathbb{P}[e_{\max}(T) \mid T]$ is smallest. For the caterpillar tree $\mathbb{P}[e_{\max}(T) \mid T]$ may be calculated exactly, and asymptotically, as $n \to \infty$,

$$\mathbb{P}\left[e_{\max}(C_n) \mid C_n\right] \sim \frac{2}{3}\sqrt{\frac{2}{\pi}}\frac{1}{\sqrt{n}} \to 0 . \tag{7}$$

7 Extending the Yule model

In the Yule model, at any time each existing species has the same probability of giving rise to a new species, and all lineages are treated exchangeably. Here we consider a simple modification of this model, in which the rate of speciation events on a given lineage is a function of the time back to the last speciation event on that lineage.

More precisely, we suppose that at time $t = 0$ there is just one species, labeled s_0, subject to a 2-state Markov process on state space $\{1, 2\}$. Under this process, s_0 is initially in state 1, and state 2 corresponds to a "speciation event", that is, the replacement of the original species by two species (either two new species, or the original species plus one new one, and we will not distinguish here between these two possibilities). Let $s(t)$ denote the rate of change from state 1 to state 2 at time t, we call this the "speciation rate". Once a speciation event occurs (say at time Λ) the two species are again assumed to be independently subject to the same Markov process, with time reset to 0 (that is, with rate function

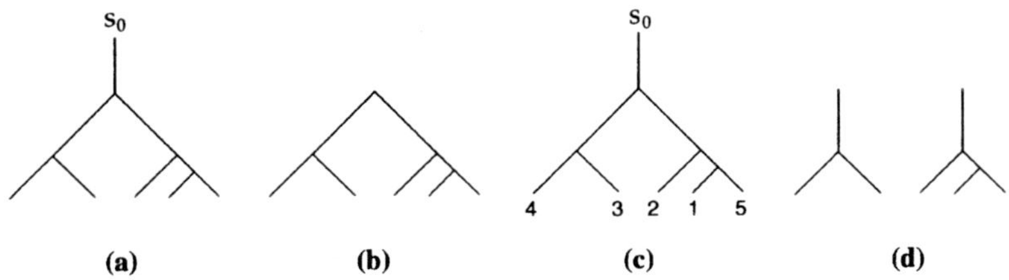

Fig. 7. Rooted tree types **(a)** An unlabeled binary tree on 5 leaves ; $\tau \in \mathrm{UB}(5)$. The root vertex is labeled s_0. **(b)** The edge-rooted unlabeled binary tree on 5 leaves obtained by removing the root leaf and its incident edge ; $\tau^* \in EUB(5)$ **(c)** A labeled binary tree on 5 leaves ; $\tau \in \mathrm{LB}(5)$ **(d)** The two subtrees, τ^1 and τ^2 of the tree τ in **(a)**

$s(t - \Lambda))$. Continuing in this way, we obtain a probability distribution on the trees of descent of species starting from s_0 up to some fixed time t which we can assume (by rescaling s if necessary) lies in the range $[0, 1]$.

The biological motivation for this model is that a recently evolved species, or the species that it has split off from, are often colonizing new regions or niches, and so may be more likely to give rise to further new species (in a given short time period) than a species that has existed for a very long time without giving rise to any new species (thus we are thinking of s being a monotone decreasing function). It would also be interesting and useful to build extinctions into such a model, however we do not pursue this here.

Kubo and Iwasa [6] consider a rate-varying model of speciation, however in their case, the speciation rate is a function of (absolute) time, rather than the lineage-specific time back to the last speciation event. Our model has more similarity to that discussed by Heard [5] who used computer simulation rather than analytical techniques in his analysis. Our general approach, which encompasses more than one model in a single analytical framework, is akin to that taken by Aldous [14]. We are interested in the probability distribution that this model induces on the tree that describes the species descendent from s_0. Since we are only interested in the "shape" of these speciation trees, we will mostly deal with trees in which the vertices are unlabeled. More detail of the work that follows may be found in [15].

In this section the following additional terminology for rooted trees is necessary.

- For $n \geq 1$, let $UB(n)$ denote the (finite) set of *u*nlabeled *b*inary trees consisting of n leaves together with an additional leaf, the *root leaf* s_0 (where the root leaf is the top-most vertex), and whose remaining internal vertices are all of degree 3 (Fig. 7a).
- For $n \geq 2$, let $EUB(n)$ denote the (finite) set of *e*dge-rooted *u*nlabeled *b*inary trees obtained from $UB(n)$ by deleting from each tree the root leaf and its incident edge. If $\tau \in UB(n)$ we will let τ^* denote the associated tree in $EUB(n)$ (Fig. 7b).

- For $\tau \in UB(n)$, let $L(\tau)$ be the set of distinct trees that can be obtained by assigning the (species) labels $\{1, \ldots, n\}$ bijectively to the n non-root leaves of τ. Let $LB(n) := \cup_{\tau \in UB(n)} L(\tau)$, the set of *l*abeled *b*inary trees (Fig. 7c).

For the model described above, the speciation tree at time $t \in [0, 1]$, $T(t)$, is the unlabeled tree of descent of the species that have evolved up to time t from the root leaf s_0. For $0 \leq t \leq 1$ and $\tau \in UB(n)$, consider the following (absolute and conditional) probabilities

$$f(\tau, t) := \mathbb{P}[T(t) = \tau]; \qquad p(\tau) := \mathbb{P}[T(1) = \tau | T(1) \text{ has } n \text{ leaves}] . \tag{8}$$

Let $\Lambda(s_0)$ denote the time until speciation of s_0, and set

$$S(x) := \mathbb{P}[\Lambda(s_0) \geq x]; \qquad \sigma(x) := s(x)S(x) , \tag{9}$$

where $s(x)$ is, as previously, the "speciation rate" at moment x.

Since the speciation of s_0 is a time-dependent Poisson process we have, from [40]

$$\mathbb{P}[\Lambda(s_0) \geq x] = exp[-\int_0^x s(\lambda) d\lambda] . \tag{10}$$

Thus, $\sigma(x) = \lim_{\delta \to 0+} \frac{\mathbb{P}[\Lambda(s_0) \in (x, x+\delta)]}{\delta}$ and so, by the assumptions that define the model, we have the following fundamental recursion:

$$f(\tau, t) = 2^{\delta(\tau)} \int_0^t f(\tau^1, t - x) f(\tau^2, t - x) \sigma(x) dx \tag{11}$$

where τ^1 and τ^2 denote the two subtrees of τ whose two vertex sets (i) intersect precisely on v and (ii) cover all vertices of τ except s_0 (Fig. 7d), and where

$$\delta(\tau) = \begin{cases} 1 & \text{if } \tau^1 \neq \tau^2 \\ 0 & \text{otherwise.} \end{cases}$$

Let $N(t)$ denote the total number of species existing at time $t \in [0, 1]$, and let

$$\nu(k, t) := \mathbb{P}[N(t) = k] .$$

For $\tau \in UB(n)$ we wish to calculate the conditional probability:

$$p(\tau) = \mathbb{P}[T(1) = \tau | N(1) = n] = \frac{f(\tau, 1)}{\nu(n, 1)} . \tag{12}$$

We may also wish to compute the probability of the induced edge-rooted tree. Thus, given $\tau \in UB(n)$ and its associated tree $\tau^* \in EUB(n)$ let

$$p(\tau^*) := \lim_{\varepsilon \to 0+} \mathbb{P}[T(1) = \tau | N(1) = n; \Lambda(s_0) < \varepsilon] .$$

The motivation for considering $p(\tau^*)$ is that one is frequently interested in the distribution on edge-rooted trees, and we can simplify matters by supposing that the first speciation event happened at time 0. We have the recursion

$$p(\tau^*) = 2^{\delta(\tau)} p(\tau^1) p(\tau^2) , \tag{13}$$

with τ^1, τ^2 as in Eq. (11).

7.1 Two classes of models

1. The simplest model has $s(x) = s > 0$, constant. This gives the Yule model as described in Sect. 3. In this case, $\sigma(x) = se^{-sx}$ and $N(t)$ models a pure birth process, so $\nu(k,t) = e^{-st}(1-e^{-st})^{k-1}$. Under this model $p(\tau) = p(\tau^*) = 2^{u(\tau)} \prod_{i>2}(i-1)^{-d_i(\tau)}$, where $d_i(\tau)$ denotes the number of internal vertices of τ which have exactly i descendant leaves, and $u(\tau)$ is the number of *unbalanced* internal vertices of τ - that is, internal vertices for which the two descendant subtrees are not identical.
2. A second class of models are those which satisfy the condition:

$$s(x) = 0 \text{ for } x > \varepsilon \,,$$

which we will call "explosive radiation" models. In these model, unless a species has undergone a speciation event within the last ε time interval, it will never do so. Thus, in this model, speciation events would tend to be clustered close together. Our last result (Theorem 6) shows that, provided epsilon is sufficiently small, then this model is precisely that induced by a *uniform distribution* on leaf-labeled trees.

Under the uniform model, on rooted trees, a tree is selected uniformly from $LB(n)$, and then it is viewed as an unlabeled tree $\tau \in UB(n)$. The probability of the tree shape τ is calculated as $p_{unif}(\tau) = |L(\tau)| \,/\, |LB(n)|$, where $L(\tau) = \{T \in LB(n) : T \text{ is a leaf-labeling of } \tau\}$. Fortunately, the numerator and denominator of this ratio can both be evaluated exactly, and so we get an explicit formula for $p_{unif}(\tau)$ as follows. We have $|L(\tau)| = n!2^{-b(\tau)}$ where $b(\tau)$ is the number of *balanced* internal vertices of τ - that is, internal vertices for which the two descendant subtrees are identical. Now, from [17], $|LB(n)| = \frac{(2n-2)!}{(n-1)!2^{n-1}}$ Therefore, under the uniform model on rooted trees,

$$p_{unif}(\tau) = \frac{|L(\tau)|}{|LB(n)|} = \binom{2n-2}{n-1}^{-1} 2^{u(\tau)} \,, \tag{14}$$

where $u(\tau)$ is the number of *unbalanced* internal vertices (and so $b(\tau) + u(\tau) = n-1$).

Theorem 6. *[15] Under an explosive radiation model, with $\varepsilon < 1/n$, the probability distribution on trees is precisely that induced by the uniform model. That is,*

$$p(\tau) = p_{unif}(\tau), \quad \forall \tau \in UB(n) \,.$$

Acknowledgement

We thank Charles Semple for some helpful comments on an earlier version of this manuscript. This research was supported by the New Zealand Marsden Fund (UOC-MIS-003).

References

1. D. Aldous: Probability distributions on cladograms, in *Random Structures*, ed. by D. Aldous, R. Permantle, Vol. 76 (Springer, 1996), pp. 1–18
2. J. Brown: Syst. Biol. **43**(1), 78–90 (1994)
3. E. Harding: Adv. Appl. Prob. **3**, 44–77 (1971)
4. S. Heard: Evolution **46**(6), 1818–1826 (1992)
5. S. Heard: Evolution **50**(6), 2141–2148 (1996)
6. T. Kubo, Y. Iwasa: Evolution **49**, 694–704 (1995)
7. A. McKenzie, M. Steel: Mathematical Biosciences **164**(1), 81–92 (2000)
8. A. Mooers, S. Heard: The Quarterly Review of Biology **72**(1), 31–54 (1997)
9. J. Slowinski: Syst. Zool. **39**(1), 89–94 (1990)
10. B. Rannala, Z. Yang: Molecular Evolution **43**, 304–311 (1996)
11. B. Mau, M. Newton, B. Larget: Biometrics **55**, 1–12 (1999)
12. S. Li, D.K. Pearl, H. Doss: Journal of the American Statistical Association **95**(450), 493–508 (2000)
13. J. Haigh: J. Appl. Prob. **7**, 79–88 (1970)
14. D. Aldous: (2000), Stochastic models and descriptive statistics for phylogenetic trees, from Yule to today [online], Available: http://www.stat.berkeley.edu/ users/ aldous/ bibliog.html [Accessed: August 25]
15. M. Steel, A. McKenzie: Math. Biosci. **170**(1), 91–112 (2001)
16. R. Page, E. Holmes: *Molecular Evolution: a phylogenetic approach* (Blackwell Science, 1998), Chap. 2, pp. 11–36
17. L. Cavalli-Sforza, A. Edwards: Evolution **21**, 550–570 (1967)
18. A. Edwards, L. Cavalli-Sforza: Reconstruction of evolutionary trees, in *Phenetic and phylogenetic classification*, ed. by W. Heywood, J. McNeill (1964), no. 6, pp. 67–76
19. J. Slowinski, C. Guyer: The American Naturalist **134**(6), 907–921 (1989)
20. S. Nee, R. May, P. Harvey: Phil. Trans. R. Soc. Lond. B **344**, 305–311 (1994)
21. J. Losos, F. Adler: The American Naturalist **145**(3), 329–342 (1995)
22. J. Kingman: J. Appl. Prob. **19A**, 27–43 (1982)
23. F. Tajima: Genetics **105**, 437–460 (1983)
24. D. Knuth: *The Art of Comuter Programming*, Vol. 3 (Addison-Wesley, Reading, Massachusetts, 1997), Chap. 5, p. 67, problem 20
25. A. Edwards: J. R. Stat. Soc. Ser. B **32**, 155–174 (1970)
26. M. Sanderson: Syst. Biol. **45**(2), 168–173 (1996)
27. W. Lynch: The Computer Journal **71**, 299–302 (1965)
28. H. Mahmoud: *Evolution of random search trees* (John Wiley and Sons Ltd, New York, 1992), p. 72
29. R. Stanley: *Enumerative Combinatorics: Vol. 1*, no. 49 in Cambridge Studies in Advanced Mathematics (Cambridge University Press, 1997), pp. 18–19
30. M. Steel, D. Penny: Syst. Biol. **42**(2), 126–141 (1993)
31. M.D. Hendy, D. Penny: Math. Biosci. **59**, 277–290 (1982)
32. M.A. Steel: SIAM J. Discr. Math. **1**(4), 541–551 (1988)
33. M. Härlin: Biological Journal of the Linnean Society **58**, 325–342 (1996)
34. J.P. Huelsenbeck, M. Kirkpatrick: Evolution **50**(4), 1418–1424 (1996)
35. M. Ridley: *Evolution* (Blackwell Science,Inc., Cambridge, Massachusetts, USA, 1993), Chap. 17, pp. 460–466
36. A. Smith: Biological Journal of the Linnean Society **51**, 279–292 (1994)
37. J. Farris: American Naturalist **106**, 646–668 (1972)

38. D.L. Swofford, G.J. Olsen, P.J. Waddell, D.M. Hillis: Phylogenetic inference, in *Molecular Systmatics*, ed. by D.M. Hillis, C. Moritz, B.K. Mable (Sinauer Associates, Inc., Sunderland, Massachusetts, U.S.A, 1996), p. 488
39. Y.V. de Peer, R.D. Wacher: Comput. Applic. Biosci. **9**, 177–182 (1993)
40. J. Medhi: *Stochastic Processes* (John Wiley and Sons Ltd, New York, 1982), Chap. 4, p. 119

Part III

The evolution of populations and species

Fitness landscapes

Peter F. Stadler[1,2]

[1] Institut für Theoretische Chemie und Molkulare Strukturbiologie, Universität Wien, Währingerstrasse 17, A-1090 Wien, Austria
[2] The Santa Fe Institute, 1399 Hyde Park Rd., Santa Fe, NM 87501

Abstract. Fitness landscapes are a valuable concept in evolutionary biology, combinatorial optimization, and the physics of disordered systems. A fitness landscape is a mapping from a configuration space that is equipped with some notion of adjacency, nearness, distance or accessibility, into the real numbers. Landscape theory has emerged as an attempt to devise suitable mathematical structures for describing the "static" properties of landscapes as well as their influence on the dynamics of adaptation. This chapter gives a brief overview on recent developments in this area, focusing on "geometrical" properties of landscapes.

1 Introduction

The concept of a *fitness landscape* originated in theoretical biology more than seventy years ago [1]. It can be thought of as a kind of "potential function" underlying the dynamics of evolutionary optimization. Implicit in this idea is both a *fitness function* f that assigns a fitness value to every possible genotype (or organism), and the arrangement of the set of genotypes in some kind of abstract space that describes how easily or frequently one genotype is reached from another one.

The same abstract construction arises in a natural way in the physics of disordered systems. Spin-glasses, for example, can be cast into the same form [2,3]. Each spin configuration is assigned an energy by virtue of the Hamiltonian that specifies the model; the dynamic properties invoke a collection of transitions between configurations. In biophysics energy landscapes govern the folding of biopolymers, including proteins [4–6] and nucleic acids [7,8]. Conceptually, there is a close connection with the *potential energy surfaces* of theoretical chemistry [9,10]: As a consequence of the validity of the Born-Oppenheimer approximation, the PES provides the potential energy $U(\boldsymbol{R})$ of a molecule with n atoms as a function of its nuclear geometry $\boldsymbol{R} \in \mathbb{R}^{3n}$. Electoral Landscapes are used to explain party platform behavior in spatial voting models [11,12].

In combinatorial optimization the fitness function is usually referred to as the *cost function*, and a *move-set* allows to inter-convert the elements of the *search space* [13]. The application of evolutionary models to combinatorial optimization problems has lead to the design of so-called *evolutionary algorithms* such as Genetic Algorithms, Evolution Strategies, and Genetic Programming [14–18].

The intuitive notion of *ruggedness* is closely related to the difficulty of optimizing (or adapting) on a given landscape. It depends obviously on both the fitness function and the geometry of the search space, which is induced by the search process. On the other hand, simulations of adaptation of biologically realistic landscapes derived from RNA folding [19] have shown that *neutrality*, that is, the occurrence of adjacent configurations with the same fitness, can play a dominating role in evolutionary dynamics as well.

One of the main topic in "landscape theory", and the focus of this contribution, is therefore a detailed understanding of the geometric features of landscapes: Mountain massives, valleys, basins, peaks, plains and ridges in multidimensional combinatorial objects may look quite different from our 3D experience and oftentimes require a mathematical description in terms of algebraic combinatorics rather than calculus.

Landscapes can also be studied from a "dynamical" point of view, focusing on the features of a dynamical system, for instance an evolving population, that uses the landscape as its substrate. The challenge for a *theory of landscapes* is therefore to link these two points of views, for instance by determining how geometric properties influence the dynamical behavior.

Given that landscapes arise naturally in many different fields, it is not surprising that the concept of a *fitness landscape* has emerged as a unifying theme in the literature on complex systems [17,20–22]. In formal terms, a landscape consists of three ingredients

1. A set X of configurations,
2. a notion $\mathcal{X}$ of neighborhood, nearness, distance, or accessibility on X, and
3. a fitness function $f : X \to \mathbb{R}$,

The set X together with the "structure" $\mathcal{X}$ forms the configuration space. The definition of $\mathcal{X}$ is purposefully left vague at this point as we will elaborate on the structure $\mathcal{X}$ in section 2. A common source of confusion is the fact that biologists like to maximize fitness on their landscapes, while physicists minimize energy on theirs. Obviously, replacing f by $-f$ maps one picture into the other.

Despite its wide range of applicability, the usefulness of fitness landscapes is limited to certain situations. Let us consider a general evolutionary process as an example for the limitations of the landscape concept. Since genetic variation is generated independently from the natural selection acting on it, the generic structure of an evolutionary model in discrete time can be written as

$$x' = S(x, \mathsf{w}) \circ T(x, \mathsf{t}) \ , \tag{1}$$

where x is e.g. the vector of haplotype frequencies [23,24]. As usual, $\circ$ denotes the Schur (Hadamard, component-wise) product of vectors. The transmission term $T(x, \mathsf{t})$ describes the probability of transforming one type into another one by mutation and/or recombination [25] and hence determines the structure $\mathcal{X}$ on the set X of all vectors of haplotype frequencies. In genetics this structure can be understood in terms of certain classes of algebraic structures [26] that depend on the details of transmission mechanism represented by the paramters t. The term $S(x, \mathsf{w})$ describes the selection forces acting on x. The parameters w form the *fitness function*, since they can be regarded as a mapping from the set of types into the real numbers. Whether or not the dynamics of equ.(1) is consistent with or determined by a *fitness landscape*, depends on the particularities of of the selection term. Setting $S(x, \mathsf{w}) = x \circ F(x, \mathsf{w})$, or in component-wise notation, $S_k(x, \mathsf{w}) = x_k F_k(x, \mathsf{w})$, we have selection proportional to a *growth-rate function* F_k for each type [27]. We suggest that one should speak about *fitness landscapes* only when $F_k(x, \mathsf{w}) \approx f(k)$, where $f(k)$ is a constant that is characteristic for the type k, since otherwise the fitness (growth rate) of type k depends on and changes with the frequencies of all other types. Models of co-evolution are sometimes viewed as "coupled dancing landscapes" [17] where a species A changes the landscape of species B, and B changes the landscape of A, at the same time scale at which both species adapt to their respective landscapes. We prefer here to limit the term fitness landscape to situations in which (1) fitness is characteristic of a type and (2) constant in time, at least approximately.

2 Configuration spaces

There appear to be three distinct approaches to organizing the configuration space.

1. In computer science one typically specifies a "move set" or "genetic operator" that inter-converts one or more configurations into a new one.
2. Sometimes transition probabilities are specified that describe how frequently a system attempts to move from one configuration to another one.
3. A rigorous mathematical analysis often starts with specifying a metric distance or a topology on X.

2.1 Move sets

In its most abstract form a *move set* assigns to a k-tuple $(x_1, \ldots, x_k) \in X^k$ of "parents" a list $N(x_1, \ldots, x_k) \subseteq X$ of "children". We will restrict our attention here to the two most commonly used move-set types, mutation and recombination.

A mutation operator simply assigns a set $N(x)$ of "accessible neighbors" or "elementary mutants" to each configuration x. This allows us to interpret X as a (possibly directed) graph with vertex set X and $N(x)$ the (out)neighbors of $x \in X$. Most commonly the move sets are constructed such that $y \in N(x)$ if and only if $x \in N(y)$, in which case the graph is symmetric, or, equivalently, undirected.

In the case of strings (i.e., sequences of characters taken from a fixed alphabet $\mathcal{A}$) typical moves consist of the replacement of a character at a single position by another one. The resulting graphs are the Hamming graphs. In particular, for a two letter alphabet such as *spin-up* ($\uparrow$), and *spin-down* ($\downarrow$), one obtains hypercubes, see Figure 1.

In some cases the configurations are naturally interpreted as algebraic objects. For instance, the tours of a Traveling Salesman Problem are permutations

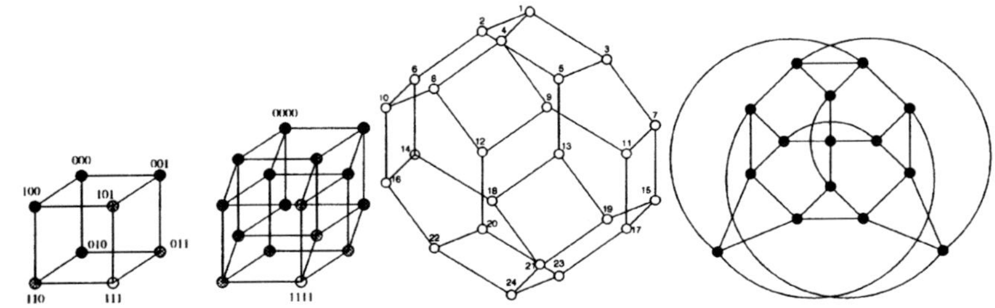

Fig. 1. Some Configuration space graphs: hypercubes $\mathcal{Q}_2^3$ and $\mathcal{Q}_2^4$, the permutohedron graph $\Gamma[\mathsf{S}_5, \mathcal{K}]$, the line graph of the Petersen graph $L[P]$ which equals the Robinson graph for $n = 5$ taxa.

of a list of cities. Configurations hence are group elements and moves become generators of the group. Let G be a permutation group acting on a finite set X. Furthermore, let $\Omega \subset \mathsf{G}$ be a set of generators of G such that (i) $\imath \notin \mathsf{G}$ and (ii) $x \in \Omega \implies x^{-1} \in \Omega$. A graph $G = G(\mathsf{G}, \Omega)$ with vertex set $V = \mathsf{G}$ and edges $\{x, y\} \in E$ if and only if $xy^{-1} \in \Omega$ is called a *Cayley graph* of the group G. In case of the symmetric group S_n suitable generator sets are e.g. transpositions, reversals, or the "canonical" transpositions of two *subsequent* cities which, for $n = 4$, lead to the permutohedron graph shown in Fig. 1. It is not hard to show that Hamming graphs are also Cayley graphs of a suitable group, see e.g. [28,29].

In biology evolutionary relationships between species or individual genes are customarily represented as phylogenetic trees. The vertices of a phylogenetic tree represent taxonomic units, the graph's topology delineates the genealogical relationships between them, and the branch lengths reflect the time of divergence. Many methods exist for the construction of phylogenetic trees, the more sophisticated among them seeking those trees in which the taxonomic units evolve with the least evolutionary change [30] (most parsimonious trees) or trees of maximum likelihood given a stochastic model of sequence evolution [31]. The search for the optimal tree is hence recast as a combinatorial optimization problem on the set of all phylogenetic trees with a given number of leaves (taxa). The basic variants of these tree reconstruction problems are all NP-complete [32,33], hence search heuristics are used in practice which employ a variety of editing operations on phylogenetic trees. So-called "nearest neighbor interchange" (swapping of subtrees separated by an inner edge of the tree), for instance, leads to a rather well-studied family graph sometimes referred to as Robinson graphs [34–36].

All move sets discussed above are symmetric and regular, i.e., any two configurations have the same number of neighbors. Of course, this is not always the case. Biological sequences, for instance, not only undergo point mutations but also insertions and deletions leading to highly irregular graphs. Other mutation operators of interest in this context include gene duplications and genomic rearrangements [37,38]. A graph is faithfully represented by its adjacency matrix $\mathbf{A}$ which has the entries $\mathbf{A}_{xy} = 1$ if $x \in N(y)$ and $\mathbf{A}_{xy} = 0$ otherwise. Obviously, $\mathbf{A}$ is symmetric if and only if the graph is undirected.

The most immediate consequence of the fact that recombination acts on two arguments is that the recombination induced configuration space can not be represented as a simple graph with the set of genotypes representing the set of vertices. This leaves two alternatives: One can change the nature of the vertex set and have pairs of types as vertices. Then one obtains again a (di-)graph, since each elementary recombination event creates up to two different strings. This approach was pioneered by Culberson [39] and Jones [40]. The alternative is to leave the vertices to represent individual genotypes and to make the edges more complex. In Gitchoff and Wagner [41] it was shown that recombination spaces can be represented as hypergraphs (which consist of a vertex set X and a collection $\mathcal{E}$ of (not necessarily) distinct subsets of X called (hyper)edges), where the hyperedges are the sets of all recombinants that can arise from the recombination of two types. With this approach it is was easy to show that string recombination spaces and point mutations spaces are homomorphic. Hypergraphs are still not completely satisfactory, since they do not indicate which pair of types produces which set of recombinants, i.e., which hyper-edge arises from which mating. This led us to invent *P-structures*, which are mappings of pairs of types to the hyperedges of the hypergraph [42,43].

Let us first consider homologous recombination on a genome consisting of n loci. For each locus k, there are α_k alleles. The set of all the $\prod_k \alpha_k$ possible genotypes will be denoted by V. For each locus k, we label the alleles using a letter from the alphabet $\mathcal{A}_k = \{0, \ldots, \alpha_k - 1\}$. Thus $V = \prod_k \mathcal{A}_k$. A particular genotype (or sequence) $x \in V$ can be regarded as a vector with components $x_k \in \mathcal{A}_k$. A particular cross-over operator χ is determined by the list $\boldsymbol{\chi}$ of loci that the child inherits from the first parent. Thus the loci in $\overline{\boldsymbol{\chi}} = \{1, \ldots, n\} \setminus \boldsymbol{\chi}$ come from the second parent. More formally, given $\boldsymbol{\chi}$, the offspring $x = \chi(y, z)$ of the two parents y and z has the component-wise representation

$$x_k = \begin{cases} y_k \text{ if } k \in \boldsymbol{\chi} \\ z_k \text{ if } k \in \overline{\boldsymbol{\chi}} \end{cases} \tag{2}$$

It will be convenient in the following to express equ.(2) by means of an "incidence operator"

$$\mathbf{H}^{\chi}_{x,(y,z)} = \begin{cases} 2 \text{ if } x = y = z \\ 1 \text{ if } y \neq z \text{ and } x = \chi(y,z) \\ 0 \quad \text{otherwise}. \end{cases} \tag{3}$$

Here we shall restrict ourselves to recombination on strings. Crossover operators for permutation, such as Traveling Salesman tours, are reviewed for instance in [44].

A *recombination operator* in the sense of most of the GA literature is then a family $\mathcal{F}$ of cross-over operators that act on $X \times X$ with probability $\pi(\chi)$. The incidence "matrix" associated with a recombination operator is simply

$$\mathbf{H}^{\mathcal{F}} = \sum_{\chi \in \mathcal{F}} \mathbf{H}^{\chi} \tag{4}$$

The two most important recombination operators are

[∞] Uniform recombination contains all 2^n possible crossover operators. In this case it is natural to include the identity $\imath$.

[1] 1-point recombination contains all cross-over operators χ for which the characteristic set is of the form $\boldsymbol{\chi} = \{1, \ldots, k\}$.

Homologous recombination (of strings) under very general conditions leads to very regular configuration spaces. In particular, one can show that the automorphism group of $\mathbf{H}^{\mathcal{F}}$ is generously transitive [42]. (A permutation group G on a set X is generously transitive if for each pair $x, y \in X$ there is a group element $\mathsf{g} \in \mathsf{G}$ such that $\mathsf{g}(x) = y$ and $\mathsf{g}(y) = x$, see e.g. [45].) This picture, however, changes radically, if unequal crossover is considered, where the number of genes on a chromosome can change [46].

2.2 Transition matrices

Regarding X as a set of "states" we may alternatively specify transition probabilities $\mathbf{T}_{xy}$ for moving from y to x. The Markov process with transition matrix $\mathbf{T}$ organizes the configuration space in this case. Typically one requires $\mathbf{T}$ to be ergodic (i.e., every state can be reached from every other state) and reversible, i.e., to satisfy

(E) $\mathbf{T}$ is irreducible, or, equivalently,
there is a unique stationary distribution p on X such that $\mathbf{T}p = p$. Furthermore $p(x) > 0$ for all $x \in X$.

(R) $\mathbf{T}_{xy}p(y) = \mathbf{T}_{yx}p(x)$. This condition is also known as "detailed balance".

In other words, $\mathbf{T}$ is self-adjoined w.r.t. to the scalar product

$$\langle f, g \rangle_p = \sum_x p(x) f(x) g(x)^* \tag{5}$$

where the star denotes complex conjugation.

A most useful observation is that the matrix $\mathbf{S}$ defined by

$$\mathbf{S}_{xy} = p(x)^{-1/2} \mathbf{T}_{xy} p(y)^{1/2} \tag{6}$$

is symmetric and similar to $\mathbf{T}$. Hence given a non-symmetric transition matrix $\mathbf{T}$ and a landscape f we may transform the model to new coordinates with the symmetric operator $\mathbf{S}$ and the transformed landscape

$$f^\sigma(x) = p(x)^{-1/2} f(x) \tag{7}$$

This allows the application of the spectral landscape theory discussed in section 4 also to the non-symmetric case.

The move sets discussed in the previous section can be translated into the Markov chain setting in a natural way. With each (directed or undirected) graph there is an associated Markov process on its vertex set [47] defined by the transition matrix

$$\mathbf{T} = \mathbf{A}\mathbf{D}^{-1} \tag{8}$$

where $\mathbf{D}$ is the so-called degree matrix, which is diagonal and $\mathbf{D}_{xx} = |N(x)|$ is the number of neighbors of x, and $\mathbf{A}$ is the adjacency matrix introduced in the previous section. This Markov process describes a *random walk* on X which has been suggested as a means to sample information about a landscape by Ed Weinberger [48,49]. We remark that in the case of undirected and symmetric directed graphs the stationary distribution is given by

$$p(x) = \frac{\mathbf{D}_{xx}}{2E} \tag{9}$$

where E is the number of undirected edges, which, for symmetric directed graphs, is of course $E = \sum_x |N(x)|/2$.

A *cross-over walk* [50,51] on X is the Markov process based on the following rule: The "father" y is mated with a randomly chosen "mother" z. The offspring is "son" x which becomes the "father" of the next mating. We regard the sequence of "fathers" as a random walk on X. It is straightforward [52] to derive the transition matrix of this Markov process for homologous recombination from the incidence "matrix" $\mathbf{H}^{\mathcal{F}}$. One obtains

$$\mathbf{S}^{\mathcal{F},\wp}_{xy} = \sum_{\chi\in\mathcal{F}} \pi(\chi)\frac{1}{2}\sum_{z\in X} \mathbf{H}^{\chi}_{x,(y,z)}\wp(z) = \sum_{\chi\in\mathcal{F}} \pi(\chi)\mathbf{S}^{\chi,\wp}_{xy} \tag{10}$$

where $\wp(z)$ denotes the frequency distribution of the genotypes in the equilibrium population.

2.3 Configuration space topologies

We shall see in the following section that finite ("discrete", or combinatorial) landscapes are treated quite differently from their manifold ("continuous") counterparts. The reason is that functions on $\mathbb{R}^n$, or more generally Riemannian manifolds can be analyzed in terms of differential operators such as gradients, while finite sets are usually discussed in terms of graph-theoretical properties. It seems desirable therefore, to find a basic framework that allows to deal with landscapes on arbitrary configurations spaces. A suitably general language is provided by the theory of *pretopological spaces.*

A pretopological space consists of an arbitrary set X and a collection $\mathcal{N}(x)$ of neighborhoods for every point $x \in X$, such that

(P1) $N \in \mathcal{N}(x)$ implies $x \in N$;
(P2) $N \in \mathcal{N}(x)$ and $N \subseteq N'$ implies $N' \in \mathcal{N}(x)$

(P2) $N, N' \in \mathcal{N}(x)$ implies $N \cap N' \in \mathcal{N}(x)$ Pretopologies are more general then the much more familiar topological spaces. In fact, $(X, \mathcal{N})$ is a topological space if and only if

(T) For each $N \in \mathcal{N}(x)$ there is an $N' \in \mathcal{N}(x)$ such that $N \in \mathcal{N}(y)$ for all $y \in N'$.

Directed graphs are exactly the finite pretopological spaces. Their neighborhood systems consists of all sets N' containing x and all vertices adjacent to x, i.e., $N(x) \cup \{x\} \subseteq N'$. Notions such as minima, maxima, or continuity of a function, connectedness, convergence, limits, etc. can be defined on pretopological spaces [53–56]. Their usefulness in the context of genotype-phenotype maps and fitness landscapes will be discussed in forthcoming manuscripts [57,58].

3 Basic properties of landscapes

3.1 Local optima

Combinatorial optimization is concerned with finding "optimal" i.e., minimal or maximal values of the cost function. Local optima thus play an important role since they might be obstacles on the way to the optimal solution. In the theory of disordered systems, local minima of the energy function are usually called metastable states. For the sake of definiteness we shall consider local minima in the following. Analogous expressions for local maxima can be obtained by replacing f with $-f$. Let us start with a formal

Definition. A configuration $\hat{x} \in X$ is a *local minimum* if there is a neighborhood $N \in \mathcal{N}(\hat{x})$ such that $f(\hat{x}) \leq f(y)$ for all $y \in N$.

Clearly, this definition makes sense on arbitrary pretopological spaces and it coincides with the usual definition in the graph case, which requires $f(\hat{x}) \leq f(y)$ for all $y \in N(\hat{x})$. A minimum $\hat{x}$ is global, of course, if $f(\hat{x}) \leq f(y)$ for all $y \in X$. Note that landscapes need not have local or even global minima unless they are defined on a compact configuration space.

The number of local optima is a measure for the "ruggedness" of landscape. Richard Palmer [59], for instance, suggested to call a landscape f *rugged* if the number M_f of local optima scales exponentially with some measure of "system size" such as the number of cities in a TSP or the number of spins in spin glass. Unfortunately, there is in general no simply way of computing M_f without exhaustively generating the landscape. Alternatively, one can of course estimate M_f by checking whether a randomly generated $x \in X$ is a local minimum. Numerical data of this kind are reported e.g. in [60–62]. Methods from statistical mechanics can be used, however, to obtain the scaling of the expected value $\mathbb{E}[\mathsf{M}]$ with the system size for a variety of disordered systems, see e.g. [63–70].

3.2 Basins

To each local minimum $\hat{x}$ there is an associated *basin* $\mathcal{B}(\hat{x})$. On manifolds it can be defined as the set of all $y \in X$ such that $\hat{x}$ is the ω-limit of the gradient dynamics $\dot{z} = -\nabla f(z)$ with initial condition y. In the graph case one can use the steepest descent algorithm instead: Starting with $z_0 = y$ we choose at each step the neighbor $z_{k+1} \in N(z_k)$, $f(z_{k+1}) < f(z_k)$ with the smallest fitness value and repeat the procedure until it terminates when $z_{k+1} = \hat{x}$ is a local minimum. The notion of a basin hence may become ambiguous when there is "local neutrality" in $N(x)$, i.e., if there are $x \in X$ and $y, y' \in N(x)$ with $f(y) = f(y')$. It is an open question how the basin should be defined in full generality, or what kind of structure on X must be required in order to properly define basins. It is not surprising that the distribution of basin sizes is crucial for the performance of simple optimization heuristics [71]. So far there does not appear to be a good method for estimating basin sizes beyond exhaustive enumeration or random sampling, however.

An important aspect is the correlation between basin size and fitness of the minimum: In general, deeper minima have larger basins. Figure 2 shows that this is not only true for well-behaved landscapes such as the Sherrington-Kirkpatrick spin glass (r.h.s.), but also for random landscapes (l.h.s.). The difference is, however, that basin sizes appear to scale exponentially with fitness in well-behaved landscapes, while they approach a constant in essentially random landscapes.

3.3 Gradient walks and adaptive walks

A simple measure for the size of a basin $\mathcal{B}(\hat{x})$ is the average length L of the steepest descent walks from $y \in \mathcal{B}(\hat{x})$ to $\hat{x}$. The average length L of a gradient walk has been investigated as a ruggedness measure in a few models, including random landscapes, Kauffman's NK landscapes [72], fitness landscapes derived from RNA folding [73].

An *adaptive walk* accepts a neighbor $x_{k+1} \in N(x_k)$ provided $f(x_{k+1}) < f(x_k)$ instead of looking for the steepest descent. Gillespie [74] suggested to use adaptive walks as models of evolutionary adaptation. They have been studied extensively in NK models [72,75,76], in particular in the context of the maturation of the immune response [77–79], in RNA folding landscapes [73] and in a model of early vascular land plants [80]. The lengths distribution of adaptive walks appears to be linked to the size distribution of the basins; the details of this connection, however, remain to be elucidated.

3.4 Barriers

The basins of local minima are separated by saddle points and fitness barriers. Let $\hat{x}$ and $\hat{y}$ be two local minima and let $\mathbf{p}$ be a path in X from $\hat{x}$ to $\hat{y}$. Then the fitness barrier separating $\hat{x}$ from $\hat{y}$ is

$$f[\hat{x}, \hat{y}] = \min \left\{ \max \left[f(z) \middle| z \in \mathbf{p} \right] \middle| \mathbf{p} : \text{path from } \hat{x} \text{ to } \hat{y} \right\}, \tag{11}$$

A point $\hat{z} \in X$ satisfying the minimax condition in equ.(11) is a *saddle point* of the landscape. It should be noted that this definition is meaningful both in the graph case and on $\mathbb{R}^n$. However, in the context of potential energy surfaces one typically defines a saddle point as a critical point $\nabla f = 0$ that is neither a minimum nor a maximum. The saddle-point energies $f[\hat{x}, \hat{y}]$ form an ultrametric distance measure on the set of local minima, see e.g. [81–83]. This hierarchical

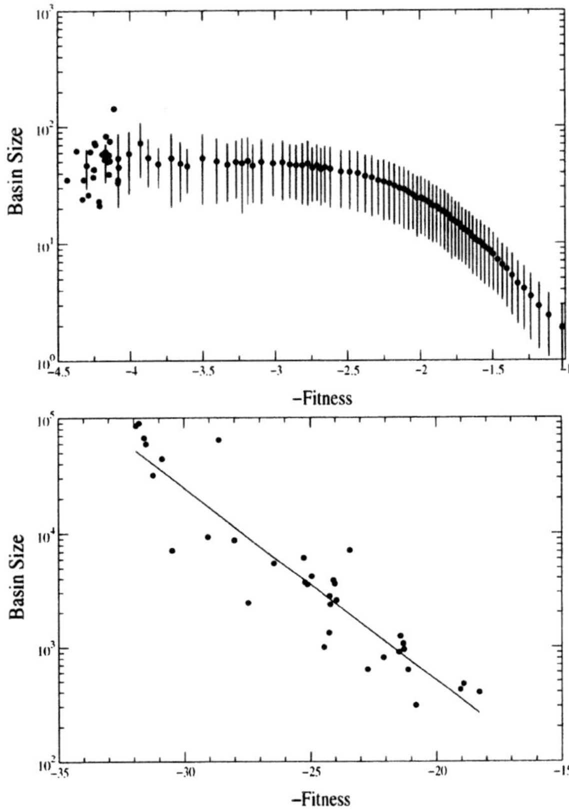

Fig. 2. Size distribution of basins of attraction for the SK-Model (quadratic Ising spin glass with i.i.d. Gaussian coefficients $a_{i_1 i_2}$ in equ.(20) and all other coefficients 0, $n = 20$), and a random assignment of Gaussian random numbers with mean 0 and variance 1 to the vertices of the 20-dimensional hypercube $\mathcal{Q}_2^{20}$. Error bars show the standard deviation of the distribution of basin sizes in a fixed fitness interval for the random Gaussian landscape, which has $\mathsf{M}_f = 49935$ local minima. The SK model on the r.h.s has only $\mathsf{M}_f = 70$ local minima and therefore much larger basins. The important observation is, however, that the basin size scales exponentially with fitness (energy) in this case.

structure can be represented by the *barrier tree* of the landscape, Figure 3. Its leaves are the local minima and its internal nodes correspond to saddle points.

The *barrier* enclosing a local minimum is the height of the lowest saddle point that gives access to a more favorable minimum. In symbols:

$$B(\hat{x}) = \min\left\{f[\hat{x}, \hat{y}] - f(\hat{x})\middle|\hat{y} : f(\hat{y}) < f(\hat{x})\right\} \tag{12}$$

If $B(s) = 0$ then the local minimum s is degenerate. It is easy to check that eq.(12) is equivalent to the definition of the depth of a local minimum in [85].

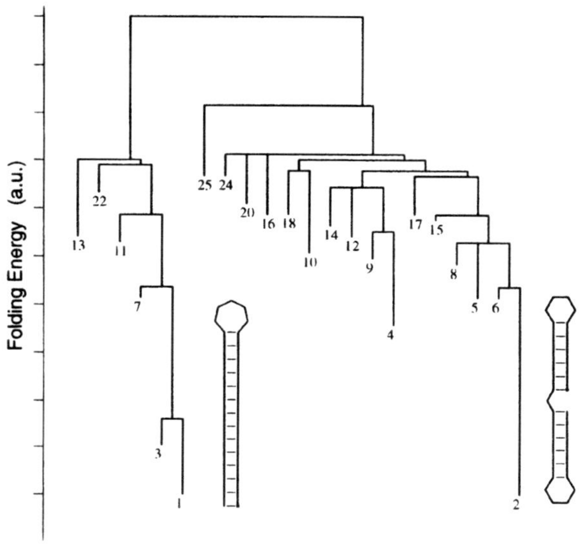

Fig. 3. Example of a barrier tree for the folding energy landscape of a bi-stable RNA molecule. The secondary structures of the two lowest energy states are indicated. For details see [84].

For metastable states it agrees with the more general definition of the depth of a "cycle" in the literature on inhomogeneous Markov chains [86–88].

3.5 Depth

The information contained in the energy barriers is conveniently summarized by two global parameters that e.g. determine the convergence behavior of Simulated Annealing and related algorithms. Let Ω_f be the set of all global minima of f. Now consider the following two quantities

$$\begin{aligned} \mathsf{D} &= \max\left\{B(s)\middle|s\notin\Omega_f\right\} \\ \psi &= \max\left\{\frac{B(s)}{f(s)-f(\min)}\middle|s\notin\Omega_f\right\} \end{aligned} \tag{13}$$

Both parameters are easily obtained from the barrier tree. The *depth* D and *difficulty* ψ [85,87–90] play a crucial role in theory of Simulated Annealing. For instance, Simulated Annealing converges almost surely to a ground state if and only if the cooling schedule T_k satisfies $\sum_{k\geq 0}\exp(-\mathsf{D}/T_k)=\infty$ [89]. The difficulty parameter is directly related to the optimal speed of convergence of Simulated Annealing.

3.6 Correlation

Correlation measures are by the far the most accessible indicators of ruggedness. Weinberger [48] considers the autocorrelation function $r(s)$ of the "time series" of fitness values $f(x^{(t)})$ sampled along a random walk $\{x^{(0)}, \dots, x^{(t)}, \dots\}$ on X with transition matrix T and initial conditions distributed as p. In [91,92] distance dependent correlation functions $\rho(d)$ are considered, where d is a metric on X. The walk correlation function $r(s)$ of a landscape can be obtained without reference to the stochastic sampling process as [29]

$$r(s) = \langle f, \mathbf{T}^s f \rangle \tag{14}$$

The relationship between the walk correlation function $r(s)$ and the distance correlation function $\rho(d)$ is described in detail in [93] for highly symmetric transition operators. In many applications the correlation length

$$\ell = \sum_{s=0}^{\infty} r(s) \tag{15}$$

is used as a convenient measure of ruggedness, see e.g. [73,94].

The correlation length ℓ, the length L of gradient walks, and the expected number M of local optima appear to be closely related in "typical" landscapes. The notion of a "typical" landscape is made precise in [95]. Denoting by $X(x_0, \ell)$ the set of configurations that can be reached in at most ℓ applications of $\mathbf{T}$ from x_0 on a graph Γ, the "correlation length conjecture" [96] states that there should be roughly $\mathsf{M} \approx |X|/|X(x_0, \ell)|$ local optima. This estimate is based on the assumption that the correlation length determines the diameter of the large mountains and valleys, and that due to the high dimensional nature of typical search spaces each mountain typically contains only a small number of local optima. The correlation length conjecture has been tested on a variety of combinatorial optimization problems and appears to be a very good approximation [62,97].

4 Spectral landscape theory

Spectral approaches to fitness landscapes start with one of the symmetric non-negative operators on X discussed in section 2.2 above. The basic idea is to interpret the adjacency matrix $\mathbf{A}$ of a symmetric graph, or the operator $\mathbf{S}$, as a representation of the configuration space and to discuss the fitness function in terms of the regularities of $\mathbf{S}$. From an algebraic point of view it appears to be more natural to start with a discrete *Laplace operator*

$$-\boldsymbol{\Delta} = \mathbf{D_S} - \mathbf{S} \qquad \text{with } (\mathbf{D_S})_{xx} = \sum_{y \in X} \mathbf{S}_{xy} \tag{16}$$

since it has number of desirable mathematical properties:

$-\Delta$ is symmetric and has non-positive off-diagonal entries.

$-\Delta$ has 0 as an eigenvalue with eigenvector $\mathit{1} = (1, \dots, 1)$. The eigenvalue 0 is unique if and only if the graph associated with the off-diagonal entries is irreducible.

$-\Delta$ is non-negative definite.

The graph Laplacian arises naturally as the discretization of the Laplacian differential operator for instance in finite element computations. For recent surveys on graph Laplacians see [98–101].

Let $\{\varphi_k\}$ be an orthonormal basis of eigenvectors of $-\Delta$. Of course, we can interpret φ_k as a fitness function on X, hence we use the "function" notation $\varphi_k(x)$ for the coordinate of φ_k indexed by x. It appears natural to expand a fitness function f into a *Fourier series*

$$f(x) = \sum_k a_k \varphi_k(x) \tag{17}$$

On so-called quasi-abelian Cayley graphs, that is, Cayley graphs for which the generator set is a union of conjugacy classes of the underlying group, this graph-theoretical Fourier series and the group theoretical Fourier transformation [102] coincide (apart from a different conventional normalization), see [103].

Since $-\Delta$ is symmetric one can of course choose the basis functions φ_k to be real valued. In many instances it is much more convenient, however, to allow for complex valued eigenfunctions. For instance, the basis functions for the Hamming graph $\mathcal{Q}_\alpha^n$ on the α-letter alphabet $\mathcal{A} = \{0, \dots, \alpha - 1\}$ can be written in the form

$$\varphi_a(x) = \alpha^{-n/2} \prod_{k=1}^{n} \exp\left(2\pi\iota \frac{a_k x_k}{\alpha}\right) \tag{18}$$

for each index $a \in \mathcal{A}^n$. A real-valued basis for this case is described e.g. in [28]. The corresponding Laplacian eigenvalue is $\Lambda_a = \alpha \operatorname{ord}(a)$, where

$$\operatorname{ord}(a) = |\{k | a_k \neq 0\}| \tag{19}$$

can be interpreted as the *interaction order* of the eigenfunction φ_a. This notion becomes more intuitive by considering Ising spin models. The most general spin glass Hamiltonian is

$$f(x) = a_0 + \sum_{p=1}^{n} \sum_{i_1 < i_2 < \dots < i_p} a_{i_1 i_2 \dots i_p} x_{i_1} x_{i_2} \dots x_{i_p} \tag{20}$$

with Ising spins $x_j = \pm 1$. In other words $f(x)$ is a superposition of p-spin models, where $p = \operatorname{ord}(i_1, \dots, i_p)$ is the interaction order. In fact, the Fourier basis on the hypercube *are* the (normalized) Walsh functions

$$\varphi_I(x) = 2^{-n/2} \prod_{i \in I} x_i = 2^{-n/2} x_{i_1} x_{i_2} \dots x_{i_p} \tag{21}$$

with the index set $I = \{i_1, i_2, \ldots, i_p\}$, and hence $p = \text{ord}(I) = |I|$. The standard way of specifying a spin glass model therefore is its Fourier expansion (17). In the following we shall assume without loosing generality that the landscape is normalized such that

$$\begin{aligned} \overline{f} &= \sum_{x\in X} p(x) f(x) = 0 \\ \text{Var}[f] &= \sum_{x\in X} p(x)(f(x) - \overline{f})^2 = \langle f, f\rangle_p - \overline{f}^2 = 1 \end{aligned} \tag{22}$$

where $p(x)$ is the stationary distribution of the transition operator in questions, see section 2.2. Thus we may assume $a_0 = 0$ in equ.(20).

Walsh functions, equ.(21), are used extensively in the analysis of Genetic Algorithms [42,105–109]. It is shown in [43,52] that the Walsh functions are also eigenvectors of the crossover transition matrices $\mathbf{S}^\chi$ defined in equ.(10) with uniform population distribution $\wp(z) = 1/|X|$. The corresponding eigenvectors are

$$\lambda_a^\chi = \begin{cases} 1 & \text{if } \tilde{a} = \emptyset \\ 1/2 & \text{if } \tilde{a} \neq \emptyset \quad \text{and } \tilde{a} \subseteq \chi \text{ or } \tilde{a} \subseteq \overline{\chi} \\ 0 & \text{otherwise} \end{cases} \tag{23}$$

where $\tilde{a} = \{k | a_k \neq 0\}$ is the set of non-zero indices for the corresponding eigenfunction φ_a. This fact allows a direct comparison of the landscapes formed by the same fitness function for a variety of crossover and mutation operators. The bottom line of such an analysis is that fitness functions with low interaction order look smoother with mutation, while recombination appears to be favorable for high interaction orders. For the details we refer to [42,43,52].

The usefulness of the spectral approach is by no means limited to Walsh functions. Its general applicability is established by the following observations:

(1) The landscapes of many of the most studied combinatorial optimization problems are *elementary*, i.e., their normalized fitness functions, equ.(22), are eigenvectors of the graph Laplacian, when X is organized according to the most natural move sets [29,110,111]. Examples include the Traveling Salesman Problem, Graph Bipartitioning, certain Satisfiability problems, Graph Coloring with a fixed number of colors, see Table 1. Furthermore, most of the examples belong to the 2nd non-zero eigenvalue of the Laplacian, see the last column of Table 1. Not all of the "classical" landscape are elementary, of course, but many of the non-elementary ones have non-zero projections to only a few eigenspaces with small eigenvalues; an example is the Quadratic Assignment Problem [103,112] or the asymmetric TSP [29].
(2) Eigenvectors of graph Laplacians have local minima and maxima that are well-separated on the fitness scale. Lov Grover [110] showed that for any local minimum $\hat{x}$ and any local maximum $\hat{y}$ one has

$$f(\hat{x}) \leq \overline{f} \leq f(\hat{y}) \tag{24}$$

Table 1. Some Elementary Landscapes. For a detailed discussion see section 4 of [29]. The *order* $\mathrm{ord}(\Lambda)$ of an eigenspace is its position in the spectrum of $-\boldsymbol{\Delta}$, not counting multiplicities and starting to count with 0 for the "flat landscape" with $\Lambda = 0$. For strings the order is given by equ.(19).

Problem	Graph	Degree	Λ	$\mathrm{ord}(\Lambda)$
p-spin glass	$\mathcal{Q}_2^n$	n	$2p$	p
NAES[(1)]	$\mathcal{Q}_2^n$	n	4	2
Weight Partitioning	$\mathcal{Q}_2^n$	n	4	2
Graph α-Coloring	$\mathcal{Q}_2^\alpha$	$(\alpha-1)n$	2α	2
XY-spin glass	$\mathcal{Q}_\alpha^n$	$(\alpha-1)n$	2α	2
for $\alpha > 2$:	$\mathcal{C}_\alpha^n$	2	$8\sin^2(\pi/\alpha)$	2
Linear Assignment	$\Gamma(\mathcal{S}_n, \mathcal{T})$	$n(n-1)/2$	n	1
TSP symmetric	$\Gamma(\mathcal{S}_n, \mathcal{T})$	$n(n-1)/2$	$2(n-1)$	2
	$\Gamma(\mathcal{S}_n, \mathcal{J})$	$n(n-1)/2$	n	2
	$\Gamma(\mathcal{A}_n, \mathcal{C}_3)$	$n(n-1)(n-2)/6$	$(n-1)(n-2)$	?
antisymmetric	$\Gamma(\mathcal{S}_n, \mathcal{T})$	$n(n-1)/2$	$2n$	3
	$\Gamma(\mathcal{S}_n, \mathcal{J})$	$n(n-1)/2$	$n(n+1)/2$	$\mathcal{O}(n)$
Graph Matching	$\Gamma(\mathcal{S}_n, \mathcal{T})$	$n(n-1)/2$	$2(n-1)$	2
Graph Bipartitioning	$J(n, n/2)$	$n^2/4$	$2(n-1)$	2

[(1)] Not-All-Equal-Satisfiability, see [104].

(3) Eigenvectors of graph Laplacians satisfy a discrete version of Courant's nodal domain theorem [113]. A nodal domain of $f : X \to \mathbb{R}$ is a maximal connected subset of X such that f does not change sign. Suppose the eigenvalues of $-\boldsymbol{\Delta}$ arranged in ascending order $0 = \Lambda_1 < \Lambda_2 \leq \cdots \leq \Lambda_{|X|}$, counting multiplicities. Note that the "ground state" Λ_1 corresponds to the *flat* landscape. On Hamming graphs we have therefore $\mathrm{ord}(\Lambda_k) = k - 1$.
The nodal theorem now states that if $-\Delta f = \Lambda_k f$, then f has at most k nodal domains, [114–118]. Thus landscapes may have more nodal domains, and hence more "mountain massives" if they belong to larger eigenvalues of the Laplacian. We remark that the nodal domain theorem holds for the class of so-called discrete Schrödinger operators which includes the symmetric transition matrices discussed in section 2.2.

Given an arbitrary landscape, we can measure the importance of a particular eigenspace of $-\Delta$ by means of the *amplitude spectrum*

$$B(\Lambda) = \sum_{k: -\Delta\varphi_k = \Lambda\varphi_k} |a_k|^2 \tag{25}$$

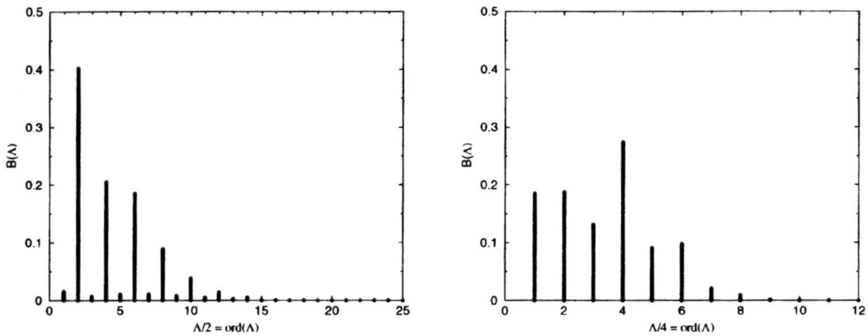

Fig. 4. Amplitude Spectra of RNA folding energy landscapes for **GC** sequences of length $n = 25$ and **GCAU**-sequences of length $n = 12$. The amplitude spectra are computed from explicit FFTs of the landscape as described in [103]. Note that the **GC** landscape has large amplitudes $B(\Lambda)$ only for even interaction orders, while the **GCAU**-landscape also contains a substantial linear, $\text{ord}(a) = 1$, component. For a discussion of the biophysical reasons see [119,120].

using $\sum_k |a_k|^2 = 1$ for normalized landscapes. Thus we have $B(\Lambda) \geq 0$ and $\sum_{\Lambda\neq 0} B(\Lambda) = 1$. For regular graphs, or bi-stochastic transition matrices, the Laplacian (16) and **T** have the same eigenvectors. In this case we can express random walk correlation functions and correlation lengths in terms of the amplitude spectrum

$$r(s) = \sum_{\Lambda\neq 0} B(\Lambda)\,(1-\Lambda/d)^s \qquad \text{and} \qquad \ell = d\sum_{\Lambda\neq 0}\frac{B(\Lambda)}{\Lambda} \tag{26}$$

where $d = (\mathbf{D_S})_{xx}$ for all $x \in X$. The amplitude spectrum, or an aggregate parameter such as the average eigenvalue $\Lambda^* = \sum_\Lambda B(\Lambda)\Lambda$, thus may serve as an alternative measure of ruggedness. It is interesting to note that Davidor's "epistasis variance" [121] corresponds to $\sum_{\Lambda>\Lambda_2} B(\Lambda)$, while $B(\Lambda_2)$ measures the linear (additive) part of a landscape defined on a set of strings. Elementary landscapes belonging to Λ_3, or equivalently, $\text{ord}(\Lambda) = 2$ on Hamming graphs, thus belong to the simplest class of landscapes with epistasis. In biology, epistasis is the interaction between genes such that the contribution of a gene to the fitness depends on the value of other genes in the chromosome [122,123].

5 Concluding remarks

The present contribution is by no means an exhaustive survey of fitness landscapes. There is a great number of topics that have not been discussed here. Most importantly, we have excluded a thorough discussion of ensembles of landscapes and all the topics associated with them such as statistical mechanics methods, see e.g. [3], the notion of isotropic ensembles of landscapes [95], a stochastic treat-

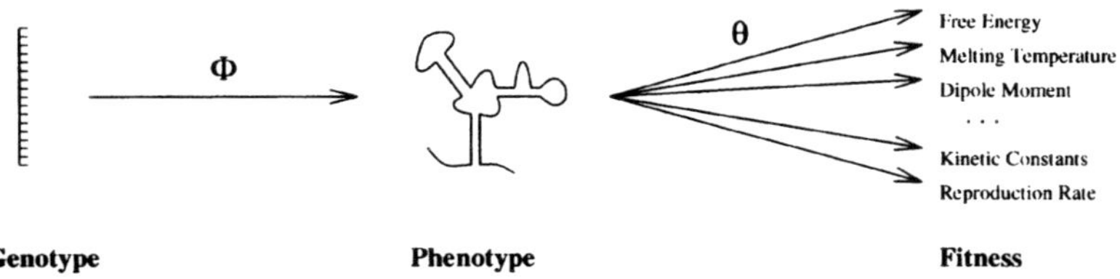

Fig. 5. Biologically realistic fitness landscapes are compositions of the genotype-phenotype map Φ and the evaluation θ of the phenotype by the environment.

ment of neutrality [124], or the random graph approach to neutral landscapes [125–129].

The second topic that we have not touched yet is the structure of *biologically* realistic fitness landscapes, which typically can be viewed as the composition of a genotype-phenotype map and the fitness evaluation of the phenotype, Figure 5. Such landscapes inherit many of their properties, including ruggedness and neutrality, essentially from the genotype-phenotype map Φ [130]. Genotype-Phenotype maps have been studied extensively for RNA molecules. In this model the RNA sequence serves as genotype, while the secondary structure approximates the phenotype [19,131,132].

Another important topic concerns the connection of Genetic Algorithm and landscape structure. Schemata, i.e., hyperplanes in $\mathcal{Q}_a^n$ play an important role here [105–109,133–135]. For a discussion of the Schema Theorem and the Building Block Hypothesis we refer to the literature [16,136–140]. The fitness function $f : X \to \mathbb{R}$ can be extended in a natural way to subsets of X by setting

$$f(A) = \frac{1}{|A|} \sum_{x \in A} f(x) \tag{27}$$

If A is a schema, then $f(A)$ is the schema-fitness. A variety of landscape classes can be defined in terms of schema fitnesses, most notably *deceptiveness* of landscape [141,142]. Some of these are compared in [143,144]. The impact of properties such as ruggedness, neutrality, deceptiveness, isotropy, etc. on the performance of particular optimization strategies is the subject of ongoing research [112,145–147].

Acknowledgements

Thanks to Ivo Hofacker and Bärbel Stadler for useful discussion. Partial support by the Austrian *Fonds zur Förderung der Wissenschaftlichen Forschung*, Proj. No. P14094-MAT is gratefully acknowledged.

References

1. S. Wright: "The roles of mutation, inbreeding, crossbreeeding and selection in evolution" in *International Proceedings of the Sixth International Congress on Genetics*, ed. by D.F. Jones, Vol. 1 (1932), pp. 356–366

2. K. Binder, A.P. Young: Rev. Mod. Phys. **58**, 801–976 (1986)
3. M. Mézard, G. Parisi, M. Virasoro: *Spin Glass Theory and Beyond* (World Scientific, Singapore, 1987)
4. H.S. Chan, K.A. Dill: J. Chem. Phys. **95**, 3775–3787 (1991)
5. K.A. Dill, S. Bromberg, K. Yue, K.M. Fiebig, D.P. Yeo, P.D. Thomas, H.S. Chan: Prot. Sci. **4**, 561–602 (1995)
6. J.N. Onuchic, H. Nymeyer, A.E. Garcia, J. Chahine, N.D. Socci: Adv. Protein Chem. **53**, 87–152 (2000)
7. C. Flamm, I.L. Hofacker, P.F. Stadler: Adv. Complex Syst. **2**, 65–90 (1999)
8. C. Flamm, W. Fontana, I. Hofacker, P. Schuster: RNA **6**, 325–338 (2000)
9. P.G. Mezey: *Potential Energy Hypersurfaces* (Elsevier, Amsterdam, 1987)
10. D. Heidrich, W. Kliesch, W. Quapp: *Properties of Chemically Interesting Potential Energy Surfaces*, Vol. 56 of Lecture Notes in Chemistry (Springer-Verlag, Berlin, 1991)
11. K. Kollman, J.H. Miller, S.E. Page: Amer. Pol. Sci. Rev. **86**, 929–937 (1992)
12. B.M. Stadler: Adv. Complex Syst. **2**, 101–116 (1999)
13. M. Garey, D. Johnson: *Computers and Intractability. A Guide to the Theory of $\mathcal{NP}$ Completeness* (Freeman, San Francisco, 1979)
14. I. Rechenberg: *Evolutionstrategie* (Frommann-Holzboog, Stuttgart, 1973)
15. J.R. Koza: *Genetic Programming: On the Programming of Computers by Means of Natural Selection* (MIT Press, Cambridge, MA, 1992)
16. J.H. Holland: *Adaptation in Natural and Artificial Systems* (MIT Press, Cambridge, MA, 1993)
17. S.A. Kauffman: *The Origin of Order* (Oxford University Press, New York, Oxford, 1993)
18. D.B. Fogel: *Evolutionary Computation* (IEEE Press, New York, 1995)
19. P. Schuster, W. Fontana, P.F. Stadler, I.L. Hofacker: Proc. Roy. Soc. Lond. B **255**, 279–284 (1994)
20. A.S. Perelson, S.A. Kauffman (Eds.): *Molecular Evolution on Rugged Landscapes: Proteins, RNA, and the Immune System* (Addison-Wesley, Reading, MA, 1991) (1991)
21. H. Frauenfelder, A.R. Bishop, A. Garcia, A. Perelson, P. Schuster, D. Sherrington, P.J. Swart (Eds.): *Landscape Paradigms in Physics and Biology: Concepts, Structures, and Dynamics* (Elsevier, Amsterdam, 1997), special Issue of *Phyica D* vol. 107(2-4)
22. J.P. Crutchfield, P. Schuster (Eds.): *Evolutionary Dynamics — Exploring the Interplay of Selection, Neutrality, Accident, and Function* (Oxford Univ. Press, Oxford UK, 2000), to appear
23. J. Hofbauer, K. Sigmund: *Dynamical Systems and the Theory of Evolution* (Cambridge University Press, Cambridge U.K., 1988)
24. R. Bürger: *The Mathematical Theory of Selection, Recombination, and Mutation* (John Wiley & Sons, Chichester UK, 2000)
25. L. Altenberg, M.W. Feldman: Genetics **117**, 559–572 (1987)
26. Y.I. Lyubich: *Mathematical structures in population genetics* (Springer-Verlag, Berlin, 1992)
27. J.W. Weibull: *Evolutionary Game Dynamics* (MIT Press, Cambridge MA, 1996)
28. A. Dress, D. Rumschitzki: Acta Appl. Math. **11**, 103–111 (1988)
29. P.F. Stadler: J. Math. Chem. **20**, 1–45 (1996)
30. J.S. Farris: "The logical basis of phylogenetic analysis" in *Advances in Cladistics*, ed. by N.I. Platnick, V.A. Funk (Columbia University Press, New York, 1983), pp. 1–36

31. J. Felsenstein: J. Mol. Evol. **17**, 368–376 (1981)
32. L.R. Foulds, R.L. Graham: Adv. Appl. Math. **3**, 43–49 (1982)
33. W.H.E. Day, D.S. Johnson, D. Sankoff: Math. Biosci. **81**, 33–42 (1986)
34. D.F. Robinson: J. Combin. Theory B **11**, 105–119 (1971)
35. M. Li, J. Tromp, L. Zhang: J. Theor. Biol. **182**, 463–467 (1996)
36. O. Bastert, D. Rockmore, P.F. Stadler, G. Tinhofer: (2000), "Landscapes on spaces of trees" submitted, SFI preprint 01-01-006
37. J. Kececioglu, D. Sankoff: Algorithmica **13**, 180–210 (1995)
38. D. Sankoff, G. Sundaram, J. Kececioglu: Internatl. J. Foundations Computer Sci. **7**, 1–9 (1996)
39. J.C. Culberson: Evol. Comp. **2**, 279–311 (1995)
40. T. Jones: (1995), "One operator, one landscape" Tech. Rep. #95-02-025, Santa Fe Institute
41. P. Gitchoff, G.P. Wagner: Complexity **2**, 47–43 (1996)
42. P.F. Stadler, G.P. Wagner: Evol. Comp. **5**, 241–275 (1998)
43. G.P. Wagner, P.F. Stadler: "Complex adaptations and the structure of recombination spaces" in *Algebraic Engineering*, ed. by C. Nehaniv, M. Ito (World Scientific, Singapore, 1999), pp. 96–115, proceedings of the Conference on Semi-Groups and Algebraic Engineering, University of Aizu, Japan
44. P. Larrañaga, C.M.H. Kuijpers, R.H. Murga, I. Inza, S. Dizdarevic: Articial Intelligence Review **13**, 129–170 (1999)
45. A. Brouwer, A. Cohen, A. Neumaier: *Distance-regular Graphs* (Springer Verlag, Berlin, New York, 1989)
46. M. Shpak, G.P. Wagner: Artificial Life **6**, 25–43 (2000)
47. L. Lovasz: "Random walks on graphs: A survey" in *Combinatorics, Paul Erdős is Eighty*Vol. 2Budapest (1996), pp. 353–398, keszthely (Hungary) 1993
48. E.D. Weinberger: Biol. Cybern. **63**, 325–336 (1990)
49. E.D. Weinberger: Biol. Cybern. **65**, 321–330 (1991)
50. W. Hordijk: Evolutionary Computation **4**(4), 335–360 (1996)
51. W. Hordijk: Physica D **107**, 255–264 (1997)
52. P.F. Stadler, R. Seitz, G.P. Wagner: Bull. Math. Biol. **62**, 399–428 (2000)
53. H.R. Fischer: Math. Annalen **137**, 269–303 (1959)
54. D.C. Kent: Fund. Math. **54**, 125–133 (1964)
55. E. Čech: *Topological Spaces* (Wiley, London, 1966)
56. D.C. Kent: Fund. Math. **62**, 95–100 (1968)
57. B.M.R. Stadler, P.F. Stadler, W. Fontana, G.P. Wagner: J. Theor. Biol. (2001). Submitted, SFI preprint 00-12-070
58. B.M.R. Stadler, P.F. Stadler, M. Shpak, G.P. Wagner: Artificial Life (2001). Submitted, SFI preprint 01-02-011
59. R. Palmer: "Optimization on rugged landscapes" in *Molecular Evolution on Rugged Landscapes: Proteins, RNA, and the Immune System*, ed. by A.S. Perelson, S.A. Kauffman (Addison Wesley, Redwood City, CA, 1991), pp. 3–25
60. P.F. Stadler, W. Schnabl: Phys. Lett. A **161**, 337–344 (1992)
61. P.F. Stadler, B. Krakhofer: Rev. Mex. Fis. **42**, 355–363 (1996)
62. R. García-Pelayo, P.F. Stadler: Physica D **107**, 240–254 (1997)
63. D.J. Thouless, P.W. Anderson, R.G. Palmer: Phil. Mag. **35**, 593–601 (1977)
64. F. Tanaka, S.F. Edwards: J. Phys. F **10**, 2769–2778 (1980)
65. A.J. Bray, M.A. Moore: J. Phys. C **14**, 1313–1327 (1981)
66. D.J. Gross, M. Mèzard: Nucl. Phys. B **240**, 431–452 (1984)
67. B. Derrida, E. Gardner: J. Physique **47**, 959–965 (1986)

68. H. Rieger: Phys. Rev. B **46**, 14 655–14 661 (1992)
69. V.M. de Oliveira, J.F. Fontanari, P.F. Stadler: J. Phys. A: Math. Gen. **32**, 8793–8802 (1999)
70. F.F. Ferreira, J.F. Fontanari, P.F. Stadler: J. Phys. A: Math. Gen. **33**, 8635–8647 (2000)
71. J. Garnier, L. Kallel: SIAM J. Discr. Math. (2000). Submitted
72. E.D. Weinberger: Phys. Rev. A **44**, 6399–6413 (1991)
73. W. Fontana, P.F. Stadler, E.G. Bornberg-Bauer, T. Griesmacher, I.L. Hofacker, M. Tacker, P. Tarazona, E.D. Weinberger, P. Schuster: Phys. Rev. E **47**, 2083 – 2099 (1993)
74. J.H. Gillespie: Evolution **38**, 1116–1129 (1984)
75. S.A. Kauffman, S. Levin: J. Theor. Biol. **128**, 11–45 (1987)
76. H. Flyvbjerg, B. Lautrup: Phys. Rev. A **46**, 6714–6723 (1992)
77. C.A. Macken, A.S. Perelson: Proc. Natl. Acad. Sci. USA **86**, 6191–6195 (1989)
78. C.A. Macken, P.S. Hagan, A.S. Perelson: SIAM J. Appl. Math. **51**, 799–827 (1991)
79. A.S. Perelson, C.A. Macken: Proc. Natl. Acad. Sci. USA **92**, 9657–9661 (1995)
80. K.J. Niklas: Amer. J. Botany **84**, 16–25 (1997)
81. R. Rammal, G. Toulouse, M.A. Virasoro: Rev. Mod. Phys. **58**, 765–788 (1986)
82. A.M. Vertechi, M.A. Virasoro: J. Phys. France **50**, 2325–2332 (1989)
83. S.R. Morgan, P.G. Higgs: J. Phys. A **31**, 3153–3170 (1998)
84. C. Flamm, I.L. Hofacker, S. Maurer-Stroh, P.F. Stadler, M. Zehl: RNA **7**, 254–265 (2001)
85. W. Kern: Discr. Appl. Math. **43**, 115–129 (1993)
86. R. Azencott: *Simulated Annealing* (John Wiley & Sons, New York, 1992)
87. O. Catoni: Ann. Probab. **20**, 1109–1146 (1992)
88. O. Catoni: "Simulated annealing algorithms and Markov chains with rate transitions" in *Seminaire de Probabilites XXXIII*, ed. by J. Azema, M. Emery, M. Ledoux, M. Yor (Springer, Berlin/Heidelberg, 1999), Vol. 709 of Lecture Notes in Mathematics, pp. 69–119
89. B. Hajek: Math. Operations Res. **13**, 311–329 (1988)
90. J. Ryan: Discr. Appl. Math. **56**, 75–82 (1995)
91. G.B. Sorkin: (1988), "Combinatorial optimization, simulated annealing, and fractals" Tech. Rep. RC13674 (No.61253), IBM Research Report
92. M. Eigen, J. McCaskill, P. Schuster: Adv. Chem. Phys. **75**, 149 – 263 (1989)
93. P.F. Stadler: Discr. Math. **145**, 229–238 (1995)
94. P. Schuster, P.F. Stadler: Computers & Chem. **18**, 295–314 (1994)
95. P.F. Stadler, R. Happel: J. Math. Biol. **38**, 435–478 (1999)
96. P.F. Stadler, W. Schnabl: Phys. Letters A **161**, 337–344 (1992)
97. B. Krakhofer, P.F. Stadler: Europhys. Lett. **34**, 85–90 (1996)
98. B. Mohar: "The Laplacian spectrum of graphs" in *Graph Theory, Combinatorics, and Applications*, ed. by Y. Alavi, G. Chartrand, O. Ollermann, A. Schwenk (John Wiley and Sons, Inc., New York, 1991), pp. 871–898
99. R. Merris: Lin. Alg. Appl. **39**, 19–31 (1995)
100. F.R.K. Chung: *Spectral Graph Theory*, Vol. 92 of CBMS (American Mathematical Society, Providence RI, 1997)
101. B. Mohar: "Some applications of Laplace eigenvalues of graphs" in *Graph Symmetry: Algebraic Methods and Applications*, ed. by G. Hahn, G. Sabidussi (Kluwer, Dordrecht, 1997), Vol. 497 of NATO ASI Series C, pp. 227–275

102. D. Rockmore: "Some applications of generalized FFTs" in *Groups and Computation II*, ed. by L. Finkelstein, W. Kantor (American Mathmatical Society, Providence, RI, 1995), Vol. 28 of DIMACS, pp. 329–370
103. D. Rockmore, P. Kostelec, W. Hordijk, P.F. Stadler: Appl. Comput. Harmonic Anal. (2000). In press Santa Fe Institute preprint 99-10-068
104. T.J. Schaefer: "The complexity of satisfiability problems" in *Proceedings of the 10th Annual ACM Symposium on Theory of Computing*, ed. by N.N. (Association for Computing Machinery, New York, 1978), pp. 216–226
105. D.E. Goldberg: Complex Systems **3**, 129–152 (1989)
106. D.E. Goldberg: Complex Systems **3**, 153–176 (1989)
107. S.E. Page, D.E. Richardson: Complex Systems **6**, 125–136 (1992)
108. M.D. Vose, A.H. Wright: Evol. Comp. **6**, 253–274 (1998)
109. M.D. Vose, A.H. Wright: Evol. Comp. **6**, 275–289 (1998)
110. L.K. Grover: Oper. Res. Lett. **12**, 235–243 (1992)
111. B. Codenotti, L. Margara: (1992), "Local properties of some np-complete problems" Tech. Rep. TR 92-021, International Computer Science Institute, Berkeley, CA
112. E. Angel, V. Zissimopoulos: Discr. Appl. Math. **99**, 261–277 (2000)
113. I. Chavel: *Eigenvalues in Riemannian Geometry* (Academic Press, Orlando Fl., 1984)
114. Y.C. de Verdière: Rendiconti di Matematica **13**, 433–460 (1993)
115. J. Friedman: Duke Math. J. **69**(3), 487–525 (1993)
116. H. van der Holst: (1996), "Topological and spectral graph characterizations" Ph.D. thesis, Universiteit van Amsterdam
117. A.M. Duval, V. Reiner: Lin. Alg. Appl. **294**, 259–268 (1999)
118. E.B. Davies, G.M.L. Gladwell, J. Leydold, P.F. Stadler: Lin. Alg. Appl. (2001). In press, see also: math.SP/0009120
119. R. Happel, P.F. Stadler: Complexity **2**, 53–58 (1996)
120. W. Hordijk, P.F. Stadler: J. Complex Systems **1**, 39–66 (1998)
121. Y. Davidor: Complex Systems **4**, 369–383 (1990)
122. M. Whitlock, P.C. Phillips, F.B.G. Moore, S. Tonsor: Ann. Review Ecol. Systematics **26**, 601–629 (1995)
123. J.B. Wolf, E.D. Brodie III, M.J. Wade (Eds.): *Epistasis and the Evolutionary Process* (Oxford Univ. Press, Oxford, UK, 2000)
124. C.M. Reidys, P.F. Stadler: Appl. Math. & Comput. **117**, 321–350 (2001)
125. C.M. Reidys: Adv. Appl. Math. **19**, 360–377 (1997)
126. S. Gavrilets, J. Gravner: J. Theor. Biol. **184**, 51–64 (1997)
127. C.M. Reidys, P.F. Stadler, P. Schuster: Bull. Math. Biol. **59**, 339–397 (1997)
128. S. Gavrilets, H. Li, M.D. Vose: Proc. Roy. Soc. London B **265**, 1483–1489 (1998)
129. S. Kopp, C.M. Reidys: Adv. Complex Syst. **2**, 283–301 (1999)
130. P.F. Stadler: J. Mol. Struct. (THEOCHEM) **463**, 7–19 (1999)
131. W. Fontana, D.A.M. Konings, P.F. Stadler, P. Schuster: Biopolymers **33**, 1389–1404 (1993)
132. W. Fontana, P. Schuster: Science **280**, 1451–1455 (1998)
133. D.E. Goldberg, M. Rudnik: Complex Systems **5**, 265–278 (1991)
134. G.E. Liepins, M.D. Vose: Complex Systems **5**, 45–61 (1991)
135. P. Field: Complex Systems **9**, 11–28 (1995)
136. L. Altenberg: "The schema theorem and the Price's theorem" in *Foundations of Genetic Algorithms 3*, ed. by L.D. Whitley, M.D. Vose (Morgan Kauffman, San Francisco CA, 1995), pp. 23–49

137. A.D. Bethke: (1991), "Genetic algorithms and function optimizers" Ph.D. thesis, University of Michigan
138. J.H. Holland: "Genetic algorithms and classifier systems: foundations and future directions" in *Proceedings of the 2nd International Conference on Genetic Algorithms* (1987), pp. 82–89
139. S. Forrest, M. Mitchell: "Relative building block fitness and the building block hypothesis" in *Foundations of Genetic Algorithms 2*, ed. by L.D. Whitley (Morgan Kaufmann, San Mateo, CA, 1993), pp. 109–126
140. C.R. Stephens, H. Waelbroeck: Phys. Rev. E **57**, 3251–3264 (1998)
141. L.D. Whitley: "Fundamental principles of deception in genetic search" in *Foundations of Genetic Algorithms*, ed. by G. Rawlins (Morgan Kaufmann, San Mateo, CA, 1991), pp. 221–241
142. K. Deb, D.E. Goldberg: "Analyzing deception in trap functions" in *Foundations of Genetic Algorithms 2*, ed. by L.D. Whitley (Morgan Kaufmann, San Mateo, CA, 1993), pp. 93–108
143. P.F. Stadler: "Spectral landscape theory" in *Evolutionary Dynamics—Exploring the Interplay of Selection, Neutrality, Accident, and Function*, ed. by J.P. Crutchfield, P. Schuster (Oxford University Press, New York, 2001), in press
144. B. Naudts, L. Kallel: IEEE Trans. Evol. Comp. (2000). To appear
145. M. Mitchell: *An Introduction to Genetic Algorithms* (MIT Press, Cambridge MA, 1996)
146. D.F. T. Baeck, Z. Michalewicz (Eds.): *Handbook of Evolutionary Computation* (Oxford University Press, New York, 1997)
147. E. Angel, V. Zissimopoulos: Theor. Computer Sci. **191**, 229–243 (1998)

Tempo and mode in quasispecies evolution

Joachim Krug[1,2,3,4]

[1] Fachbereich Physik, Universität Essen, D-45117 Essen, Germany[‡]
[2] CAMP and Department of Physics, DTU, DK-2800 Kongens Lyngby, Denmark
[3] Niels Bohr Institute, Blegdamsvej 17, DK-2100 Copenhagen Ø, Denmark
[4] Institute for Theoretical Physics, UCSB, Santa Barbara, CA 93106-4030, USA

Abstract. Evolutionary dynamics in an uncorrelated rugged fitness landscape is studied in the framework of Eigen's molecular quasispecies model. We consider the case of strong selection, which is analogous to the zero temperature limit in the equivalent problem of directed polymers in random media. In this limit the population is always localized at a single temporary master sequence $\sigma^*(t)$, and we study the statistical properties of the evolutionary trajectory which $\sigma^*(t)$ traces out in sequence space. Numerical results for binary sequences of length $N = 10$ and exponential and uniform fitness distributions are presented. Evolution proceeds by intermittent jumps between local fitness maxima, where high lying maxima are visited more frequently by the trajectories. The probability distribution for the total time T required to reach the global maximum shows a T^{-2}-tail, which is argued to be universal and to derive from near-degenerate fitness maxima. The total number of jumps along any given trajectory is always small, much smaller than predicted by the statistics of records for random long-ranged evolutionary jumps.

"The concept of quasispecies is not just a model that involves any odd assumption; it shows how to view the darwinian world of replicating and mutating species from a physical viewpoint." M. Eigen [1]

1 Introduction and motivation

Eigen's quasispecies theory of molecular evolution is the simplest mathematical model that incorporates the central Darwinian paradigm of natural selection acting on variability created by random mutations. The model was originally developed to understand the conditions for the maintenance of information in systems of self-instructive replicating macromolecules [2]. Such systems can be realized in the laboratory in the form of populations of RNA strands which replicate *in vitro* in the presence of RNA replicase, displaying a wide range of evolutionary phenomena [3–6]. The notion of a *quasispecies* [7] refers to the structure of self-replicating populations, which typically consist of a distribution of related mutants centered around a most abundant *master* genotype (see below). The quasispecies structure plays an important role in the evolution of RNA viruses, where the presence of a wide range of mutants allows the virus to adapt rapidly to environmental changes [8,9]. On the other hand, the existence

[‡] Permanent address.

of an error threshold beyond which no localized quasispecies can be maintained (see Eq.(4)) places an upper bound on the genome length of RNA viruses [10].

The mathematical structure of the quasispecies model has made it a favored entrance way for statistical physicists into the field of biological evolution[1]. It was first observed by Leuthäusser [14] that the discrete time dynamics (3) can be interpreted as a transfer matrix of a two-dimensional Ising model, where the genotype sequences become one-dimensional spin configurations that are coupled in the time direction through the mutation matrix (1) [15]. A similar relation can be established between (3) and the transfer matrix of a polymer directed along the time axis [16,17]. In addition, Baake and coworkers have recently exploited the equivalence between quantum spin chains and a class of kinetic evolution equations closely related to the quasispecies model, in which mutation and selection occur in parallel [12,18].

In its most basic version, the quasispecies model is formulated in terms of standard chemical reaction equations[2] written for the concentrations $n_\sigma(t)$ of sequences $\sigma = (\sigma_1, ..., \sigma_N)$, each of which is composed of N symbols drawn from an alphabet of K letters; the usual choice is a binary alphabet ($K = 2$), so that σ_i takes the values 0 and 1. The resulting *sequence space* consists then of $S = 2^N$ points arranged on the vertices of an N-dimensional hypercube. Each sequence σ reproduces at a rate $W(\sigma)$, which may be taken as a measure of its fitness [11]. In the reproduction process errors occur with a mutation probability μ per site. The probability of creating a sequence σ' when attempting to copy sequence σ is therefore equal to

$$Q_\mu(\sigma \to \sigma') = \mu^{d_H(\sigma,\sigma')}(1-\mu)^{N-d_H(\sigma,\sigma')} \tag{1}$$

where

$$d_H(\sigma,\sigma') = \sum_{i=1}^{N}(\sigma_i - \sigma_i')^2 \tag{2}$$

is the Hamming distance between the two sequences, i.e. the number of digits in which the two differ. The dynamical evolution in discrete time is then given by

$$n_\sigma(t+1) = \sum_{\sigma'} W(\sigma')Q_\mu(\sigma' \to \sigma)n_{\sigma'}(t). \tag{3}$$

Introducing the constraint of a fixed number of molecules leads to nonlinear loss terms on the right hand side of (3) [7]. However since these can generally be transformed away, we ignore this complication here, at the expense of dealing with exponentially growing population numbers.

The linear form of the evolution equation (3) makes it plain that, for long times, the concentrations will approach that eigenvector of the evolution matrix $W(\sigma)Q_\mu(\sigma \to \sigma')$ which corresponds to the largest eigenvalue. Provided this eigenvector is *localized* in sequence space, it defines the quasispecies: A distribution of related mutants centered around the master sequence, which usually

[1] Recent articles which review this connection are [11–13].

[2] Similar equations arise in classical population genetics [12].

is the sequence with the maximum replication rate $W(\sigma)$. The most celebrated property of the model is its prediction of a sharp *error threshold*, where the quasispecies delocalizes, and the population spreads uniformly over sequence space. In terms of the sequence length N and the mutation probability μ, the condition for a localized quasispecies takes the form [2,7]

$$N < N_{\max} = \frac{\ln A}{\mu}, \tag{4}$$

where A denotes the *selective advantage*, a measure for the superiority of the master sequence compared to the other sequences. For the simplest case of a single peak fitness landscape, where the master sequence replicates at rate W_0 and all other sequences replicate at rate $W_1 < W_0$, the selective advantage is $A = W_0/W_1$, while for randomly distributed replication rates it is a functional of the rate distribution [19,20]. In terms of the physical analogies described above, the error threshold phenomenon is equivalent to the thermal phase transition in the Ising model [14,15,18,20,21] and to the thermal unbinding of a directed polymer bound to an attractive columnar defect along the time direction [16,17].

Much less appears to be known about the evolutionary dynamics of the model, that is, the approach to the final quasispecies distribution from an initial localized or delocalized state. It was first pointed out by McCaskill [19,22] that this dynamics should take the form of a "slowing optimization walk" through a succession of metastable states which correspond, in some sense, to local maxima in the fitness landscape. The separation of time scales between the (long) residence time in a metastable state and the (brief) transition time to the next maximum implies a *punctuated* pattern of evolution [23–25], which can be analyzed in analogy to variable range hopping in condensed matter physics [26–28].

We should concede from the outset that the deterministic rate equations (3) are not an entirely appropriate description for this evolutionary regime, since the transition between two local maxima proceeds through the tails of the localized, metastable quasispecies, where the population numbers are small and fluctuations due to the finite number of molecules cannot be ignored [22,28,29]. It seemed nevertheless worthwhile to explore these questions within the most basic, deterministic model, before turning to more sophisticated approaches.

The present paper reports on some preliminary results from such an investigation. To avoid the complications due to a finite error threshold, we consider a strong selection limit (to be described in the next section), in which the population is localized at a single site in sequence space at all times. This allows a direct comparison with simple schematic models of evolutionary dynamics, such as adaptive walks [30–33] and record dynamics [34–36]. Adaptive walks describe the evolution of a genetically homogeneous population under the assumption that deleterious mutations (which decrease the fitness) are eliminated, while advantageous mutations spread instantaneously. The population then performs an uphill walk in the fitness landscape, which terminates at a local maximum where no fitter one-step neighbors are available. In contrast, in the quasispecies model the population evolves through a chain of local fitness maxima which progresses all the way to the *global* optimum. Some qualitative properties of these *evolutionary trajectories* are described in Section 3, while Section 4 focuses on a specific statistical feature, the total evolution time. A comparison with record dynamics is provided in Section 5, and some open questions are formulated in Section 6.

Throughout the paper we will consider maximally rugged fitness landscapes, in which the reproduction rate $W(\sigma)$ is chosen independently and randomly for each sequence. In contrast to simpler permutation invariant landscapes [12], this makes it necessary to store all 2^N sequences during the iteration of the evolution equations[3], restricting our numerical treatment to rather short sequences; the results shown here are for $N = 10$. A systematic analysis of the dependence on sequence length will be presented elsewhere [38].

2 The strong selection limit

The form of the strong selection limit is motivated by the analogy with the zero temperature limit in the associated problems of statistical physics. Following Peliti [11,21] we introduce an inverse "selective temperature" $k > 0$ by writing the reproduction rates in the form

$$W(\sigma) = e^{kF(\sigma)}. \tag{5}$$

We want to take the strong selection limit $k \to \infty$ in such a way that only a few mutations occur in each time step. This requires to scale the mutation rate as

$$\mu = e^{-k\gamma} \tag{6}$$

where $\gamma > 0$ is a constant. Inserting (5) and (6) into (3) it is clear that the sequence concentrations will grow for large k as

$$n_\sigma(t) = e^{kE(\sigma,t)}. \tag{7}$$

In the limit $k \to \infty$ the evolution equation (3) then reduces to the recursion

$$E(\sigma, t+1) = \max_{\sigma'}[E(\sigma', t) + F(\sigma') - \gamma d_H(\sigma, \sigma')]. \tag{8}$$

Since the term $-\gamma d_H$ suppresses mutations to far away sequences, we expect similar behavior for a model in which only nearest neighbor mutations are allowed,

$$E(\sigma, t+1) = \max_{d_H(\sigma,\sigma')\leq 1}[E(\sigma', t) + F(\sigma') - \gamma d_H(\sigma, \sigma')]. \tag{9}$$

All results shown in this paper were obtained using the nearest neighbor rule (9), with the parameter γ set to unity[4].

We still need to specify the probability distribution $p(F)$ of the fitnesses $F(\sigma)$. Two choices will be considered: The exponential distribution $p(F) = e^{-F}$, $F \geq 0$, and a uniform distribution on the interval $[0, S]$, where $S = 2^N$. The reason for this particular scaling of the width of the uniform distribution will become clear below in Section 4.

[3] An approximate scheme which reduces the storage requirement from 2^N to N is described in [38].

[4] For a discussion of the differences between the rules (8) and (9), see [38].

3 Evolutionary trajectories

It is evident from (7) that, in the strong selection limit $k \to \infty$, the entire population resides at the global maximum of the function $E(\sigma, t)$. The position of this maximum in sequence space will be referred to as the master sequence at time t, and denoted by $\sigma^*(t)$. At time $t = 0$ the master sequence is placed at a randomly chosen point $\sigma^{(i)}$ by setting $E(\sigma^{(i)}, t) = 0$ and $E(\sigma, t) = -\infty$ for $\sigma \neq \sigma^{(i)}$. The subsequent time evolution $\sigma^*(t)$ defines an evolutionary trajectory.

Inspection shows that, after one or two time steps, such a trajectory passes exclusively through local fitness maxima, and eventually, after a total evolution time T, it invariably reaches the global fitness maximum. During the evolution, the master sequence spends increasingly long time intervals at local maxima of increasing fitness, with a few abrupt transitions in between (Figure 1). The number of transitions is small (see Figure 5), much smaller than the number of local fitness maxima, which is on average equal to $2^N/(N+1) \approx 93$ [30]. This implies that most local maxima are bypassed by a typical trajectory.

A quantification of this statement is shown in Figure 2, in which the fitness F of local maxima is plotted against the number of times it is visited by an evolutionary trajectory. These data were generated by going through all possible starting points $\sigma^{(i)}$ in a fixed fitness landscape. The figure shows a roughly linear correlation between the fitness of a maximum and the logarithm of the number of visits.

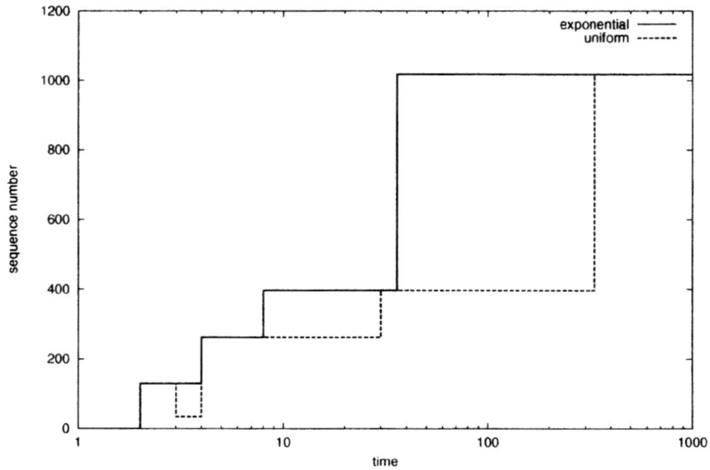

Fig. 1. Two evolutionary trajectories generated in two fitness landscapes with identical ordering of fitnesses but different fitness distributions. The monotonic increase in sequence number for the exponential distribution is fortuitous - there is no correspondence between the position of a sequence on the y-axis and its fitness

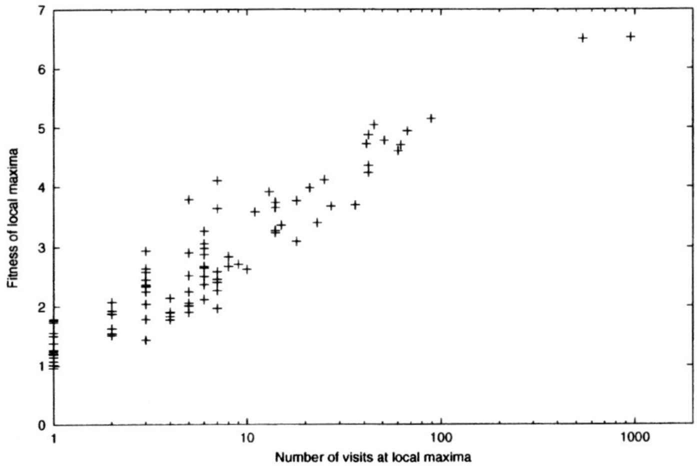

Fig. 2. Fitness $F(\sigma)$ vs. the number of visits for all local maxima in a fitness landscape with exponential fitness distribution $p(F)$. The particular landscape used here is near degenerate (gap size $\epsilon \approx 0.02$)

In relation to adaptive walks [30–33], which respond only to the relative ordering of fitnesses and not to their actual values, it is of interest to ask to what extent the set of maxima visited by a given trajectory is determined by the ordering of fitnesses. For this reason Figure 1 shows two trajectories evolving in landscapes which were generated using the same random numbers – thus having identical ordering of fitnesses – but with different fitness distributions. It can be seen that the set of local maxima visited by the two trajectories is almost identical, apart from a small detour taken by the "uniform" trajectory, but the timing of the evolutionary transitions is markedly different in the two cases. With reference to G.G. Simpson's classic treatise [23], we may say that the fitness distribution affects only the *tempo*, but not the *mode* of quasispecies evolution. For a quantitative analysis of the temporal aspects we next turn to the distribution of evolution times.

4 Distribution of evolution times

Figure 3 shows the distribution $P(T)$ of the number of time steps T required to reach the global fitness maximum, obtained by averaging over 500000 landscapes with exponential and uniform fitness distributions. The time distribution for the exponential case displays a distinct maximum around $T = 7$, followed by a slowly decaying tail which is well described by the power law

$$P(T) \sim T^{-2} \tag{10}$$

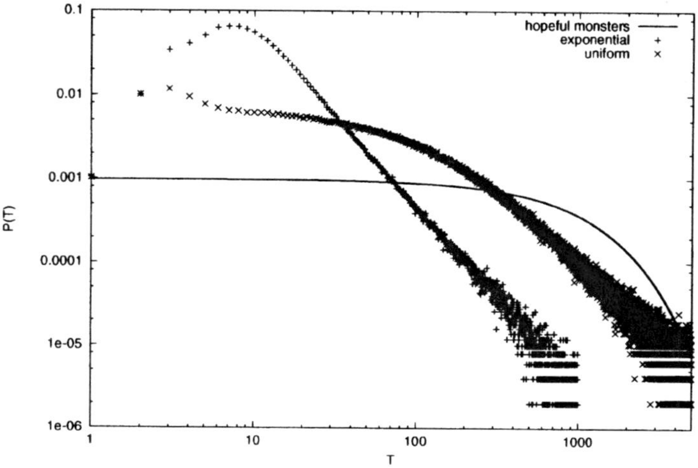

Fig. 3. Distribution of the total time T required to reach the global fitness maximum. Symbols show data obtained by averaging over 500000 uncorrelated fitness landscapes with exponential (+) and uniform ($\times$) fitness distributions, while the full line shows the distribution (17) obtained for record dynamics. The simulations with exponential fitness distribution were stopped after 1000 time steps

over roughly two decades. The distribution for the uniform case is much broader, but a similar power law tail can be seen for times $T \geq 500$.

The power law (10) appears to be a simple consequence of the order statistics of uncorrelated fitness landscapes. Let $F^{(1)} > F^{(2)} > ... > F^{(S)}$ be a realization of fitnesses arranged in decreasing order. As a measure of the spread in fitnesses among the most fit sequences we introduce the *fitness gap*

$$\epsilon = F^{(1)} - F^{(2)} > 0, \tag{11}$$

which is a random variable characteristic of each fitness landscape. In the late stage of evolution the population will typically make a a transition from the second best sequence $\sigma^{(2)}$ (or some local fitness maximum with comparable fitness[5]) to the globally optimal sequence $\sigma^{(1)}$. From the evolution rule (9) it is easy to see that, for small ϵ, this transition will require a time of the order of

$$T \approx a(N)/\epsilon, \tag{12}$$

where the coefficient a is determined by the early stages of the evolution process [38]. Thus given the gap distribution $P_{\rm g}(\epsilon)$ the tail of the distribution of evolution times can be estimated to be

$$P(T) \approx aT^{-2}P_{\rm g}(a/T), \tag{13}$$

[5] In fact $\sigma^{(2)}$ *is* a local fitness maximum with high probability $1 - N/(2^N - 1) \approx 0.990$.

and a T^{-2} power law follows for $T \gg a$, provided that $0 < P_g(0) < \infty$. The gap distribution is given by $P_g(\epsilon) = e^{-\epsilon}$ *both* for exponentially distributed fitnesses, and for fitnesses distributed uniformly between 0 and S, when S is large [39]. The striking difference between the two evolution time distributions seen in Figure 3 is related to the different scaling of the coefficient $a(N)$ in (12) with sequence length: In the exponential case $a(N) \sim N^{3/2}$, while in the uniform case $a(N) \sim \sqrt{N}S$ [38].

To compute $P_g(0)$ for general fitness distributions, note first that the joint distribution of $F^{(1)}$ and $F^{(2)}$ is given by [40]

$$P_2(F^{(1)}, F^{(2)}) = S(S-1)p_c(F^{(2)})^{S-2}p(F^{(1)})p(F^{(2)}) \tag{14}$$

where $p_c(F) = \int_0^F dF'\ p(F')$ denotes the cumulative fitness distribution. The cumulative gap distribution is obtained by integration,

$$\text{Prob}[F^{(1)} - F^{(2)} < \epsilon] =$$

$$S(S-1)\int_0^\infty dF^{(2)}\ p(F^{(2)})p_c(F^{(2)})^{S-2}\int_{F^{(2)}}^{F^{(2)}+\epsilon} dF^{(1)}\ p(F^{(1)}), \tag{15}$$

which tends to $P_g(0)\epsilon$ for $\epsilon \to 0$. Thus we conclude that

$$P_g(0) = S(S-1)\int_0^\infty dF p(F)^2 p_c(F)^{S-2} \tag{16}$$

which is clearly finite and nonzero.

The relationship (13) implies that the near-degenerate fitness landscapes, which have very small gaps, are the ones that give rise to anomalously long evolution times. Figure 4 illustrates this connection. The data shown as crosses were obtained from an average over exponential fitness landscapes, for which the fitness gap ϵ was increased artificially by increasing the global fitness maximum according to $F^{(1)} \to F^{(1)} + 1$. This is seen to immediately remove the power law tail (10).

5 Comparison to record dynamics

A simple schematic analogue of the nonstationary (ever slowing) evolution process found in the quasispecies model is provided by the dynamics of records [34–36,39], which is equivalent to evolution by long-ranged random mutations known in the classical literature as the theory of "hopeful monsters" [23,30]. In the present context it reduces to the following rule for the motion of the master sequence $\sigma^*(t)$ in sequence space: At each time step, the population attempts a jump to another, randomly chosen sequence $\sigma' \neq \sigma^*(t)$. The move is accepted, and $\sigma^*(t+1) = \sigma'$, if $F(\sigma') > F(\sigma^*(t))$; otherwise it is discarded and $\sigma^*(t+1) = \sigma^*(t)$. Thus the current sequence $\sigma^*(t)$ represents the *fitness record* among the sequences which the population has encountered so far.

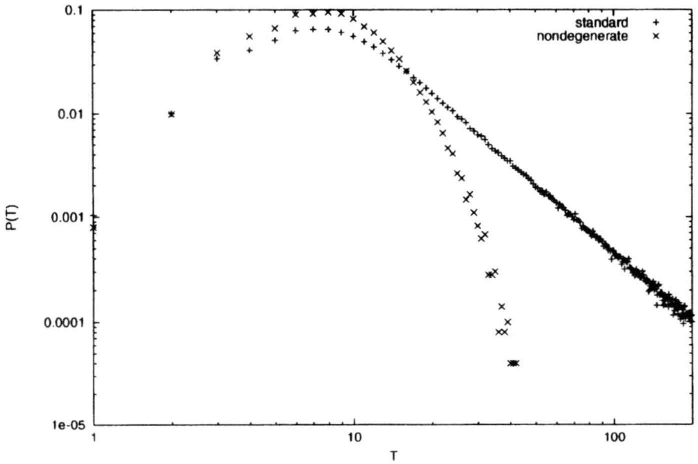

Fig. 4. Comparison of the distribution of evolution times for exponentially distributed fitnesses (+) with an ensemble of landscapes for which near-degeneracies (small gaps) have been removed (×). The latter data were averaged over 50000 realizations

Clearly this process gives rise to a step-like, punctuated pattern which is *qualitatively* similar to that shown above in Figure 1. Here we are concerned with a *quantitative* comparison of statistical properties. Let us first compute the probability distribution of the total evolution time T for the record dynamics. Since the probability of finding the global fitness maximum in any jump is $1/S$, the probability that it has not been found up to time t is $(1 - 1/S)^t \approx e^{-t/S}$ for large S and t. Taking the derivative one obtains

$$P(T) = S^{-1}e^{-T/S}. \tag{17}$$

The typical evolution times are of the order of the number of sequences, much larger than in the quasipecies model. This demonstrates impressively the "guided" nature of quasispecies evolution [7], which is much more efficient than a random search. Figure 3 shows how broad the distribution (17) is compared to that of the quasispecies model. Note, however, that for very long times (longer than S) Eq.(17) decays exponentially, faster than the degeneracy-induced power law (10). Taken literally, Eq.(10) implies that the mean evolution time is infinite.

Next we consider the distribution P_n of the total number n of evolutionary jumps which occur on the way to the global fitness maximum. Adapting the results of Sibani and collaborators [36,37], for the case of record dynamics we find that P_n is a Poisson distribution with parameter $\ln S$,

$$P_n \approx S^{-1}\frac{(\ln S)^{n-1}}{(n-1)!}. \tag{18}$$

In Figure 5 this is compared to numerical data obtained for the quasispecies model. Again the distributions for the quasispecies dynamics are much narrower,

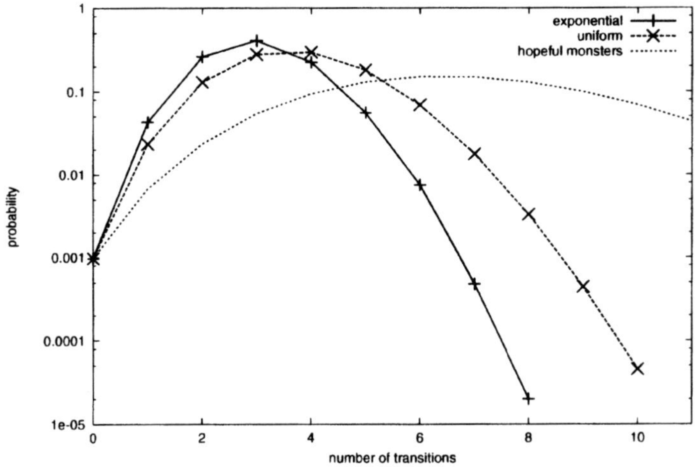

Fig. 5. Probability distributions for the total number of evolutionary jumps required to reach the global fitness maximum. Symbols show data obtained by averaging over 500000 fitness landscapes with exponential (+) and uniform (×) fitness distributions, while the dotted line is the log-Poisson distribution (18) predicted by record dynamics

showing that less transitions are required to reach the global maximum. Simulations for longer sequences show that the mean number of transitions increases sublinearly in N, more slowly than the linear behavior predicted by (18) [38,41].

6 Outlook

The simplicity of the strong selection dynamics (8,9) suggests to use it for a dynamical characterization of different kinds of fitness landscapes. In contrast to the random landscapes considered here, realistic fitness landscapes obtained e.g. from RNA folding contain extended neutral networks in sequence space, in which the fitness (defined in terms of the RNA secondary structure) does not change [42]. Central concepts of quasispecies theory have been extended to such landscapes [43]. Extended neutrality provides a distinct mechanism for the appearance of punctuation patterns in evolution, since changes in the genotype do not show up in the phenotype, as long as the former moves within a neutral network [44].

Another interesting direction for further research inspired by the analogy with directed polymers is to include effects of environmental fluctuations, which amounts to making the fitness landscape time-dependent [45]. In the directed polymer analogy, the issue is the interplay between *columnar* disorder, which is provided by the time-independent part of the landscape, and *point* disorder modeling the time-dependent variations [46]. It is well known that point disorder can depin a polymer from an attractive columnar defect in much the same way as thermal fluctuations [47]. This suggests the intriguing possibility of an error threshold delocalization phenomenon induced by environmental fluctuations.

Acknowledgements

I would like to thank T. Halpin-Healy and C. Karl for their contributions to this project, and K. Sneppen, H. Flyvbjerg and L. Peliti for useful discussions. This work has been supported in part by NATO within CRG.960662, and by NSF under Grant No. PHY99-07949.

References

1. M. Eigen: Trends in Microbiology **4**, 216 (1996)
2. M. Eigen: Naturwissenschaften **58**, 465 (1971)
3. D.R. Mills, R.L. Peterson, S. Spiegelman: Proc. Natl. Acad. Sci. USA **58**, 217 (1967).
4. C.K. Biebricher: 'Darwinian Selection of Self-Replicating RNA Molecules'. In: *Evolutionary Biology, Vol. 16*, ed. by M.K. Hecht, B. Wallace, G.T. Prance (Plenum, New York 1983) pp. 1-52
5. C.K. Biebricher, W.C. Gardiner: Biophys. Chem. **66**, 179 (1997)
6. J.S. McCaskill, G.J. Bauer: Proc. Natl. Acad. Sci. USA **90**, 4191 (1993)
7. M. Eigen, J. McCaskill, P. Schuster: Adv. Chem. Phys. **75**, 149 (1989)
8. E. Domingo: Clinical and Diagnostic Virology **10**, 97 (1998)
9. R.V. Solé, R. Ferrer, I. González-García, J. Quer, E. Domingo: J. theor. Biol. **198**, 47 (1999)
10. M. Eigen and C.K. Biebricher: 'Role of Genome Variation in Virus Evolution'. In: *RNA Genetics. Vol. III: Variability of Virus Genomes*, ed. by E. Domingo, J.J. Holland, P. Ahlquist (CRC Press, Boca Raton, FL, 1988) pp. 211-245
11. L. Peliti: 'Introduction to the statistical theory of Darwinian evolution'. `cond-mat/9712027`
12. E. Baake, W. Gabriel: 'Biological evolution through mutation, selection, and drift: An introductory review'. In: *Annual Reviews of Computational Physics VII*, ed. by D. Stauffer (World Scientific, Singapore, 2000) pp. 203-264
13. B. Drossel: 'Biological Evolution and Statistical Physics'. `cond-mat/0101409` (to appear in Adv. Phys.)
14. I. Leuthäusser: J. Stat. Phys. **48**, 343 (1987)
15. P. Tarazona: Phys. Rev. A **45**, 6038 (1992)
16. S. Galluccio, R. Graber, Y.-C. Zhang: J. Phys. A **29**, L249 (1996)
17. S. Galluccio: Phys. Rev. E **56**, 4526 (1997)
18. E. Baake, M. Baake, H. Wagner: Phys. Rev. Lett. **78**, 559 (1997)
19. J.S. McCaskill: J. Chem. Phys. **80**, 5194 (1984)
20. S. Franz, L. Peliti, M. Sellitto: J. Phys. A **26**, L1195 (1993)
21. S. Franz, L. Peliti: J. Phys. A **30**, 4481 (1997)
22. J.S. McCaskill: Biol. Cybern. **50**, 63 (1984)
23. G.G. Simpson: *Tempo and Mode in Evolution* (Columbia University Press, New York, 1944)
24. C.M. Newman, J.E. Cohen, C. Kipnis: Nature **315**, 400 (1985)
25. S.J. Gould, N. Eldredge: Nature **366**, 223 (1993)

26. W. Ebeling, A. Engel, B. Esser, R. Feistel: J. Stat. Phys. **37**, 369 (1984)
27. J. Krug, T. Halpin-Healy: J. Phys. I France **3**, 2179 (1993)
28. Y.C. Zhang: Phys. Rev. E **55**, R3817 (1997)
29. D. Alves, J.F. Fontanari: Phys. Rev. E **57**, 7008 (1998)
30. S. Kauffman, S. Levin: J. theor. Biol. **128**, 11 (1987)
31. C.A. Macken, A.S. Perelson: Proc. Natl. Acad. Sci. USA **86**, 6191 (1989)
32. C.A. Macken, P.S. Hagan, A.S. Perelson: SIAM J. Appl. Math. **51**, 799 (1991)
33. H. Flyvbjerg, B. Lautrup: Phys. Rev. A **46**, 6714 (1992)
34. P. Sibani, M.R. Schmidt, P. Alstrøm: Phys. Rev. Lett. **75**, 2055 (1995)
35. P. Sibani: Phys. Rev. Lett. **79**, 1413 (1997)
36. P. Sibani, M. Brandt, P. Alstrøm: Int. J. Mod. Phys. B **12**, 361 (1998)
37. P. Sibani, P.B. Littlewood: Phys. Rev. Lett. **71**, 1482 (1993).
38. J. Krug and C. Karl (in preparation)
39. W. Feller, *Introduction to Probability Theory and Its Applications*, Vol. 2 (Wiley, New York 1971)
40. H.A. David, *Order Statistics* (Wiley, New York 1970)
41. C. Karl: Diploma thesis (University of Essen, 2001)
42. P. Schuster: Biophys. Chem. **66**, 75 (1997)
43. C. Reidys, C.V. Forst, P. Schuster: Bull. Math. Biol. **63**, 57 (2001)
44. P. Schuster, W. Fontana: Physica D **133**, 427 (1999)
45. M. Nilsson, N. Snoad: Phys. Rev. Lett. **84**, 191 (2000)
46. I. Arsenin, T. Halpin-Healy, J. Krug: Phys. Rev. E **49**, R3561 (1994)
47. L. Balents, M. Kardar: Phys. Rev. B **49**, 13030 (1994)

Multilevel processes in evolution and development: Computational models and biological insights

Paulien Hogeweg

Theoretical Biology and Bioinformatics Group, Utrecht University Padualaan 8, 3584CH Utrecht, the Netherlands. Email P.Hogeweg@bio.uu.nl

Abstract. We argue that it is profitable (and necessary) to study biotic systems as multilevel systems in which the behavior of 'higher levels' is not only determined by the lower level processes but where the reverse is true as well: the higher level processes determine the structure of the lower level entities In this paper we will discuss two rather different examples of this phenomenon.

The first example discusses how the properties of (mutating) reproducing entities are, over evolutionary time, shaped by the dynamics of the larger scale spatial patterns which they generate. Also the dynamics of the evolutionary process, as represented in the shape of the phylogeny, is determined by these patterns. A novel type of 'punctuated evolution', i.e. periods of rapid change alternated by periods of stasis *at the genotypic level* is shown to be the result of unstable pattern formation. We will also show that the dynamics of the large scale spatial patterns, in its turn, shows features which result from the evolutionary dynamics of the micro entities.

The second example examines morphogenesis as multilevel process. Here, unlike the previous example, the basic model formulation incorporates several levels, and morphogenesis is the result of intricate coordination between these levels. The modeling of such coordination is achieved through modeling an evolutionary process. We will discuss both an example of an resulting morphogenetic process, and some properties of the evolutionary paths which leads to it. The re-occurrence of similar morphogenetic 'innovations' is one feature which our model evolutionary process shares with an ever increasing database of examples of biotic molecular phylogenies in which such re-inventions at the morphological level also seem to occur frequently.

1 Introduction

The extensive formal theory developed in population genetics has shaped our insight in the process of Darwinian evolution. Curiously most of these theoretical population genetics models fail to deal in a realistic manner with the process of mutation, they focus on selection only (e.g. [1]). More recently statistical physics approaches have been applied to biological evolution. Here the basic dynamics is formulated in microscopic equations, including mutations etc. and macroscopic properties, are derived.

"A theory should be as simple as possible, but no simpler" is an often quoted slogan of Einstein. In my opinion many of the evolutionary models, although yielding useful baseline insights, are in the 'too simple' category, because they ignore the multilevel nature of biotic systems. The aim of this paper is to develop models which are 'as simple as possible' but in which we can explore and exploit the interplay between levels in biotic systems Level refers here to a process at a particular space and time scale.

I will argue and demonstrate that 'clean' micro-macro transitions are not sufficient to describe biological evolution satisfactorily Instead I argue that because of the very process of evolution, a two-way information flow among levels occurs. In figure 1 of the next session this proposition is illustrated in a eco-evolutionary setting. Interacting, replicating entities living in space are prone to the generation large scale spatial temporal patterns. These macro-scale properties of the micro-scale interactions between the entities constitute the 'environment' in which these entities live. Thus, it is rather obvious that, when the entities are subjected to mutations and selection, the environment and hence the spatial temporal pattern generated by the entities determine the evolutionary fate of the mutants and ultimately the micro-scale interactions which define the system.

The aim of this paper is not only to argue that such two way interactions between levels do occur, but also to argue that it is a good research strategy to examine these two or multiple way processes explicitly: it allows us to make our models, and the analysis of the models, simpler.

An important reason for the usefulness is that the macro-scale spatial-temporal dynamics may be relatively well understood even when it is beyond present day analysis tools to derive the macro-scale dynamics from the micro-scale interactions. This is true for 'generic' patterns which occur for a large set of micro-scale' dynamics. In the example we discuss in section 2 spiral wave and turbulent patterns arise from a stochastic model for host-virus interactions. This is easily seen in movies of monte-carlo simulations of the system. Once the wave patterns are recognized, we can use their well studied properties to try to understand the evolutionary dynamics at the micro-scale level. We demonstrate that the evolutionary dynamics of the viruses can best be explained in terms of generic properties of spiral waves. On the other hand 'not so generic properties' of the spiral waves in the studied system, can best be explained in terms of the mutational dynamics of the viruses.

The entanglement of levels in the products of biological evolution, i.e. present day biotic systems, is mostly seen as simply an unwanted complication in both experimental and theoretical studies. Experimental and theoretical studies therefore try to isolate processes at one space/time scale, or to study simple one way micro-macro transitions. An example of the latter is the use of spatial pattern formation in e.g. Turing systems as theories of development. I think this approach is in the 'too simple category', as indeed was acknowledged by Turing himself when he allegedly said: "the stripes are simple, but what about the horse part?". In my opinion entanglement of levels is such a basic feature of biotic systems, that we should address it explicitly in our models. I think it is the very process we want to understand about biotic systems. In the second part of this paper I will explain a modeling approach by which we can explore and exploit the entanglement of various processes at various space and time scales. By using an evolutionary process. and by taking into account the entanglement of levels our models (and the parameter space to be explored) can in fact become simpler/smaller because the levels constrain each other. We will apply this approach to study biological development We will show that by using a simple evolutionary process we can derive interesting simple models for morphogenesis, i.e. we can at least begin to address "the horse part". In these models the interplay between differential adhesion, cell signaling and cell growth, cell death and cell differentiation, is shaped by evolution to produce intricate morphologies.

In short, in this paper I discuss how entanglement between levels arise through Darwinian evolution, and how a Darwinian evolutionary process can be used as tool to study (evolved) complex entangled systems. Moreover in both contexts I will study the evolution process itself - and how it is shaped by the entanglement of levels. I will stress in both context the biological insights we have obtained. The insights, of course, are what validates the modeling approach!

2 Multilevel evolution: Pattern formation, phylogenetic trees and punctuated equilibria

2.1 Dynamics of meso-scale patterns shape of micro-scale replicators through evolution: A review

In explicit spatial ecological models of interacting replicators, pattern formation will occur in many circumstances. Minimally local replication will result in clumping of lineages and larger scale clumping of species is also often observed. Moreover many ecological interactions, as modeled in classical ODE models, give rise to oscillatory population densities. Examples are Host-Parasitoid, Predator-Prey, Hypercycles, In-transient competition etc . In space such oscillations lead to waves - which may organize in e.g. spiral waves. The formation of such spatial patterns is crucial for the qualitative outcome of the interactions. For example coexistence of mutualistic species and so called 'cheaters', which take extra benefits and do not reciprocate, is impossible in the classical ODE models, but is possible in spatial models over a large range of parameters. Such coexistence is also observed in nature, and may persist over geological time scales.

Here we discuss how such patterns formed by ecological interactions of replicating entities, influence the evolutionary 'fate' of these replicators. Thus we consider a feedback of the larger scale patterns to the micro-scale entities which generate these patterns (Fig. 1).

We can best analyze this feedback in cases where the dynamics of the the meso-scale patterns is generic and well understood in systems other than the eco-evolutionary system under consideration. This is pre-eminently the case for spiral waves, and the patterns formed by the complex Ginsberg-Landau equation, in which regions of spirals and regions of turbulence occur. We have demonstrated the feedback of spiral wave on the evolutionary dynamics of the replicators which generate the waves, in a variety of cases and a variety of aspects. For example we have shown:

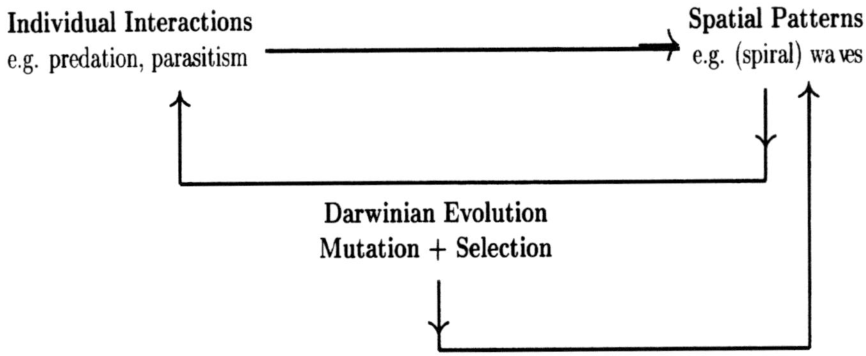

Fig. 1. Relation between local interactions and spatial pattern formation in eco-evolutionary models

- **Positive selection for fast decay**
  Boerlijst and Hogeweg [2] [3] have shown that very counter-intuitive selection pressures occur in an explicit spatial version of the classical ODE Hypercycle model of Eigen and Schuster [4]. The model is formulated as a stochastic cellular automata (CA) model. The states represent the presence of one molecule of a particular type at that position; state 0 represents an empty position. Surrounding molecules can replicate into an empty position with a probability depending on their type and the catalysis they get. Non-zero states have a probability to become zero, representing the decay of molecules. For Hypercycles of length 5 and longer stable spiral waves, mostly organized in pairs, form from initial random distribution of the molecules. It is well known that faster rotating spiral waves overtake the domain of slower rotating ones. Increase of decay rate increases the rotation speed and hence spirals with faster decaying molecules will overtake the domain of spirals consisting of species with greater longevity - a feature which is entirely incomprehensible from the standpoint of the individual.
- **Self-reinforcing spatial patterns**
  Savill et al. [5] have shown that selection pressures in an explicit spatial version of the classical Nicholson Bailey Host-Parasitoid model [1] are such that the spatial pattern in which the evolving parasitoids happen to find themselves are reinforced. The model is formulated as a Lattice Map model. At each Lattice site Host and Parasitoids interact as in the N-B model, i.e. the within site dynamics is an unstable spiral. The hosts diffuse randomly to neighboring sites, whereas the parasitoids have biased diffusion towards the Hosts: $\beta_{A->B} = cH_B/{H_{tot}}^{\mu}$, where H_{tot} is the total number of hosts surrounding the local parasitoid population in A , and $\beta_{A->B}$ the fraction thereof which will migrate to B. The parameter on which the mutation selection process operates is μ; The patterns which form in a large parameter

[1] Host: $H_{t+1} = \lambda H_t \exp(-aP_t)$ Parasitoid: $P_{t+1} = bH_t(1 - \exp(-aP_t))$

region resemble those studied extensively in the Complex Ginsburg-Landau equation: regions of spiral waves and regions of turbulence occur in models where all hosts/parasitoids are identical. Selection pressure will be towards 'overexploitation (large μ) in turbulent regions, leading to μ values for which only turbulence occurs. Conversely selection pressure within the spiral regions is towards under-exploitation, but alternate between increasing and decreasing μ within a range for which spiral patterns are reinforced. Within a spiral selection is towards low μ, i.e. towards remaining in the spiral core (from which all offspring descend in the long run), whereas competition between spirals favor higher μ values, as this leads to faster rotation spirals. New spirals are formed at the border with turbulent regions and have high μ .
From the standpoint of the parasitoids neither outcome seems 'optimal'. The outcome is determined by the spatial patterns and indeed reinforces them.

- **Long term information integration**
Pagie and Hogeweg [6] [7] have shown that in explicit spatial models of co-evolving antagonistic populations long term evolution leads to entities which can cope with a full set of circumstances, although each generation experiences only a very small subset of these (i.e. experiences sparse fitness evaluation). We have even shown that sparse fitness evaluation leads to better performance of individual entities towards all circumstances than full fitness evaluation [7]. We studied this in stochastic CA models in which an external fitness criterion is given. One species represent instances of this fitness function (e.g. coordinates of a function) and the other population should generate the value of the function at these coordinates. Information integration depends crucially on spatial pattern formation, although the patterns formed in these circumstances are not characterized (characterizable) independently, so that we cannot point at the dynamical features of the patterns which explain the effects seen at the level of the individuals, as in the cases mentioned above. However, as shown in [6] , mixing the populations after each step leads to so called red queen dynamics: no information integration occurs but the populations "have to run very fast to remain at the same place" i.e. the individuals only adapt to the current circumstances.

In this paper we extend these studies by looking at "tempo and mode" of an evolutionary process resulting from the interplay between evolution and pattern formation, rather than at selection pressures and the resulting the properties of the evolved individuals, as done above.

2.2 Interplay between pattern formation and the shape of phylogenetic trees

Reconstruction of phylogenetic trees on the basis of DNA sequence similarity is widely used to study the evolutionary history of species. Traditionally the main interest was "who was more closely related to whom". More recently there are "New uses for new phylogenies" [8], e.g. (1) assessment of population dynamics [9] from phylogenetic trees, using simple models of evolution (see [10] for an application to virus evolution) and (2) assessment of 'novelty' and 'reinvention' of morphological/behavioral features ([11] and many other examples in the same volume). In this paper we will use reconstructed phylogenetic trees of *simulated* evolutionary processes to investigate the phylogenetic signatures one might expect with respect to these latter uses. In this section we focus on the first type of issue, i.e. we study how the shape of phylogenetic trees is influenced by spatial patterns generated by the ecological interactions of the species under consideration.

We examine a virus host situation, in which the host becomes immune to viruses which have infected it. The study is inspired by data on influenza virus NP protein. Sequence data are available of human, pig and bird infecting strains from 1910 onwards. Phylogenetic reconstruction of the human and pig strains shows a very skewed, almost linear tree shape, with older strains at the bottom and younger strains at the top. Thus, although some strain diversification occurs, in the longer run it is only one of the strains whose offspring produce the next generation [12]. In contrast, the bird phylogeny is rake-shaped. The time of isolation and position in the tree are not related. The length of the tree from the common ancestor to present day strains is an order of magnitude less than that of the human and the pig trees [13]. It is well know that influenza epidemics sweep as waves over the world. We study a generic spatial model of host virus interaction, which is not particularly moulded to influenza. We want to study which general properties might lead to such different tree shapes.

A model of host virus interactions. We model long lived hosts which become (temporarily) immune for the virus which has infected it. The virus is represented by a amino-acid sequence, which is subject to mutation. Specifically the assumptions of the model are:

- Immunity declines exponentially in time
  $I = m * exp(-z * t)$ $(m = 120; z = .04)$
- sensitivity: saturated function of protein distances
  $s = d/(K + d)$ $(K = 15; sizeprotein = 60)$
- probability of infection: Hill function of immunity and sensitivity
  $P = s^n/(I^n + s^n)$ $(n = 3)$ (see Fig. 2)
- Decay probability of viral infectivity: (zet=.2 - .4)
  Host decay probability: (decay=.01 - .0025)
  Mutation rate (mut=.0001 - .0004)
  Field size (200x200 or 300x300 patches)

The populations are modeled in a multi-layer stochastic CA. State 0 represents empty in all layers; all other states represent a virus type or immune memory for a virus type or I, strength of immunity. I is coded as integer and exponential decay is simulated in terms of probability of unit decay of memory. Initial only 1 virus type is present.

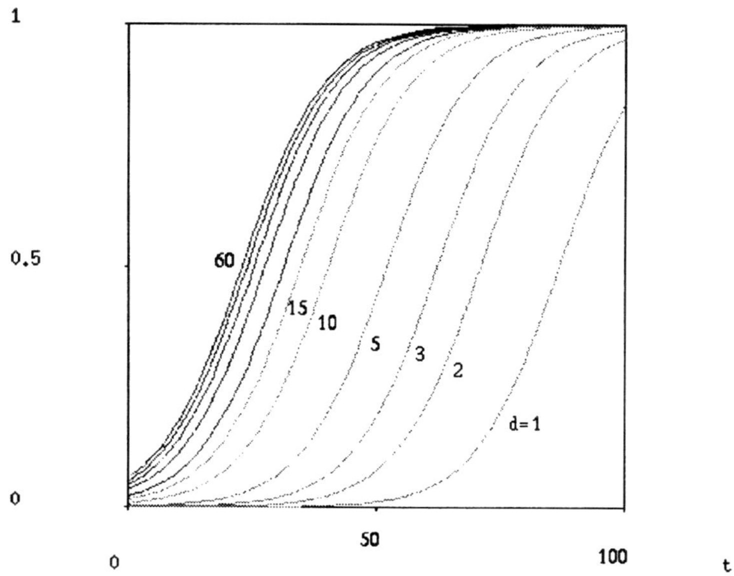

Fig. 2. Infectivity as a function of time after previous infection for indicated difference between current and previous infecting virus

Virus evolution and shape of phylogenetic trees. Fig. 3 shows the phylogenetic tree of an initial identical viral population evolved in two different hosts. The difference of the host is their longevity. In the long-lived host the phylogenetic pattern is similar to that reported for influenza in humans and pigs, whereas for a host which lives half as long phylogeny is rake-like, like observed of influenza in birds. Indeed figure 3 is strikingly similar to the similarly composed figure 6 in [13].

The difference in shape of phylogeny coincides, and is indeed caused by (see next section), a difference in pattern formation. For the longer lived hosts, no stable spatial structures emerge. Infection spreads as target waves, which sometimes convert temporarily into spiral waves. New mutations can invade somewhat before the ancestral types and are therefore strongly selected. Temporarily several strains can coexist, in different regions, but regularly a new strain sweeps away all other existing strains, as none maintains a stable 'source'. For the shorter lived hosts, in contrast, pattern formation does occur, so that hosts are alternately infected by virus strains which have diverged significantly.

When the duration of infectivity is relatively long (e.g. $zet = .2$), two- or three-armed spirals develop of the different strains New mutants are not strongly selected because the other strains will always be more different from the parent strain than the mutant is. In spiral waves only mutants in the spiral core can re-

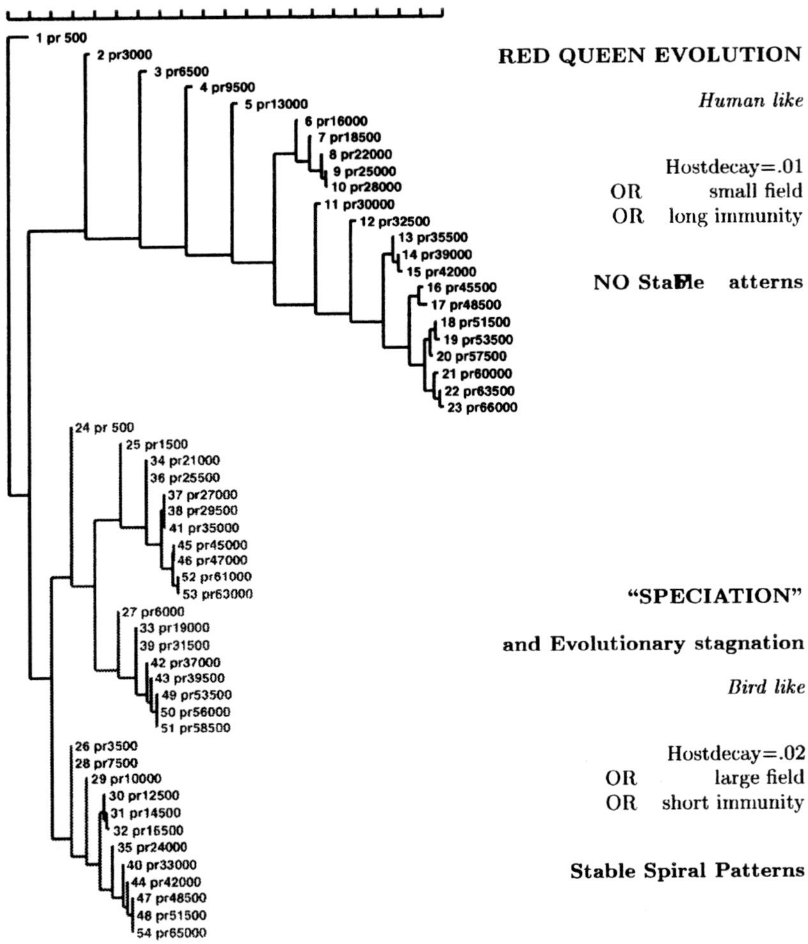

Fig. 3. Phylogeny of viruses evolved in two different hosts: the differences in evolutionary dynamics are caused by the presence/absence of patterns

place their ancestor because in the long run all offspring originates from the core. However, mutants in the core tend to destabilize the core leading to slower rotation speeds (and therefore, in a slow process, to loss of spiral domain). However the de-stabilization will also increase selection on mutants of the other strains in that core, eventually leading to more divergence of the strains coexisting in the core, and thereby to larger rotation speeds. The result is that several strains per spiral domain and several different spiral domains, which alternately gain domain on the other, coexist. The strains are stably diverged (i.e. 'radiation' or 'speciation' has occurred) and few new mutants can establish themselves. The resulting phylogeny is 'rake-shaped'.

In the complex wave patterns which occur when infectivity is short ($zet = .4$), so that only a subset of the hosts are infected in an infection wave, a similar host longevity related difference in phylogeny occurs, although somewhat less strong. Also this case the organization of the waves is such that very few mutants are 'at the right moment at the right place' to be able to invade, and, if they are, their domain remains restricted.

Other parameter differences also lead to these two types of evolutionary dynamics. They even occur with identical parameter values, with as only difference the size of the 'world'. Putting the long-lived hosts and their viruses in a world of 300*300 patches, instead of the 200*200 in the simulation above, there is enough space to develop e.g. stable and multi-armed spirals, the strains can diverge, and the evolutionary change will stagnate, just as in the shorter lived ones described above. In the next section we see that it is indeed the occurrence of pattern formation which determines the evolutionary dynamics and the shape of the evolutionary tree.

Genotypic punctuated equilibria. For intermediate parameter values the two evolutionary modes can alternate, as shown in Fig. 4. This leads to periods of rapid evolutionary change of the viral strains alternated by periods of very little change but the concurrence of several significantly diverged strains. In the figure the lowest panel shows the (cumulative) number of mutations of the strains which are present in more than 200 individuals in the 'world', over time. The spatial patterns are shown over the shaded period spanning from a period of stasis, through a period of fast change to a period of stasis. Snapshots of the world, and a space-time plot of a horizontal section are shown. For clarity of presentation the type of the last infecting strain is shown: the thin viral waves are hard to see. The snapshot is taken at the last time of the corresponding space-time panel. The two-armed spiral with slowly changing strains persist to t=23000, i.e. just before the end of panel 2. The target wave patterns continue to ca. t=26000, when again two-armed spirals appear - now as coupled pair; they last to ca t=29000; after a short unstable period, with fast changing strains, one single two-armed spiral reappears which remains stable for a longer period.

Alternation of periods of fast and slow change have been observed at the phenotypic level in many evolutionary models, in long term evolutionary experiments and in the fossil record. It is known as 'punctuated equilibria' or 'epochal evolution'. The phenomenon is well studied in models with redundant genotypic coding, where it is a consequence of diffusion on of neutral networks and occasional shifts to other (fitter) neutral nets [14], [15] Thus the epochal evolution on the phenotypic level is accompanied with a constant rate of change at the genotypic level. In the fossil record punctuated equilibria may be caused by external environmental changes. However, neither of these scenario's is true in the present model. The epochal dynamics occurs here at the genotypic level as well as on the phenotypic level. It is caused by self induced 'environmental'

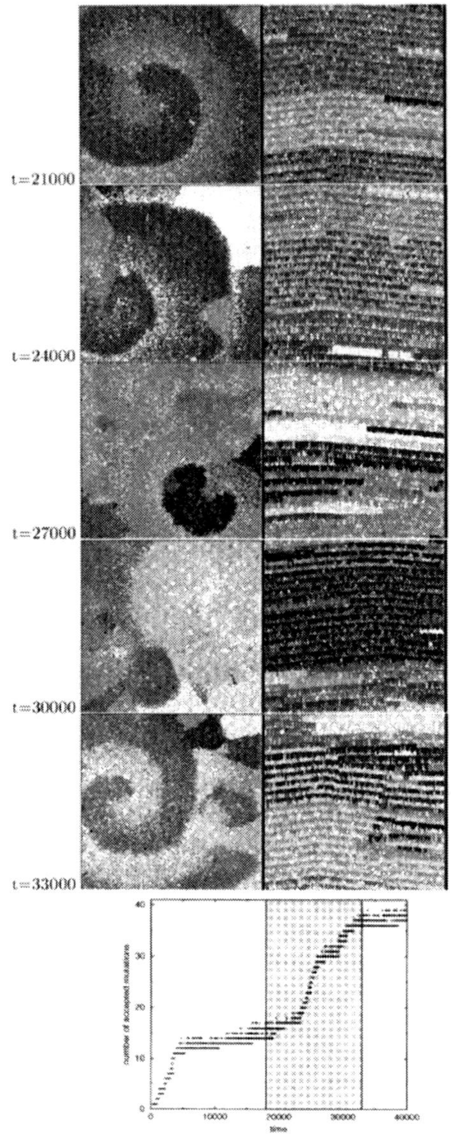

Fig. 4. Genotypic Punctuated equilibria due to unstable spatial pattern formation. The bottom panel shows the cumulative number of accepted point mutations in the dominant strains which coexist in the population over time: periods of rapid change and slow change alternate. The snapshots and space time plot above show that the periods of slow change correspond with periods in which the dominant species occur in stable two-armed spiral patterns, while during the period of fast change the spatio-temporal dynamics is chaotic. For more detail, see text

conditions, i.e. spatial pattern formation, by which periods of 'radiation' and subsequent neutrality alternate with periods of strong selection.

Mutation induced pattern dynamics. The dynamics of the spatial patterns shows interesting features induced by the mutations as well. For example:

- Mutations in a spiral arm produce inhomogeneities which, long after the mutants have died out, can generate new spiral cores "straddling' spiral arms of other cores (see e.g. Fig. 5-t17940)
- Transition from single spirals to two-armed spirals to three-armed spirals is caused by mutations close to the core and interference by neighboring domains. An example is shown in Fig. 5-a. Mutants cause destabilization of the core. The unstable core favors mutants. This way sufficient divergence of the strains can build up rapidly and stabilize the core with more members. Once it is stabilized it spreads over the entire worlds
- When one spiral domain is overtaking an other spiral domain the former may 'infect' the core of the latter: a member of the 'winning' spiral may become member of the loosing spiral. An example is shown in Fig. 5-b The infection will breed a faster rotating spiral because an immigrating 'foreign' member is more dissimilar to the remaining 'native' strains than the replaced one. Moreover in a short period of core de-stabilization following the infection fixation of new mutations are favored. Therefore such infection 'saves' the spiral, which than starts to gain domain on the 'donor' spiral. Thus, 'horizontal transfer' of strains takes place, by which diversity in the world remains limited, not withstanding very long lived separate spiral domains.
- On the border between spiral domains which have at least two members which are very dissimilar (not withstanding the homogenizing effect mentioned above) the domain can start to intercalate and chaotic dynamics slowly spreads (see Fig. 5-c). Phylogenetically this leads to a somewhat faster rate of accepted mutations, but the shape of the phylogeny remains rake-shaped. We studied the dynamics of the chaos - spiral 'competition' without the interference of new mutations: mutation is stopped after a small chaotic zone has developed. The chaos first spreads, like in the full evolutionary run. However, after it has spread over half of the world it is subsequently suppressed - leaving a multi-domain pattern of three-armed spirals, all composed of the same strains. The mechanism of chaos suppression is through the occurrence of spirals of the same members but in reversed order. These turn out to be more stable relative to the 'green' lineage (see Fig. 5-c) and effectively annihilate that lineage. Collision of spirals in which the strains occur in reversed order, leads to regions which are less frequently 'visited' by viral infection waves, as clearly seen in the space time plot of Fig. 5-c.

Thus summarizing: host-virus interactions lead to formation of wave patterns. These can be complex and chaotic, or can form (multi-armed) spirals, or target waves. The pattern formation leads to two different modes of evolution - which are seen in phylogenetic trees as 'narrow-and skewed' and 'rake-like'.

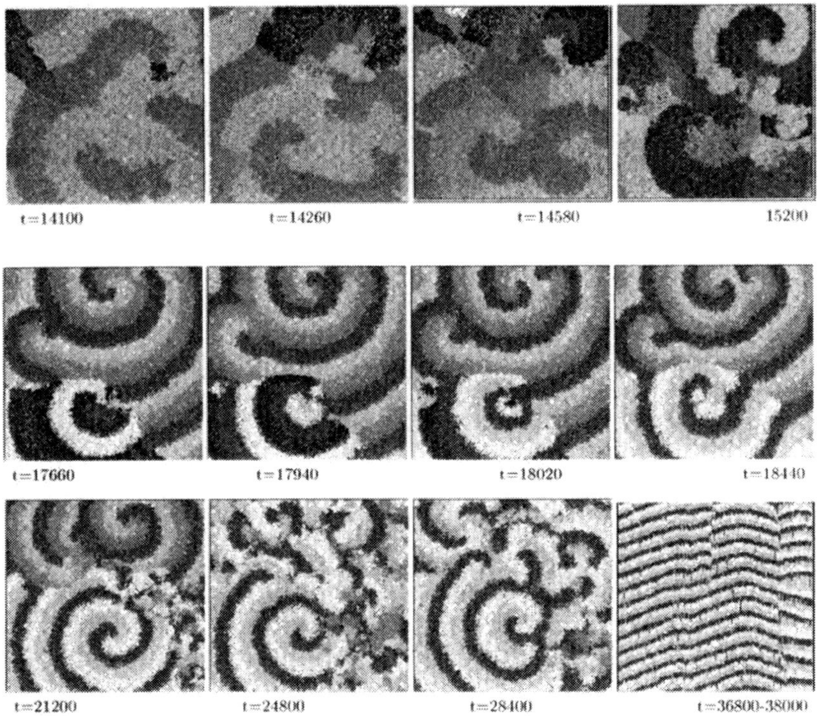

Fig. 5. Mutation induced features of spatial patterns. (a) transition from two-armed to three-armed spiral (b) 'infection' of weaker spiral by species of the stronger spiral. This infection induces new mutations, and a faster rotating spiral, which gains domain from the 'donor' spiral. Shading denotes lineages, not individual strains for clarity of presentation (c) induction and suppression of chaotic waves (later in the same run as (b); same shading)

They are similar to evolutionary modes verbally known as 'red queen dynamics' and 'radiation' respectively. In the former case there is a strong selection for being 'different' so as to (partially) escape the immunity of the hosts. Thus, many mutations are 'accepted', i.e. remain in the population. In the latter case strains diverge and waves organize themselves so that hosts are alternately infected by different (lineages of) strains. No (strong) positive selection on mutants occurs, because the diverged strains can infect the hosts better than the newly arising mutant. We have shown how the spiral patterns determine the evolutionary dynamics, and how 'genotypic punctuated equilibria' result from alternating periods of stable spirals and target waves. Moreover we have shown how the occurrence of mutations affects the spiral dynamics. Characterization of the global organization of the chaotic waves is a challenging task for the future.

3 Multilevel morphogenesis: Coordination, evolution and re-inventions along phylogenetic trees

In the previous section we have studied minimal conditions in which patterns generated at a larger scale can determine the structure of the interacting entities which form these patterns. It is shown that mutual interaction between 'levels of organization' is 'generic' in locally interacting systems subjected to mutation and selection (Darwinian evolution. In this section we study (and study how we can study) the products of long term multilevel evolution. We do this in the context of morphogenesis, i.e. the generation of a multi-cellular critter from a single cell. This process involves cell division, cell differentiation (i.e. alternative gene expression and therewith alternative function of the cells), differential growth and death of cells , and often (in animals) cell movement. Intricate coordination between intra- and inter-cellular dynamics is a prerequisite for the formation of 'well shaped' critters. Such a process does not only involve pattern formation, but also the reaction of the cells on the pattern formation.

A striking example of an interplay between pattern formation and morphogenesis in the full sense of the word is demonstrated in a recent study on the development of the slime mold *Dictyostelium discoideum* [16]. The last phase of the life cycle fruiting bodies are formed. This involves an intricate developmental process which is referred to in the experimental literature as the 'reverse fountain' because the top cells are 'pushed' down in the middle, while the bottom cells raise to the top at the periphery. We have shown how the *interaction* of waves of the signalling molecule cAMP (cyclic andenosine mono phosphate), (which are initiated by cells which produce cAMP oscillatory, and which are relayed by the other cells), differential adhesion between cells, and cell differentiation into a type of cells which produce an extracellular stiff slime layer and which do not produce or relay cAMP, generates such a pattern of cell movement and thus turns a 'blob' of cells into a fruiting body, i.e. a slender stalk with spores on top.

The *Dictyostelium* model successfully mimics a specific well studied morphogenetic process, and uses known biochemical processes. The question we examine here (and in [17],[18]) is how we can study the 'generic' behavior of the interplay between the various processes which play a role in morphogenesis. One should note, however, that here (and in evolved systems in general) the term 'generic' behavior should be understood in a special way. It is not so that from arbitrary initial conditions and for a large range of parameter values we should expect intricate morphologies to appear: this is not true for organisms either. Generic behavior should be understood in the sense of the common properties of the rare cases which do form interesting morphologies (see [19] for a discussion on such 'generic non generic properties').

We can study those 'rare' cases by formulating an evolutionary process, using an artificial fitness criterion. This fitness criterion should be chosen in such a way that maximizing it *enable* the phenomena of interest to occur *as side-effect.* The fitness criterion should not (and usually cannot) incorporate the phenomena of interest directly. Here we take 'cell differentiation' (i.e. number of gene-expression patterns and their Hamming distance) as fitness criterion to study morphogenesis.

3.1 A model for multilevel morphogenesis

An overview of the model is given in Fig. 6. Two 'tricks' enable us to formulate a very *minimal* model of multilevel development. The first is using an evolutionary process to zoom in on interesting cases as explained above. The second is the use of a 'two scale CA model' as introduced by Glazier and Graner [20]. In this

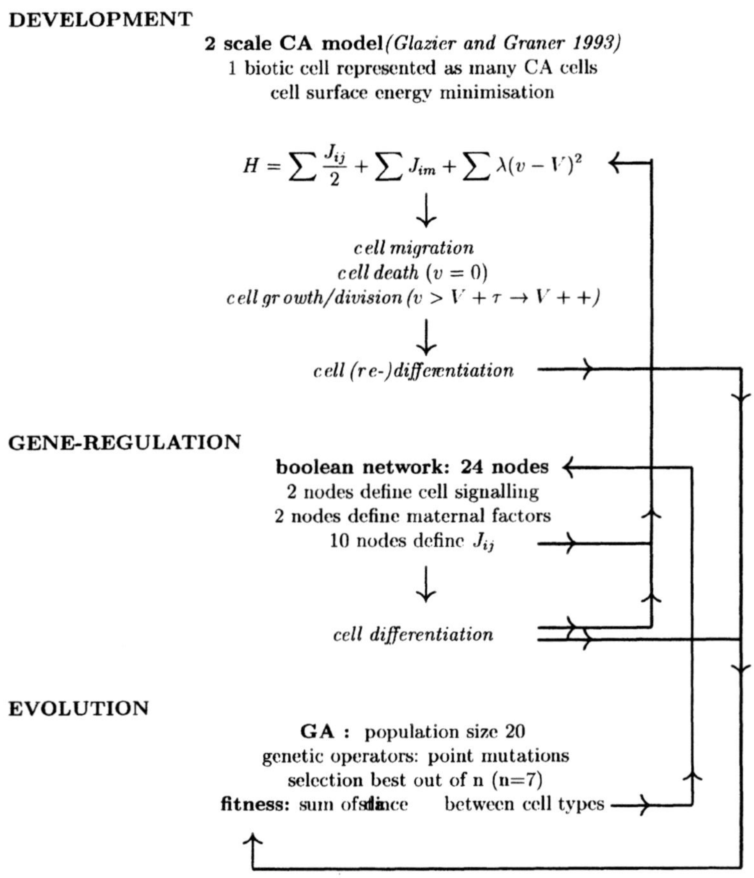

Fig. 6. Overview of the model: entanglement between gene regulation, development and evolution

model a biological cell is represented as many (here ca 40) cellular automaton cells, with the same state (= cell identification number). The model can be seen as an extension of the large Q Potts model. The cellular automaton transition rules represent a surface energy minimization process which is conditional on properties of the cell, notably its volume, which is conserved. Thus, this larger scale feeds back on the micro-scale. Glazier and Graner [20] have shown that cell sorting is a generic property of this model when cell surface tensions differ.

The basic model can be easily extended/interfaced with other processes, like chemotaxis (as was done in the slime mold model mentioned above, but not here), cell growth (i.e. increase of target volume V), cell death (which occurs automatically for small λ), cell differentiation (cell identification refers to cell type which defines surface energy $J_{i,j}$). In the model studied here the biological cells contain a boolean gene-regulation network. Some of the nodes of the network define (through a bit-matching 'mask') the surface energies with neighboring cells. Two nodes define signaling molecules, which may (dependent on the gene regulation network) influence gene expression in neighboring cells.[2] Cell growth is modeled as a reaction on stretching of the cell. [21] has shown that stretching can indeed trigger cell growth in experiments.

The development starts with one large cell, representing a fertilized egg. The first 7 divisions are pre-scheduled for all cells simultaneously (such initial cell divisions are called 'cleavage' divisions in developmental biology). Further cell growth and division (when cell size is twice a reference size) and cell death is governed by the dynamics of the development.

Evolution shapes the gene-regulation networks. The development defines the genotype to fitness mapping: cell differentiation depends on cell movement, which co-determines the cell-neighborhood and the received signals.

The full model is defined by very few parameters. Apart from the parameters defining the evolutionary process, and the gene regulation network (number of nodes, number of connections, and mapping of nodes to surface energy and cell signaling) - which are all held constant in all experiments done so far, there are only four: the strength of the volume conservation λ, the cell growth threshold τ and the dissipation constant and 'temperature' of the Boltzmann equation which defines the probability of copying the state of a neighboring cell into the current cell given a change in surface energy. These were varied, but (over the range tried) did not systematically influence the results. The initial (randomly generated) gene regulation networks, and initial stochastic differences in the evolutionary process produce over evolutionary time very different morphemes It is the properties of these evolved morphogenetic processes which we study as 'generic rare cases' as discussed above.

[2] All nodes have 2 inputs coming from nodes in the same cell, of from neighboring cells (denoted by negative numbers). Because there are only 2 signalling nodes defined, negative inputs from other nodes are always 0. This defines a variable connection network.

3.2 Mechanisms of Morphogenesis

About 20 % of the evolutionary runs produce (more or less) intricate morphogenesis (a small sample is shown in Fig. 7). One example is described in detail below (see Fig. 8). Overall the resulting morphogenetic processes as have been described in [17] lead to the following conclusions:

- Morphogenesis results as 'sustained transient' from surface energy minimization and 'intrinsic conflict' which is maintained by cell differentiation, cell growth and cell death. Without continued 'interference', the initial, high energy state would change, through shape changes to, at the end a 'blob' like low energy shape. The development shown in Fig. 8 without cell growth is an example of such 'transient to a blob'. Growth and cell division supply a continued interference which sustains zones where there is a conflict between energy minimization and internal cell changes which maintain higher energy states and intricate shapes.
- These intrinsic conflicts lead to automatic orchestration of adhesion, migration, differentiation, cell growth/division and death. It results in "pseudo-isomorphic outgrowth". Although the shapes do change during 'maturation' a 'critter' preserves its general appearance.
- Many different morphemes result from few mechanisms. Mechanisms found are:
  - engulfing: one cell type surrounds an other cell type. An example is Fig. 7-a
  - meristematic growth: a layer of dividing cells which differentiate into non-dividing (or rarely dividing) cells of several types. The zone is maintained because cell types depend on location. An example is the blue cap in Fig. 7-c
  - elongation by 'budding': a small group of differentiated cells is pushed outwards because an other cell type on the one hand tends to engulf them, but on the other stick together more firmly than to the 'bud' (Fig. 7-b) The situation is maintained because cells which do, nevertheless engulf, differentiate into bud-type cells.

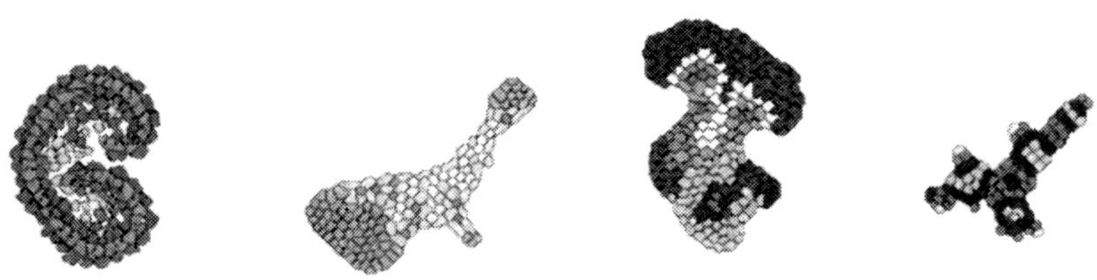

Fig. 7. Examples of morphogenetic mechanisms; (a) engulfing combined with apoptosis (b) elongation by budding (c) meristematic growth (d) dynamic growth-death and redifferentiation

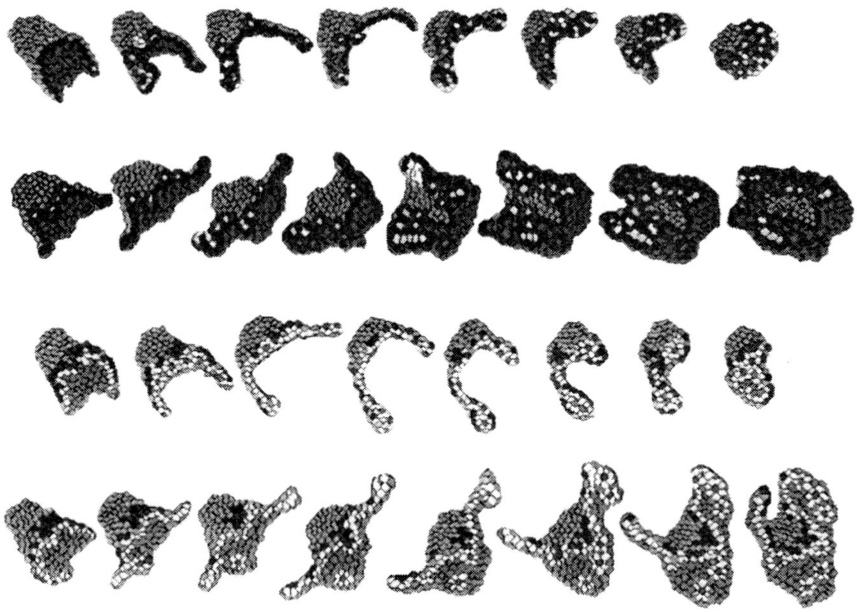

Fig. 8. Developmental histories of 'reinvented' morphologies; (a) development of ET2854 without cell growth frame=f=20,50,100,200,400,600,700,800; (1 frame equals 250 timesteps) (b) development of ET2854 with cell growth f= 35,70,140,200,273,400,440,480 (c) reinvention in an alternative evolutionary run restarted from ET=2488 The morpheme occurs at ET=1809 of the new run; Without cell growth f=20,50,100,200,280,300,600 With cell growth f=20,50,100,150,200,300,350,400

- elongation by 'convergence extension' like process. Two layers of cells maximize their border. Often a other cell types outside these layers contribute by intercalation in the boundary cell types and differentiating into boundary cell types themselves when they touch the other boundary cell type (see Fig. 8)
- elongation by intercalation of stably differentiated cell types (not shown)
- dynamic re-differentiation, cell division and cell death (Fig. 7-d).

3.3 A case study of the development and evolution of a morphotype

Here we examine a particular case of morphogenesis and its evolution. The morphogenesis is shown in Fig. 8. It is an example in which morphogenesis is mainly caused by the 'convergence extension' mechanism mentioned above, and clearly demonstrates 'conflict induced morphogenesis'. As shown in the upper panel after the cleavage divisions a two-armed structure develops. However in the long run the arms 'retract' and finally only a blob consisting of a number of cell types remains, and is stable; many cells have died. This is how the development unfolds without cell growth and division (large τ), i.e. for the parameter setting used during the evolutionary runs (developmental time up to t=250000).

Gene regulation and cell differentiation. The 'functional' gene regulation network is shown in the left (lower) panel of Fig. 9 (network was drawn with daVinci graph drawing software). The functional gene regulation network was extracted from the full gene regulation network, by an iterative procedure which determines per node whether its state is invariant or dependent on one of its inputs only. If so, non-functional inputs are deleted, and invariant states are used as such in the next iteration. As seen in the figure, although the network is defined with 2 inputs per node, the majority of nodes have zero or one inputs. The structure of the network is very hierarchical, i.e. the terminology 'upstream' and 'downstream' genes makes sense. This in contrast to the situation in random boolean networks. There is some crosstalk between pathways. Genes 1 and 2 code for signalling molecules. The nodes -1 and -2 refer to the expression of the gene in neighboring cells.

Cell differentiation is initiated at the first cleavage division, when the state of gene 21 is set to 0 in one of the daughter cells for one time step. This signals causes stable differentiation between the two cell lineages: gene 7, which is autoregulatory, preserves the signal. Further cell differentiation is through induction. Most importantly the dark-brown cell lineage differentiates by having the light-brown cell lineage in its neighborhood (and becomes blue). Extensive further history dependent cell differentiation occurs, but is of minor importance for the formation of the arms.

Morphogenesis. The mechanism of the development of the two-armed shape is as follows. The intermediate cell layer is very 'hydrophobic', i.e.the cell surface energy with the medium is very high (22) while that of the other cell types is 5 (dark brown) and 10 (light brown). Therefore the latter cell types engulf the middle layer. However when they do so they will touch each other, causing the dark-brown cells to differentiate into the blue cell type of the intermediate layer. This causes the extension of the tips of the arms. This process is aided by the fact that the dark-brown cells also intercalate into the intermediate layer: their surface energy with blue cells is only 7 whereas the surface energy among blue cells is 12. Also here re-differentiation occurs when they do intercalate and touch the light-brown cells. Thus the intermediate layer is extended by this process as well. Extension of the arms stops because of a 'shortage' of cells. Therefore the tip of the arms are 'closed off' by 'abnormal' intermediate cells (abnormal due to changing neighborhoods), which do have high surface energy toward the medium. This causes the arms to retract.

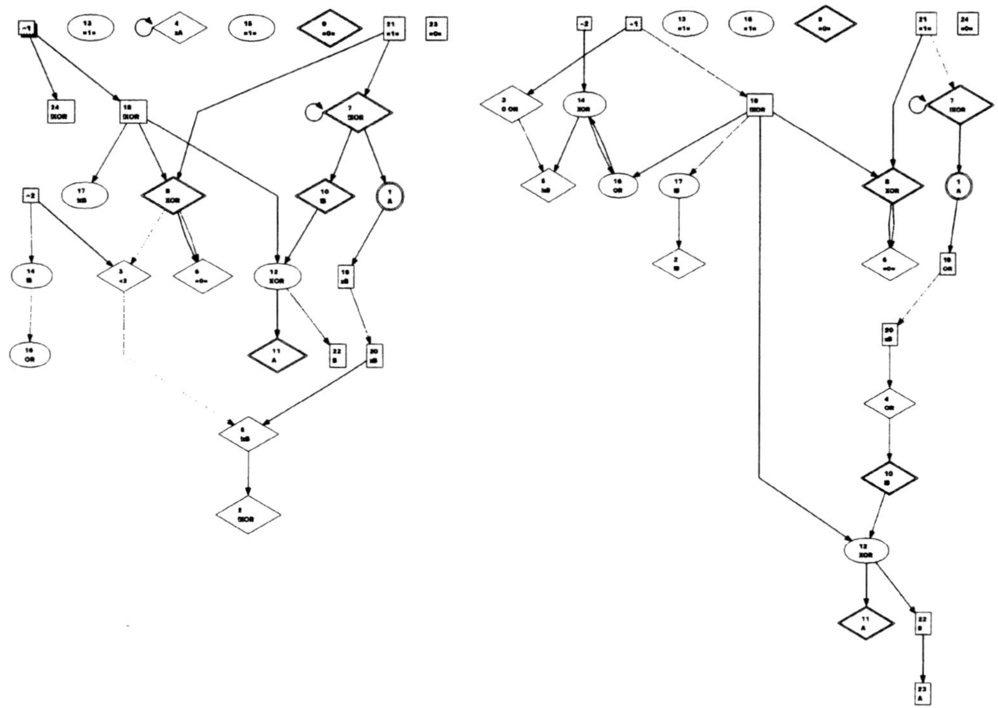

Fig. 9. Gene regulation networks of 2-armed morpheme as evolved in two parallel evolutionary histories, starting from a differentiated common ancestor

As mentioned, this is the development without cell growth and division. When cell growth is incorporated ($\tau = 4$) the morphogenesis of the 'critter' unfolds as shown in the second panel of Fig. 8: relative to the non-growing case it turns itself 'inside out'. The mechanism is as follows. The dark-brown cells have low surface energy among themselves (6) and, as described above, are 'pulled' because of intercalation and engulfing tendency. This causes them to be stretched, and thereby they grow, and eventually divide. Therefore the 'shortage' of cells mentioned above does not arise at the side which engulfs most efficiently. The net result is to push the intermediate layer backwards. When finally the two arms meet each other at the back the engulfing stops, and the growth rate of the critter drops. The last depicted stage is stable.

Evolution. From all the de novo initiated evolution runs, this particular morphogenesis was only observed once. However, during the evolutionary run similar critters arise repeatedly anew at wide intervals of time (data not shown). This is after full cell differentiation has evolved. The further evolution is along the 'neutral' path on which cell differentiation is maintained. The gene-regulation networks continue to change (at fixed rate), but the rate of change of the 'functional' gene regulation network slows down, but continues as well [18]. Within these constraints the morphogenesis as described remains a 'likely' mutant of the dominant evolutionary type, which forms rather undistinguished shapes (see Fig. 10). In these morphemes one cell type engulfs all the other cell types. This is true in the same evolutionary run, but also when evolution is restarted (i.e. 'when the tape were played twice', [22]) after the full cell differentiation pattern has arisen. In Fig. 8 an example is shown from such a parallel evolutionary history; it is 'critter' 1809 to arise in that run. Both without and with cell growth its

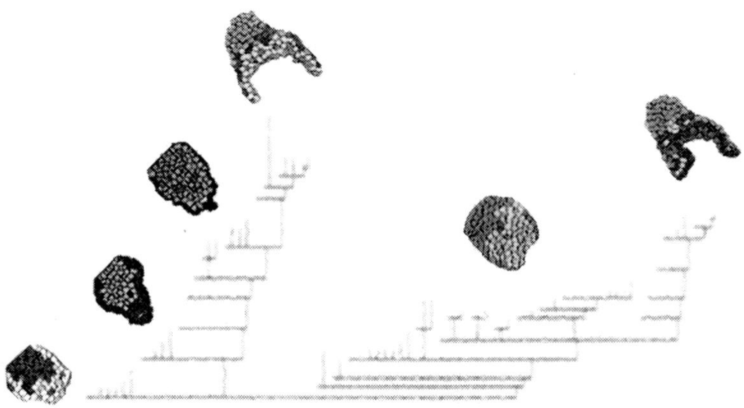

Fig. 10. Phylogenetic tree depicting evolutionary divergence and convergence from one common ancestor (left most picture).

development qualitatively mimics the one described above. Nevertheless the general shape of the functional gene regulation network has clearly changed (Fig. 9), although closer inspection reveals quite some similarities: 9 of the 24 genes are identically regulated. However the surface energies of the main cell types are identical. The observed differences between the two morphogenetic processes are partly due to noise (no 2 critters are identical, also not when they have identical genomes because of randomness in cell movement) and partly due to differences in the history dependent cell types, which have slightly different surface energies, and arise in different circumstances. The latter differences are dominant over the former: it is possible to recognize the critters in a 'color-independent way' when the full morphogenesis is observed (the 're-evolved' critter sticks its arms out more during growth, and becomes thinner without growth). Note, however, that the morphogenetic process is quite sensitive to the ratio of the value of the Boltzmann 'temperature' T and the stretch induced growth parameter τ. Together they define the growth rate of the dark brown cells.

The phylogenetic tree of the two independent evolutionary histories is shown in Fig. 10. The phylogenetic tree was calculated on the basis of the full gene regulation networks (using the neighbor-joining method). A sample of critters is shown. The evolutionary situation is comparable to that of independently evolving population on islands. Recent research (for a number of striking examples see [23] has shown that on islands repeatedly similar morphotypes and ecological adaptation evolve independently. This discovery has upset many traditional classifications, which assumed that the elaborate morphological or physiological adaptations surely would suggest a common origin. Within our experimental setup, the described two-armed critter would surely be judged as 'special'. And indeed, in a way it is, as it occurs only in one of the de novo runs. However, several samples of de novo evolution is not what we observe in our biosphere, rather we see alternative lineages evolved from a common ancestor. Our experiments suggest that given conserved basic cell differentiation pattern, the *propensity* to evolve certain seemingly complex adaptations is conserved as well. Such a 'preadaptation' to potential evolutionary change, is a 'heresy' current evolutionary theory. Our experiments suggest, however. that it is a 'generic' property of Darwinian evolution of critters evolved by Darwinian evolution when entangling of several levels of organization is not ruled out.

4 Conclusions

We have discussed models of multilevel evolution and of multilevel morphogenesis. In both cases we have seen that it is feasible to do so in quite simple models. In fact including several levels explicitly 'tunes' the processes at the various levels so as that it in fact modeling becomes easier. Moreover we have shown that two puzzling features of observed evolutionary processes can be explained in terms of the multilevel models presented.

In eco-evolutionary context we have seen that the same virus in different hosts, or even simply in a different sized world can evolve in very different modes. The modes of evolution depend on spatial pattern formation: a rake-like phylogeny, and radiation of strains occurs when wave patterns are such that hosts are infected by alternately by different strains. This is the case in multi-armed spirals. When this is not the case there is strong selection on escape of the immune system and a skewed, almost linear phylogeny arises.

In the context of morphogenesis we have seen that the 're-invention' of intricate morphologies appears to be a generic feature in multilevel evolution: some things are and remain 'easy' to evolve, given conservation of some basic features, e.g. (early) cell differentiation.

In conclusion, we have demonstrated that in a multilevel setting Darwinian evolution is an even more versatile mechanism than previously recognized. There is still much work to do to unravel its full potential.

Acknowledgements

I am much indebted to Roeland Merks for the great amount of work he did in setting up and programming the combined developmental and evolutionary model. I also thank my former students Maarten Boerlijst, Nick Savill, Ludo Pagie and Stan Mare'e for their contributions to studying evolution and development as multilevel processes. I thank Ben Hesper for long term inspiration and support.

References

1. D.L. Hartl and C.H. Taubes. Towards a theory of evolutionary adaptation. *Genetica*, 102-103:525–533, 1998.
2. M. C. Boerlijst and P. Hogeweg. Self-structuring and selection: spiral waves as a substrate for evolution. In C. G. Langton, editor, *Artificial Life II*, pages 255–276, Redwood City, CA, 1991. Addison-Wesley.
3. M. C. Boerlijst and P. Hogeweg. Spiral wave structure in pre-biotic evolution: hypercycles stable against parasites. *Physica D*, 48:17–28, 1991.
4. M. Eigen and P. Schuster. *The hypercycle: a principle of natural self-organization.* Springer, Berlin, Heidelberg, New York, 1979.
5. N. J. Savill, P. Rohani, and P. Hogeweg. Self-reinforcing spatial patterns enslave evolution in a host-parasitoid system [published erratum appears in J Theor Biol 1997 Oct 21;188(4):525-6]. *J. theor. Biol.*, 188:11–20, 1997.
6. L. W. P. Pagie. *Information Integration in evolutionary processes.* PhD thesis, Utrecht University, 1999.
7. L. Pagie and P. Hogeweg. Evolutionary consequences of coevolving targets. *Evol. Comput.*, 5:401–418, 1997.
8. P. H. Harvey, A. J. L. Brown, J. Maynard Smith, and S. Nee. *New Uses for New Phylogenies.* Oxford University Press, Oxford, 1996.
9. S. Nee, E. C. Holmes, A. Rambaut, and P. H. Harvey. Inferring population history from molecular phylogenies. *Philos. Trans. R. Soc. Lond. B. Biol. Sci.*, 349:25–31, 1995.
10. C. K. Ong, S. Nee, A. Rambaut, H. U. Bernard, and P. H. Harvey. Elucidating the population histories and transmission dynamics of papillomaviruses using phylogenetic trees. *J. Mol. Evol.*, 44:199–206, 1997.

11. T. Jackman, J. B. Losos, A. Larson, and K. de Wueiroz. Phylogenetic studies of convergent adaptive radiations in caribbeab anolis lizards. In T.J. Givnish and J. Sytsma, editors, *Molecular Evolution and Adaptive Radiation*, pages 535–557. Cambridge University Press, Cambridge, 2000.
12. W. M. Fitch, J. M. Leiter, X. Q. Li, and P. Palese. Positive Darwinian evolution in human influenza A viruses. *Proc. Natl. Acad. Sci. U.S.A.*, 88:4270–4274, 1991.
13. W. M. Fitch. Uses for evolutionary trees. *Philos. Trans. R. Soc. Lond. B. Biol. Sci.*, 349:93–102, 1995.
14. M. A. Huynen, P. F. Stadler, and W. Fontana. Smoothness within ruggedness: the role of neutrality in adaptation. *Proc. Natl. Acad. Sci. U.S.A.*, 93:397–401, 1996.
15. E. Van Nimwegen and J. P. Crutchfield. Metastable evolutionary dynamics: crossing fitness barriers or escaping via neutral paths? *Bull. Math. Biol.*, 62:799–848, 2000.
16. A. F. M. Marée. *From Pattern Formation to Morphogenesis: Multicellular Coordination in Dictyostelium discoideum.* PhD thesis, Utrecht University, 2000.
17. P. Hogeweg. Evolving mechanisms of morphogenesis: on the interplay between differential adhesion and cell differentiation. *J. theor. Biol.*, 203:317–333, 2000.
18. P. Hogeweg. Shapes in the shadow: evolutionary dynamics of morphogenesis. *Artif. Life.*, 6:85–101, 2000.
19. P. Hogeweg. On searching generic properties of non-generic phenomena, an approach to bioinformatic patter analysis. In C. Adami, R. K. Belew, H. Kitano, and C.E. Taylor, editors, *Artificial Life VI*, pages 285–294. MIT Press, Cambridge, Mass., 1998.
20. J. A. Glazier and F. Graner. Simulation of the differential adhesion driven rearrangement of bio logical cells. *Phys. Rev. E*, 47:2128–2154, 1993.
21. C. S. Chen, M. Mrksich, S. Huang, G. M. Whitesides, and D. E. Ingber. Geometric control of cell life and death. *Science*, 276:1425–1428, 1997.
22. W. Fontana and L.W. Buss. What would be conserved 'if the tape were played twice'. *PNAS*, 91:757–761, 1994.
23. T.J. Givnish and J. Sytsma. *Molecular Evolution and Adaptive Radiation.* Cambridge University Press, Cambridge, 2000.

Evolutionary strategies for solving optimization problems

Werner Ebeling, Axel Reimann, and Lutz Molgedey

Institute of Physics, Humboldt–University Berlin,
Invalidenstr.110, D-10115 Berlin, Germany

Abstract. We will give a survey of applications of thermodynamically and biologically oriented evolutionary strategies for optimization problems. Primarily, we investigate the solution of discrete optimization problems, most of combinatorial type, using a certain class of coupled differential equations. The problem is to find the minimum on a large set of real numbers (the potential) U_i, defined on the integer set $i = 1...s$, where s is an extremely large number. The stationary states of the system correspond to relative optima on the discrete set. First, several elementary evolutionary strategies are described by simple deterministic equations, leading to a high-dimensional system of coupled differential equations. The known equations for thermodynamic search processes and for simple models of biological evolution are unified by defining a two-parameter family of equations which embed both cases. The unified equations model mixed Boltzmann/Darwin- strategies including basic elements of thermodynamical and biological evolution as well.

In a next step a master equation model in the occupation number space is defined. We investigate the transition probabilities and the convergence properties using tools from the theory of stochastic processes. Several examples are analyzed. In particular we study the optimization of theoretical model sequences with simple valuation rules. In order to demonstrate that the strategies developed here may also be used to investigate realistic problems we present an example application to RNA folding (search for a minimum free energy configuration).

1 Introduction

Natural evolution is connected to optimization problems [1–5]. As in earlier works, Evolution is understood here in a rather abstract sense as an unlimited sequence of steps of self-organization [3,4]. In the course of real evolution many strategies have been used beginning from simple thermodynamic optimization steps up to higher biologically oriented strategies. Thermodynamical strategies are always based on individual search. In order to give an example of a thermodynamical search process let us consider the motion of a Brownian particle in a one-dimensional potential $U(x)$ which is represented symbolically in Fig. 1.

We start the motion at the coordinate x_0 outside the minimum. Due to the influence of motion in the force field, external friction and thermal agitation the particle tends to approach the next minimum. Possibly it could cross also potential barriers due to thermal activation. In the course of a stochastic process it will finally approach a Boltzmann distribution for the coordinates and momenta. At fixed temperature, the free energy is minimized during this process.

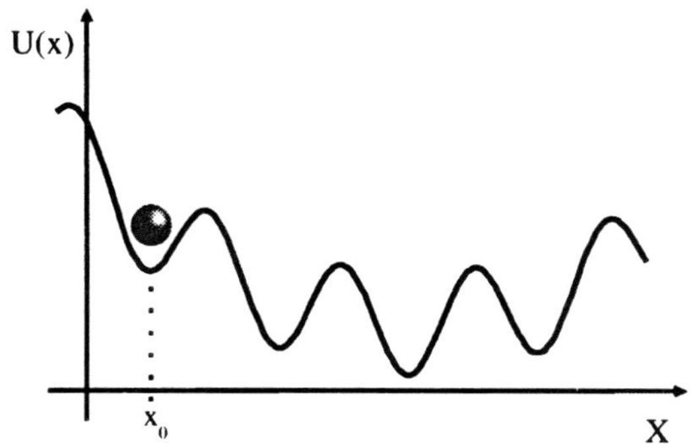

Fig. 1. Thermodynamical search of a Brownian particle

By lowering the temperature (so-called annealing procedure) the distribution will concentrate more and more around the lowest minimum of the potential. At low temperature the particle tends to assume the lowest minimum. This simple example demonstrates the essence of a thermodynamic strategy. Processes of this type are all subject to the second law.

On the other hand, biologically oriented strategies introduce several new elements connected to the biological dynamics as e.g. self-reproduction, mutation and competition. Biological entities are in general also searching for something, e.g. the individual is looking for food, good living conditions etc, the population tries to optimize a combination of biological properties, in abstract terms we may speak about fitness. Following the work of Wright, Conrad and other pioneers we model fitness here in a most abstract way by a landscape with hills and valleys. Since biological entities in general live in populations, biological strategies typically are ensemble strategies. A population, represented here as a swarm of points on a fitness landscape are moving collectively. This corresponds to a swarm dynamics. A demonstration of an ensemble search is given in Fig. 2. We have represented here symbolically 9 seekers which are searching for a minimum (the black point in Fig. 2) of a two-dimensional potential landscape. The swarm of particles approaches a distribution concentrated around the minimum. How this can be realized by dynamical processes like selection, mutation and competition will be demonstrated in the second part. The picture which we have drawn above does not claim to model any realistic evolution process (of thermodynamical or biological type). Our aim is merely to develop a caricature of real evolution processes which models at least a few of the relevant features of the strategies, which are performed.

The understanding and modeling of evolutionary strategies is an important part of any research program dealing with problems of evolution. One of the most important problems is the understanding of the advantages of a collective search in comparison to individual search.

In reality, evolution has to deal with very complex optimization problems. Here we restrict ourselves to problems which allow an elementary mathemati-

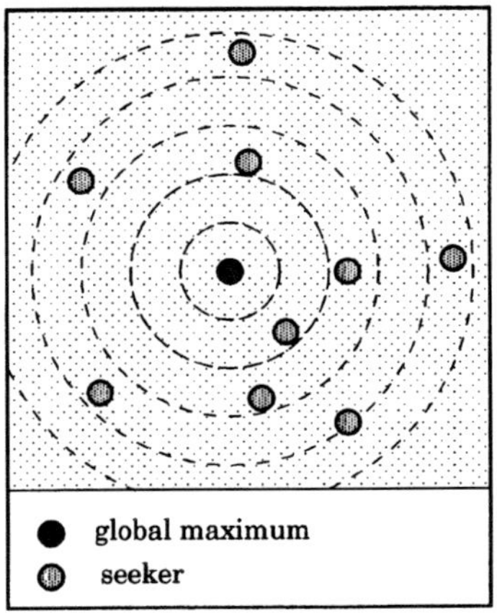

Fig. 2. Collective search for an optimum

cal description. The simplest one is the solution of an extremum in a Gaussian landscape. In more detail we will consider a model which may be considered as a caricature of the folding of polynucleotides. This optimization task is represented by a frustrated sequence game as proposed by Engel: a sequence is constructed on an alphabet consisting of 4 letters, e.g. $\{A, B, C, D\}$. The sequence is then credited a fitness point if any two letters appear in alphabetical order. (We also consider the wrapped order $\{D, A\}$ to be alphabetical.) This rule tends to create sequences with the period 4. Additionally, the periodic occurrence of a letter with period 5 is credited by a credit point (or more general by a credit b with $b < 1$) [4]. Enumerating all possible sequences and arranging them in a sequence space (*Gödel coding*) gives an impression of the corresponding fitness landscape. A complete representation requires an exhaustive search which is applicable to short sequences only. This way one gets an impression of the model landscape's difficulty. Fig. 3 lists the absolute number of local and global maxima for different sequence lengths, obtained by an exhaustive search. A different optimization problem, which is closely related to the group of spin glass models is the so-called MERIT- or LABS- problem (LABS - low autocorrelated binary sequences). The problem is to find binary sequences consisting of $+, -$ with a minimum of autocorrelations. In an earlier work we have tested several pure and mixed strategies with respect to their effectiveness to solve optimization problems of the type described above [6]. We mention also the models of evolutionary landscapes of biomolecules, based on free energies, and replication rate constants of RNA-molecules [7]. Here we concentrate on the rather simple Engel problem. In Fig. 4 we present a comparison of the autocorrelation function for a random path on Engel sequences and the corresponding result for LABS-sequences. We see that the Engel problem shows a rather long correlation tail which is well approximated by an exponential function. On the other hand the MERIT-problem shows a correlation function with a very fast decay, this shows that the parts

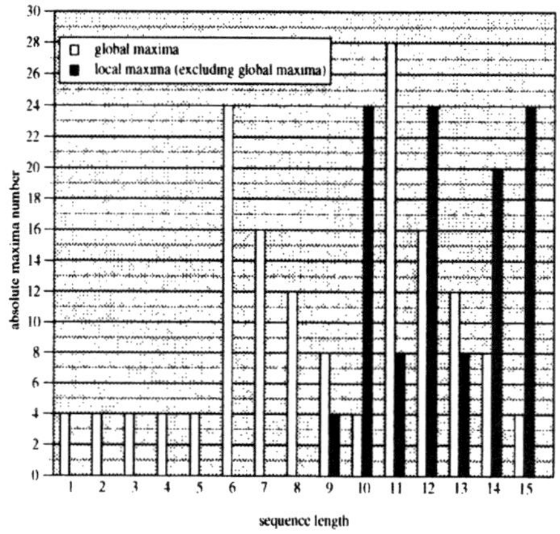

Fig. 3. Number of local and global maxima depending on the sequence's length. Up to the length $L = 8$, there are no local maxima. Longer sequences are more demanding to any search strategy, introducing local maxima where every neighboring sequence has an inferior fitness

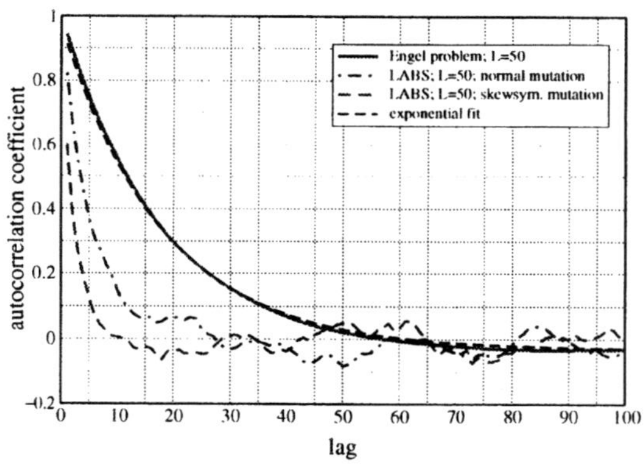

Fig. 4. Autocorrelation function for Engel sequences and for the LABS problem

of the landscapes subsequently visited by the path are nearly uncorrelated. We concentrate here on methodological studies of strategies and will use the Engel problem as an optimization task which is on one hand not trivial but also not too difficult on the other hand.

We will discuss the application of different strategies borrowed from nature to the Engel problem. Modern research has shown that the strategies developed in the process of natural evolution might also be of interest for the design and construction of technical systems. Pioneering work in this direction has been done by BREMERMANN, CONRAD, HOLLAND, KIRKPATRICK, RECHENBERG, SCHWEFEL and others [1,5,2].

2 Models of basic evolutionary strategies

Analyzing the mechanisms of natural evolution we find several basic strategies [4,8–12,6] The most important natural strategies are:

2.1 Darwin strategy

Biological strategies appear in the universe in the process of biogenesis only, i.e. about 3 – 4 billions of years ago. The basic elements of a Darwin type strategy are [2,8,5,10,11,6]

- Self-reproduction of good species that show maximum fitness.
- Mutation processes due to error reproductions that change the phenotype's properties of the species.

Consider for example the dynamics of a population distributed on s states: The population's task is to find the maximum on a non-negative set of real numbers (the potential) V_i defined on an integer set $i = 1 \ldots s$, where s is extremely large. Now, we identify V_i with the growth rate of the i-th member of the population. A simple model for a problem solving dynamics is the Fisher-Eigen equation:

$$\partial_t x_i = (V_i - \langle V \rangle) x_i + \sum_{j=1}^{s} (A_{ij} x_j - A_{ji} x_i) \tag{1}$$

Here, $x_i(t)$ denotes the occupation number of state i; we set

$$\sum_{i=1}^{s} x_i(t) = x_0 = \text{const.} \tag{2}$$

In our model , V_i is the fitness of species i in the population (expressed by the self-reproduction rate). The population average is denoted by $\langle V \rangle$ and A_{ij} describes mutation processes. Mutations are modeled here as transitions between the states i and j , i.e. the A_{ij} are transition rates from state j to state i. Usually one assumes symmetrical mutation rates $A_{ij} = A^0_{ji}$ where A_{ij} is a symmetrical matrix $A^0_{ji} = A^0_{ij}$. With the ansatz

$$x_i(t) = \exp\Big[- \int_0^t \langle V_{t'} \rangle \, \mathrm{d}t' \Big] y_i(t) \tag{3}$$

we get some kind of discrete Bloch equation

$$\partial_t y_i(t) = - \sum_{j=1}^{s} H^D_{ij} y_i(t) \tag{4}$$

where the Heisenberg matrix for the Darwin strategy H^D is defined by

$$H^D_{ij} = -A^0_{ij} + \delta_{ij} \Big(\sum_{k=1}^{s} A^0_{ki} - V_i \Big) \tag{5}$$

Thus we get the eigenvalue problem

$$\sum_{j=1}^{s} H_{ij}^{D} y_j^n = \epsilon_n^D y_i^n \qquad n = 1 \dots s \tag{6}$$

The solution may be expressed in terms of the eigenvalues ϵ_n and eigenfunctions y^n [3,4]

$$y_i(t) = \sum_{n=1}^{s} \exp[-\epsilon_n^D t]\, a_i^n y_i^n \tag{7}$$

For the time-dependent occupation number of the states i follows:

$$x_i(t) = \frac{y_i(t)}{\sum_j y_j(t)}. \tag{8}$$

2.2 Boltzmann strategy

Another fundamental strategy observed in nature is connected to the second law. Macroscopic physical systems optimize certain thermodynamic functions in their course of evolution. Simplifying, we call this the Boltzmann strategy. The Boltzmann strategy again has two basic elements:

- Motion along gradients to reach steepest ascent/descent of thermodynamic functions.
- Stochastic processes including thermal and hydrodynamic fluctuations leading to random changes. Hence, locking in local maxima is avoided.

In order to formulate a simple dynamic model, let us consider a numbered set of states $i = 1, 2, \dots, s$ again – each characterized by a potential energy $U_i = -V_i$ and a relative frequency in the population $x_i(t)$ at time t. Then, the simplest model of a Boltzmann process which tends to find minima of U_i is described by the following master equation (Pauli equation):

$$\partial_t x_i(t) = \sum_{j=1}^{s} (A_{ij} x_j(t) - A_{ji} x_i(t)) \tag{9}$$

with the transition rates

$$A_{ij} = A_{ij}^0 \begin{cases} 1 & \text{if } U_i \le U_j \\ \exp[-\beta(U_i - U_j)] & \text{otherwise.} \end{cases} \tag{10}$$

In other words, a downhill transition is always carried out whereas an uphill transition occurs only with a small probability decreasing exponentially with the threshold's height. Here, β is a parameter with the meaning of a 'reciprocal temperature'. By using a linear transformation we get a Bloch-like equation again:

$$\partial_t x_i(t) = -\sum H_{ij}^B x_j(t) \tag{11}$$

with the characteristic matrix for the Boltzmann strategy H^B being defined as

$$H_{ij}^B = -A_{ij} + \delta_{ij} \sum_k A_{ki}. \tag{12}$$

Accordingly, we get the eigenvalue problem

$$H_{ij}^B y_i^n = \epsilon_n^B y_i^n \tag{13}$$

with the lowest eigenvalue $\epsilon_1 = 0$ and the corresponding eigenvector

$$y_i^1 = \frac{e^{-\beta V_i}}{\sum_j e^{-\beta V_j}}$$

Again, the solution may be expressed in terms of these eigenfunctions

$$y_i(t) = \sum_{n=1}^{s} \exp[-\epsilon_n^t] a_i^n y_i^n \tag{14}$$

The two strategies discussed so far will be denoted in the following as elementary strategies. For the sake of completeness, let us mention only a few more complex strategies which have evolved, as e.g. Volterra strategies [3,4], Haeckel strategies [13,4] etc., leading to nonlinear equations.

2.3 Mixed Boltzmann-Darwin strategy

Boltzmann and Darwin strategies show several parallels, but also essential differences [8,9,4]. Both strategies are well suited to find the extrema in landscapes of potential functions. In general, it will depend on the structure of this landscape which search strategy is the better one. The qualitative analysis carried out in an earlier work [8] suggests that in case no knowledge about the structure of the landscape is available, it will be advantageous to apply the Boltzmann strategy combined with annealing. This strategy seems to be more universal. However, thermodynamic processes have the tendency to be locked in relative extrema surrounded by high thresholds. On the other hand, Darwin processes are able to cross high barriers by tunneling if the next minimum is in the vicinity. Anyway, we have seen that both strategies are quite different and we might expect that there exists a class of problems where the Boltzmann strategy is better and another class of problems where the Darwin strategy is more appropriate. In a situation like this it seems to be a good idea to develop a strategy which possesses components from both elementary strategies. In order to model mixed strategies let us consider a numbered set of states $i = 1, 2, \ldots, s$, again, each characterized by a fitness V_i and a population $x_i(t)$ at time t. Then, our model of a mixed strategy with the property to find maxima of V_i is:

$$\partial_t x_i(t) = \gamma \left(V_i - \langle V \rangle\right) x_i(t) + \sum_j (B_{ij} x_j(t) - B_{ji} x_i(t)) \tag{15}$$

The new two-parameter family given by eq. (15) contains the Boltzmann strategy for $\gamma = 0$ as a special case. The Darwin strategy is obtained for:

$$\gamma = 1, \ \beta \to 0 \tag{16}$$

With the transformation

$$x_i(t) = e^{-\gamma \int\limits_0^t \langle V \rangle dt'} \cdot y_i(t) \tag{17}$$

the quasi-linear differential equation (15) transforms into a linear Bloch like equation for $y_i(t)$:

$$\partial_t y_i(t) = -\sum_j H_{ij}^M y_j(t) \tag{18}$$

with the effective "Hamiltonian"

$$H_{ij}^M = -B_{ij} + \delta_{ij}\Big(\sum_{k=1}^{s} B_{ki} - \gamma V_i\Big) \tag{19}$$

$$\sum_j H_{ij}^M y_j^n = \epsilon_n^M y_i^n \tag{20}$$

Here, the eigenvalues ϵ_n and the eigenfunctions y_i^n are determined by a stationary Heisenberg problem for an effective potential. Thus we get the complete explicit solution in the form of equations (7, 8).

2.4 Tournament – nonlinear strategies

In all the previous cases the competition between the seekers was based on a linear comparison. We now introduce the following nonlinear generalization [11,6]

$$\begin{aligned} \partial_t x_i(t) &= \gamma \sum_j F(U_j - U_i) x_j(t) x_i(t) \\ &-\gamma \sum_j F(U_i - U_j) x_i(t) x_j(t) \\ &+\sum (B_{ij} x_i(t) - B_{ji} x_i(t)) \end{aligned} \tag{21}$$

Here, $F(x)$ is some monotonically decreasing function. The most interesting special case is:

$$F(x) = \text{const} - \theta(x) \tag{22}$$

where $\theta(x)$ is the step function. Due to the step character the worst partner of the game is always thrown out like in a tournament, i.e. we model a hard competition. Therefore we will use the term tournament selection. Our equation models a tournament of two competitors. We will later on in our simulations also use tournaments with more than two competitors in a selection event.

The new nonlinear two-parameter family given by eqs. (21 - 22) contains the Boltzmann strategy for $\gamma = 0$ as a special case again. The Darwin strategy is obtained for the special case of linear selection functions

$$F(x) = \text{const} - x; \quad \gamma = 1, \ \beta \to 0 \tag{23}$$

In the general case $\gamma > 0$, $\beta > 0$ we restrict ourselves – for the analysis – to a step function as competition function, or a smooth version as e.g.:

$$F(x) = \text{const}\,(1 - \arctan(x))$$

3 Stochastic modeling of mixed strategies

The mathematical problem regarding the solution of the coupled nonlinear differential equations (15 - 22) is extremely difficult. On the other hand, simulations and realistic search problems are mostly based on a finite number of seekers. Therefore we transform the problem onto an occupation number space which is often used in physics. We introduce a corresponding stochastic model leading to one partial differential equation which often is called a master equation. First, we replace the concentration vector x_i by a set of integer numbers $N_i(t)$ forming a lattice: the occupation number space. Any changes are restricted to one-step processes [1,3,4]

$$N_i \longrightarrow N_i \pm 1 \tag{24}$$

With these assumptions we get the transition probabilities for the general mixed strategy:

$$W(N_i + 1, N_j - 1 | N_i, N_j) = A_{ij} N_j + F(V_j - V_i) \frac{N_i N_j}{N} \tag{25}$$

where

$$N = \sum N_i \tag{26}$$

We see that the transition algorithm (25) which in fact depends only on differences of the fitness before and after the transition conserves the total number of seekers. Let us now introduce the probability to find N_1 seekers using the strategy 1 and N_i seekers using the strategy i at time t in the game: P($N_1 \ldots N_i \ldots N_s$,t). The corresponding Master equation reads

$$\partial_t P(N_1 \ldots N_i \ldots N_s, t) = \mathbf{W} P(N_1 \ldots N_i \ldots N_s, t) \tag{27}$$

Let us now take a look at the dynamics's averages

$$x_i = \frac{\langle N_i \rangle}{N}$$

in the simplest case of linear competition $F(x) = \text{const} - x$. Then for $\beta = 0$ ($T \to \infty$) our Master equation leads to the stochastic version of the Fisher-Eigen equation [3,4] which converges for $t \to \infty$ to a distribution centered around the absolute maximum of V_i. The case $\gamma = 0$ yields the stochastic version of the thermodynamic strategy equation which converges for $t \to \infty$ to the Boltzmann distribution possessing a maximum around the lowest minimum of U_i. Another interesting case $F(x) = \text{const} - \theta(x)$ corresponds to the so-called tournament competition. This is a hard competition in the sense that the loser is thrown out of the game. We have shown in earlier works [11,6] that this strategy is often quite effective in solving difficult optimization tasks.

4 Simulations – the dependence on search parameters

In this section we will test the effectiveness of the stochastic strategy versions introduced above by applying them to the simple model problem (Engel sequences). In particular we will study the influence of mixing and the dependence of the search on the total number of seekers, temperature and mutation/selection rates.

We have shown above that the dynamics of each member of the two - parameter family of mixed strategies is in a weak sense problem-solving. That means the dynamics will find the absolute maximum of V_i which is the target of the search, or at least a value very close to it. Since the Boltzmann factor has a maximum at the highest minimum of the potential vector U_i we see that the target vector $x_i(t \to \infty)$ and the corresponding $N_i(t \to \infty)$ center close to it. In other words, the dynamics of these processes converges to distributions near the minimum of U_i (the maximum of V_i) which is searched for. In other words the strategy is indeed problem solving. In order to guarantee the exact convergence to the minimum of U_i we may use some kind of annealing: $\gamma \to 0,\ \beta \to \infty$. Since for $\gamma = 0$ the ground state solution of equation (15) is

$$\psi_i^0 = \exp\left[\frac{-\beta U_i}{2}\right] \tag{28}$$

we see that $x_i(t \to \infty)$ degenerates in the limit of the annealing process to a δ - function centered around the deepest minimum of U_i. This shows that the dynamic process converges in the limit

$$\gamma \to 0,\ \beta \to \infty \text{ and } t \to \infty$$

indeed to the absolute minimum of U_i. In earlier works we considered the application of mixed Boltzmann-Darwin-strategies based on the discrete form of mixed strategies given by equations (15 - 21) to the traveling salesman problem (TSP) and to related street network problems [8,14]. It could be shown by simulations that in this respect mixed Boltzmann-Darwin-strategies have very good search properties.

We will study now the frustrated sequence game which we explained in the introduction: a sequence is constructed using for example the letters $\{A, B, C, D\}$. The sequence is then credited a fitness point if any two letters appear in alphabetical order. Additionally, the appearance of a letter with the period 5 is credited. We simulated sequences of length 15; the maximum possible fitness in this configuration is 22. The ratio of local to global maxima is 24 : 4 or 6 : 1. The number of representatives (seekers) N in the ensemble has been varied between 1 and 100. Let us underline that $N = 1$ always corresponds to a pure Boltzmann strategy and $N > 1$ to a mixed Boltzmann-Darwin strategy. In this case we are simulating a population of N seekers searching simultaneously and being coupled by competition for best results. We always calculated the mean values and the best results obtained for a fixed maximum number of fitness function evaluations.

The simulation results for a mixed strategy are shown in Fig. 5. We used a *mixed Boltzmann-Darwin strategy* using a tournament selection involving 2 seekers respectively. In tournament the loser disappears and the better one replaces it. In order to control the coupling (selection strength) the frequency of selection operations as compared to mutation steps was varied. The corresponding probability P_{mut} is plotted along the ordinate axes. A value of 0% means that mutation steps were never carried out. Accordingly, 50% means that *in average* every mutation step was followed by a selection step; whereas 100% indicates that there were mutation steps only. Visibly, the optimal number of seekers depends on the chosen mutation / selection ratio. The best results, however, are achieved in the region of about 25% mutation rate and with relatively small ensembles $N < 20$. In recent work we used also a tournaments of 4 seekers: the best one always replaces the most inferior. From our experience with simulations it seems that hard competition is exceedingly more efficient than linear competiti

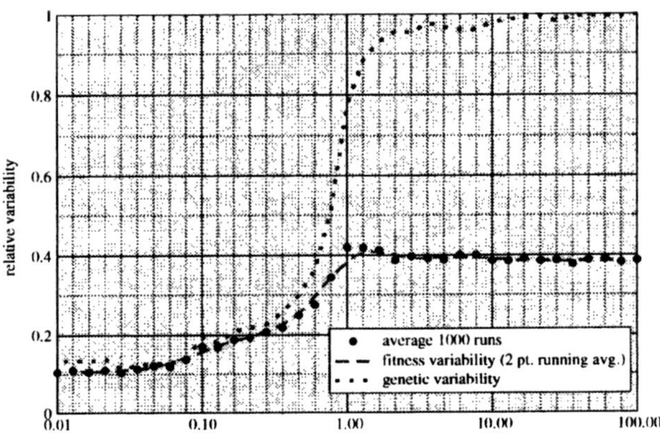

Fig. 5. Boltzmann-Darwin strategy, tournament selection: Fitness of the best seeker in the ensemble, averaged over 1000 runs. Used parameters, string length $L = 15$, temperature $T = 1$, calculation time $t_{max} = 1000$.

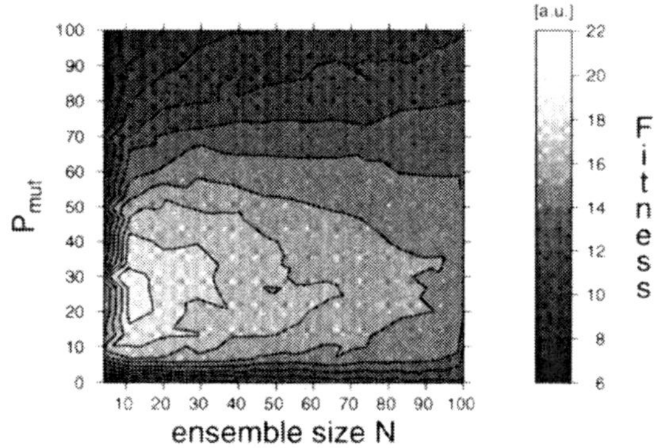

Fig. 6. Ensemble variability for the Boltzmann-Darwin strategy with tournament selection

5 Control of the mutation/selection rate

As seen in Fig. 5, the mixed strategies' effectiveness crucially depends on the mutation vs. selection ratio. Very low mutation rates cause the strategy to resemble gradient search methods with higher likelihood of getting stuck. Too high mutation rates on the other hand basically turn the search into a random walk, destroying information already gathered while searching. The optimal mutation rate occurs at the edge of a kind of phase transition. We demonstrate this by an investigation of the variability of the searchers (see Fig. 5). This fact was already pointed out by EIGEN AND SCHUSTER et al. [15]. An optimal mutation rate lies in the proximity of the phase transition, being just a little bit smaller – thus granting a quick search without information destruction. In order to automatically control the mutation/selection rates one only needs to know where the phase transition occurs. Disappointingly, one can only observe these transitions for long simulation times, i.e. when the search approaches a stationary state. Therefore one needs a different, truly dynamic indicator. An easily obtainable value is the seeker ensembles' variability, both in terms of fitness and genotype: In an ensemble of N seekers one can observe any number from 1 to N different seekers. Normalizing the variability one obtains an indicator varying from $1/N$ to 1. Our Fig. 5 shows that the variability redraws the phase transition. As shown in Fig. 5 it does not even matter for the relevant variability values below 0.25 whether we refer to the fitness- or genotype based variability. A problem is, however, that the variability's variance scales with the ensemble size:

$$\sigma_v^2 = \mathcal{O}\left(\frac{1}{N}\right) \tag{29}$$

rendering it useless as control parameter for small ensembles $N \leq 10$. To circumvent this problem we can define some 'control free' $\varepsilon \simeq \sigma_v^2$ around the optimum variability we are aiming for, thus softly blending in control with increasing ensemble size. The adaption actually implemented works as follows: The ε - interval around the optimal variability value 0.2 was set to $\varepsilon = 1/N$. The variability's deviation

$$\Delta V = V - V_{opt} \tag{30}$$

from the optimum was measured in terms of ε to exponentially adapt the mutation probability: $P_{mut}(t+1) = \alpha\, P_{mut}(t)$. In some earlier work [6] we proposed the estimates $V_{opt} \simeq 1/\sqrt(N)$ and

$$\alpha = \frac{\Delta V}{\epsilon} \quad \text{if } (\Delta V > 0) \tag{31}$$

and $\alpha = 0$ elsewhere. Mutation probabilities outside the interval $[0,1]$ were clumped back to the interval's boundaries. We have shown that this control applied to the mixed Boltzmann-Darwin strategy with tournament selection improves the results in an effective way. As expected, simulation results are relatively bad for small ensembles with less than $N = 10$ seekers. For ensemble sizes above this 'threshold' the automatic mutation rate adaption eliminates the dependence on the initially chosen rate. The optimal ensemble size ranges from $N = 10$ to more than $N = 30$ members.

6 Discussion

Mixed strategies with appropriately set parameters do not always give the best results with respect to mean and best fitness value. However a certain amount of parallelism introduced by a Darwin term ($\gamma > 0$) is in general a useful element in good search strategies. The computation time which is lost simulating a number of parallel seekers is gained by certain advantages of parallel search as e.g. the possibility of experience exchange introduced by selection processes in which inferior seekers get replaced by better ones.

In our mixed strategy this is modeled by the Darwin elements: competition between seekers, survival of the fittest. However, when the number of seekers working in parallel is too high the parallelism costs more then one gains from it. Theoretical investigations regarding this subject have been carried out by many workers [16]. It is easy to see that the computation overhead necessary for parallel strategies also cannot pay off if the fitness landscape is rather simple. According to our experience, a successful search (optimization) requires to choose the proper strategy and a fine balance between parallelism and individualism, that is between Darwin and Boltzmann elements. Similar results were obtained in another work analyzing street network problems [14]. The results summarized here show a certain advantage of including Darwin elements into the search strategy – especially if these elements get tuned appropriately. As far as the tuning parameter is concerned, it is still an open question whether the variability's optimum value of about 0.2 found in our work is model independent or has to be adapted to the particular problem.

Possibly the real power of mixed strategies including Darwin elements will show up on parallel computers with more than some 4 processors. Here, simultaneous search was always simulated on a single sequential computer. Of course,

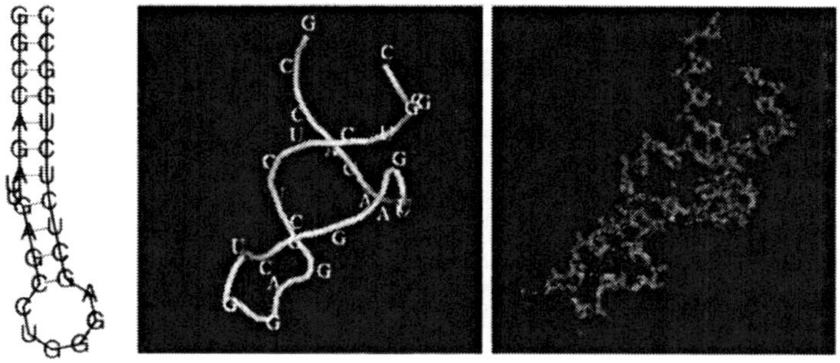

Fig. 7. Example for an RNA folding problem; a mixed Boltzmann/Darwin strategy was quickly able to not only find the best folding possible ($E = -16.8\,\mathrm{kcal/mol}$, left picture) but also suboptimal foldings along the search path. The results coincide with results obtained by NMR measurements (protein data base entry: 1AJU, middle and left picture).

it could in principle also be carried out on a net of parallel processors. Since coupling (selection process) between the elements of a Darwin ensemble is a rather seldom event, the speed up by using such an N-processor system is expected to be higher than of order $\mathcal{O}(\log N)$, thus scaling better than simulated annealing techniques [16].

In order to demonstrate that the strategies developed here may also be used to investigate realistic problems we present an example application to RNA secondary structure optimization. The problem here is to find a sequence folding which minimizes the free energy of a given RNA strand. The example presented in Fig. 7 shows the optimal secondary structure for the HIV-2 Tar-Argininamide Complex (Protein Data Base entry: 1AJU) found by our mixed strategy and compares it to experimental data. Fitness evaluations were done using the RNAeval routine contained in the Vienna RNA Software package[1].

Acknowledgment

We wish to thank the *Deutsche Forschungsgemeinschaft* for providing financial support in the framework of the 'Sonderforschungsbereich 555' and the Theoretical Biochemistry Group *TBG* of the University of Vienna for making their Vienna RNA package publicly available.

References

1. M. Conrad: *Adaptability* (Plenum Press, New York, 1983)
2. H.P. Schwefel: *Numerical Optimization of Computer Models* (Wiley, New York, 1981)

[1] http://www.tbi.univie.ac.at/~ivo/RNA

3. R. Feistel, W. Ebeling: *Evolution of Complex Systems* (Kluwer Publ., Dordrecht, 1989)
4. W. Ebeling, A. Engel, R. Feistel: *Physik der Evolutionsprozesse* (Akademie-Verlag, Berlin, 1990)
5. L. Rechenberg: *Evolutionsstrategie 94. Optimierung technischer Systeme nach Prinzipien der biologischen Evolution.* (Fromman-Verlag, Stuttgart, 1994)
6. W. Ebeling, L. Molgedey, A. Reimann: Physica A **287**, 599 (2000)
7. W. Fontana, W. Schnabl, P. Schuster: Phys. Rev. A **40**, 3301 (1989)
8. T. Boseniuk, W. Ebeling, A.Engel: Phys. Lett. **125**, 307 (1987)
9. T. Boseniuk, W. Ebeling: Europhys. Lett. **6**, 107 (1988)
10. T. Asselmeyer, W. Ebeling, H. Rosé: Biosystems **39**, 63 (1996)
11. T. Asselmeyer, W. Ebeling, H. Rosé: Phys. Rev E **56**, 1171 (1997)
12. T. Asselmeyer, W. Ebeling: Biosystems **41**, 167 (1997)
13. W. Ebeling, A. Engel, V.G. Mazenko: BioSystems **19**, 213 (1986)
14. F. Schweitzer, W. Ebeling, H. Rosé, O. Weiss: Evol. Computation **5**, 419 (1998)
15. M. Eigen, J. McCaskill, P. Schuster: Adv. Chem. Phys. **75**, 149 (1989)
16. H.P.S.Y. Davidor, R. Männer: *Parallel Problem Solving from Nature* (Springer, Berlin, Heidelberg, New York, 1994)

Review of biological ageing on the computer

Dietrich Stauffer

[1] Instituto de Física, Universidade Federal Fluminense,
Av. Litorânea s/n, Boa Viagem, Niterói 24210-340, RJ, Brazil
[2] Institute for Theoretical Physics, Cologne University,
D-50923 Köln, Germany

Abstract. The asexual Penna model of 1994/5 is the most widespread model for computer simulations of biological ageing. It is based on the half-century old mutation-accumulation hypothesis, though it does not necessarily contradict alternative explanations of ageing through oxygen radicals or specific longevity genes. It represents the genome by a bit-string such that bit = 0 means a healthy and bit = 1 a sick gene, acting on the body from that age on which corresponds to the position of the bit in the bit-string. One of the questions is: Do old men die like flies?

1 Introduction

Why do we get older and older [1]? Why don't we run 100 meters in about ten seconds even up to the age of our death? To contribute to this question we first need a clear definition of ageing, and we define it as the increase of the mortality function

$$q(a) = -d \ln S(a)/da \qquad (1)$$

with age a, where $S(a)$ is the number of survivors at age a, from an initial cohort $S(0)$ of newly born babies. Fig. 1 shows recent yearly mortality functions from Sweden, and we see that mature men obey reasonably an exponential increase, $q(a) \propto \exp(ba)$, known since the 19th century as the Gompertz law. Since q and b have the dimension 1/time, Azbel [2] wrote this law as

$$q(a)/b = A \cdot e^{b(a-X)} \qquad (2)$$

and found that for different countries and different centuries the characteristic age $X \simeq 103$ years and the proportionality factor A are roughly the same for humans, while the Gompertz slope b increases with increasing medical progress. In recent decades this trend may have been stopped and replaced by a slow increase with time of the age of the oldest person in a country [3], beyond the age X, instead of following eq(2). A better agreement, also for young ages where Eqs. (1,2) fail, was found [2] when plotting the survival probability $S(a)$ at some fixed age a versus the one at 40 years or versus the life expectancy at birth.

Women are less law abiding, Fig. 1, and mostly ignored here. Thus this paper deals mainly with asexual models, while sexual reproduction and the reasons why women live longer than men are reviewed [4] in the "Anais" of the Brazilian Academy of Sciences, since Brazil is the leading country for ageing simulations. For a longer presentation of both questions see [1].

Fig. 1. Recent yearly mortality functions of men (+) and women (x) from the Statistical Central Bureau of Sweden.

Theories about ageing [5] can be grouped as: mutation accumulation, telomeres, oxygen radicals, longevity genes. Here we deal with the first one, according to which the mutations (i.e. the errors in the replication of our genome in the DNA, accumulated in the course of biological evolution) mainly affect the old ages after childbirth. Inherited diseases which kill us before we have any children cannot be given on to the offspring and vanish from the population, while inherited mutations killing old university professors have no detrimental effects (on the number of viable offspring) and thus can stay in the population for many generations. (Most mutations have no or bad effects; those which distinguish homo sapiens coloniensis from homo sapiens neanderthalis are quite rare. The mutations we deal with are hereditary; we ignore here the somatic mutations which are not inherited.) In this evolutionary theory, due to Nobel laureate Medawar [6], ageing is therefore caused by accidental errors, i.e. by bad mutations.

Telomeres are sections at the end of the DNA strands, and some are lost at each duplication of the DNA. Thus after many replications, normal cells in vitro can no longer duplicate (Hayflick limit). In living organisms, the enzyme telomerase can restore the lost telomeres. If loss of telomeres is responsible for our death and the same for the whole species, we should all die at about the same age, in contrast to Fig. 1. Thus Hayflick [7] himself suggested that his limit for humans corresponds to an age above that ($\simeq$ 125 years) of the oldest humans: The death of an organism can occur before the maximum life time of its cells. I read only few expert opinions [8] according to which telomer loss is the main reason for normal ageing of large animals [9]. But the Hayflick limit has been connected [10] with Werner's syndrome, where young people soon look like old people: There only 20 instead of the usual 60 replications are possible.

Oxygen radicals are chemicals arising from natural metabolism, and damage our molecules in our brain [11], in our chromosomes, or elswehere in our bodies. They can be the microscopic cause for the mutations, for both the inherited mutations of evolutionary theories and the somatic mutations occuring during our lifetime [12]. In the special case of progeria, a form of accelerated ageing, not enough of an anti-oxidant enzyme is produced, allowing for more damaging oxygen radicals [13]. From very small to very large animals, about a dozen oxygen molecules per body atom are consumed in the average life [14]; this order-of-magnitude observation shows the importance of oxidation reactions for the life span. If I would eat less I would live longer [15].

These competing theories are not mutually exclusive. If genetic engineering by the change of one gene [16] changes drastically the life expectation of mice, (or nematodes and yeast [17]) this longevity gene may act through a change in the body's self-defence against oxygen radicals. In this way it may change the mutation rate which in turn is the basis of mutation-accumulation theory [18,19]. P.M.C de Oliveira therefore compared mutation-accumulation theory with the software, while the computer on which the software is running corresponds to oxygen radicals, longevity genes, etc.

In the mutation-accumulation theory, without mutations we would live youthfully forever, apart from accidents or predators, and we would never experience senescence. Thus we simulate mutation-accumulation theory, since it is most suited for simulations, and does not exclude all alternative theories.

An experimental distinction for or against evolutionary theories like mutation-accumulation can be made if the number of observed generations is varied: If the life span is changed for the same animals with which the experiment started, this change cannot be attributed to mutation accumulation; if the change is seen ten or more generations later [19], it may be due to mutation accumulation.

We now explain shortly the Penna model, then present its main results, followed by many applications, and at the end discuss future research. An appendix presents a possible alternative model.

2 Penna model

The most widespread ageing model for computer simulations is today the Penna model [20], but alternatives exist [21]. In this Penna model, the genome stored in the DNA is represented by, typically, 32 bits in one computer word. The position of the bit corresponds to a certain age; thus for humans one bit may correspond to about 5 years. A zero bit means no life-threatening inherited disease; a bit set to one means that from this age on until death a dangerous disease affects the health. Typically but not always [22], three such diseases kill the individual; also, the birth rate is diminished by these diseases in some modifications [23]. At birth, the child gets the bit-string of the parent, except for, typically, one mutation at a randomly selected bit position. After reaching a minimum reproduction age of, typically, 8, each individual produces at every time step (in units of bit positions) a few offspring. Restrictions of food and space are taken into account by a Verhulst factor with a probability $N(t)/N_{max}$ to die, with $N(t)$ the actual population at time t, and N_{max} some large parameter, typically ten times the initial population. These Verhulst deaths act on babies only [24] or, in most publications, on all ages. (See [25] for a discussion of antagonistic pleiotropy.) For parameters near these typical values, the results do not depend strongly on the parameters or the number of bits (16 to 64) in the bitstring: The Gompertz law is always roughly recovered, except for the youngest and the oldest ages. A complete Fortran program is listed in [1].

In this way the Penna model is one of many bit-string models of biology like the Eigen quasispecies model presented at this conference [26], where bit zero is good (no mutation) , and bit one means a bad mutation. In contrast to other bit-string models, the position of the bit is important and is connected with the age at which this mutation diminishes the fitness. All our mutations shorten the lifespan, none increase it. Positive mutations, so crucial for the development of life in the last 3×10^9 years, are ignored here as well as in many other ageing theories since they are quite rare. These rare cases are important for the changes of species involving millions of generations, but usually do not happen within the life of one individual. (Somatic mutations, which are not given on to the offspring, happen more often, but play no role within the mutation-accumulation theory [6] implemented in the Penna model; they might explain the difference between the life expectation of men and women [27].)

This model can be compared with the minimization of the free energy in thermodynamics, where entropy and energy counteract each other: We start with the ideal population which has no mutations, analogous to the ground state. Then due to random mutations, corresponding to positive temperatures, this state of eternal youth is changed into a realistic mortality function. Darwinian selection minimizes the number of bad mutations, like minimizing the energy, while replication errors produce such mutations and create a disordered genome corresponding to entropy. (This analogy is different from the one in [28].)

3 Main results

The main success is the agreement of simulations, Fig. 2, with real mortalities in Fig. 1: The Penna model obeys the Gompertz law. The deviations from this law at young ages due to child mortality seen in Fig. 1 and missing in Fig. 2 have been modelled by adding "housekeeping" genes to the Penna model [29]. Fig. 1 refers to humans. To find a stationary total population, about 10^3 iterations are sufficient, but this criterion is misleading: At old age, the equilibrium mortalities are found only after 10^4 iterations, while before, q at old age is smaller. Examples for the time dependence are published in [30].

The rapid death of Pacific Salmon after giving birth can be explained by assuming that all individuals give birth only once, and all at the same age [31]. This extreme example also shows nicely how the equilibrium of mutations and

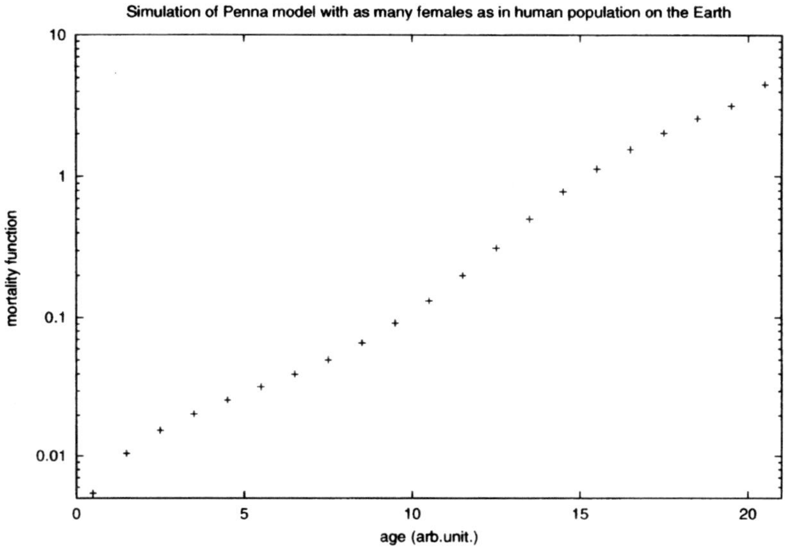

Fig. 2. Simulated mortality function in asexual Penna model with 3×10^9 individuals. The Gompertz law would give a straight line in this semilogarithmic plot and also in Fig. 1.

Darwinian selection of the fittest works: Survival after giving birth is unimportant in this model without child care, and thus mutations accumulate in the later age positions killing all individuals above the maximum reproduction age, if it exists.

For genetically completely homogeneous populations like some inbred laboratory animals, the Penna model would let everybody die at the same age. Modifications were published [32] avoiding this unrealistic effect and restoring roughly the Gompertz law: At each time interval, also a probabilistic influence of the mutations is included, in addition to the above-mentioned deterministic killing after three mutations have been reached.

As in many branching processes of this type, if we wait long enough (a number of iterations proportional to the initial population), all survivors have one common ancestor Eve (and Adam, if males and females are simulated) [1,35]. Also in reality, all present humans may be offspring of one individual women living 10^5 years ago in Africa [33] (or Australia [34]?). While for humans this effect may coincide with speciation and the formation of homo sapiens, within the model it also applies to a stationary species: The larger the population is the longer do we have to go back to find the most recent common ancestor, and this common ancestor may change in time when the population lives on for many generations.

Among other results we mention a justification of female menopause and its analogs, as found in many animals, not only humans [36]. Menopause was found to self-organize from a combination of two effects: a risk of dying at birth for the mother which increases with the number of active mutations and thus with age; and child care in the sense that young children die if their mother dies. These complications of sexual reproduction are discussed in [4]. Alternative ideas [37] restricting this effect to humans and explaining it by the help of grandmothers seem to contradict reality: [36], Austad in Wachter and Finch [5].

4 Applications

Overfishing may have been the cause for the vanishing of the Northern Cod off the Atlantic cost of Canada, and indeed simulations [30] were compatible with this possibility. They also showed, that not fishing the younger animals helps to increase the stock and thus the catch for fishing. (The fish distribution as a function of the depth below the ocean surface was simulated in [38].) Lobsters have a fertility increasing with age, similar to trees, and therefore [38] the old ones should also be saved from being eaten if the total catch is to be optimized. Hunting of Alaskan wolves was simulated by several authors [39].

Trees seldomly wander around and thus it may be more appropriate to put them on a lattice [40]. A new tree can grow only on an empty neighbour site of the parent tree. Still, the ageing curves look good [40], similar to the standard case. But Makowiec showed that the controversial [24] Verhulst factor can be avoided completely if one is using this lattice restriction [40].

Also on a lattice, an asexual bitstring model similar to Penna's gives self-organized criticality similar to sandpile automata [41]; such critical phenomena may be crucial for evolution [42].

In Germany and elsewhere, the future of retirement pensions is actively discussed. Do we have to work until we are 68 or 70 years old ? California has already shifted upwards the age for medicare. At first this solution looks obvious since life expectancy at birth has roughly doubled within the 20th century, at a roughly constant retirement age. However, neither did everybody die at age 40 a century ago, nor does everybody die at age 80 now; see Fig. 1. Much but not all of the increase of the average life expectancy comes from an enormous decrease of child mortality (Fig. 1.1 in [1]). Thus life expectancy at birth still increases in Germany every year, but appreciably weaker than in the first half of the 20th century. To fund my retirement, the crucial quantity is the average number of years which people at 65 years still can expect to live. This time increased only slightly during the first half of the 20th century in Germany, but stronger in the second half; see [3] for a similar change in Sweden. Thus careful extrapolations are needed to estimate the number of retired people in future decades. Nevertheless I do not see how the present German system can survive if in the year 2030 the strongest age cohort has age 70. Countries like Brazil seem to follow the trends of Western Europe with some delay, only perhaps with changes occuring more rapidly. Brazil made a 20-year plan to increase the retirement age by 6 months every year.

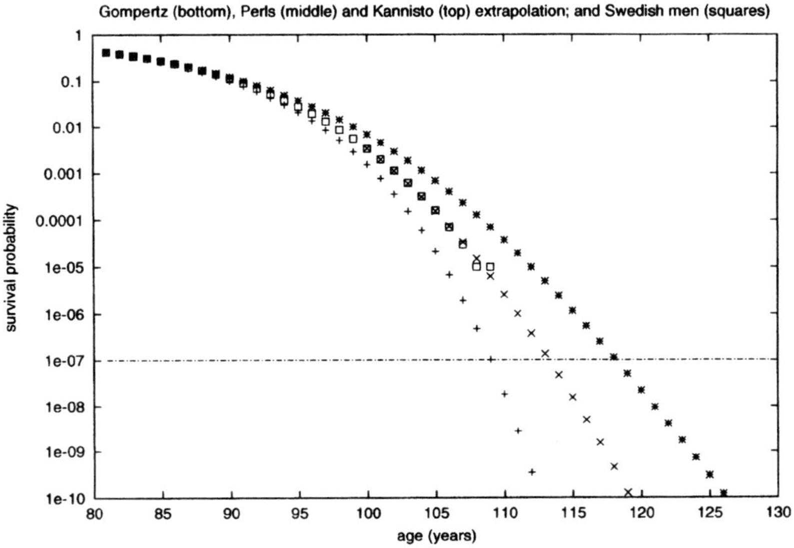

Fig. 3. Extrapolated survival probabilities for an exponential (+) or linear (x, age ≥ 100 only) or Kannisto increase of mortality function with age. The squares represent Swedish male life tables. We fit $q \propto \exp(0.1 \cdot \text{age})$ onto middle age, and for Kannisto [45] use a mortality function $q/(1+q)$. The typical maximum age is reached when the reciprocal of this survival probability becomes of the order of the population size.

A crucial question in this connection is whether humans die like flies. For medflies, and less accurately for other animals, it was established in the 1990's [43] that the mortality for the oldest old no longer increases at the same exponential rate as at middle age, and even has a maximum. Lots of exaggerations (about humans having similar properties) were published, as summarized in [44], but recent male mortalities in the USA, from Wilmoth's Berkeley Mortality Database, show no such effect. Recent Swedish life tables, Fig. 1, show downward deviations from the exponential increase of $q(a)$, but these deviations are weaker than in England and other large rich countries [45]; similar deviations were seen in the Swedish data used for [45] and communicated to me by Thatcher. They are compatible with the Perls [46] suggestion of a mortality function increasing above 100 years linearly instead of exponentially: no more acceleration but not yet deceleration of mortality function. Maybe, the better the data are, the smaller are the deviations from Gompertz. If indeed the oldest human being was a Brazilian woman who died at age 129 in 2000, and not a French woman who died at age 122 in 1997, then this single person would for me be the best indication for a deceleration of human mortality, $d^2q/da^2 < 0$, above 110 years of age. The oldest Swedish men are about 108 years old [3], and Fig. 3 shows that this age is compatible with a Gompertz plot. (See [47] for a similar USA analysis.) Predictions for retirement funding in 2050 should take into account these problems and not just extrapolate existing exponential trends. Azbel [2] has analyzed breaks in trends for many mortality tables in ways which may be useful for better extrapolations. A crucial question is: Can human death be postponed without limits, or is there some limiting maximum age, independent of medical care?

Four technical points:

i) Often the downward deviations from the Gompertz law for humans are exaggerated in the literature because the authors look at female mortalities and ignore middle age. At middle age, women compared with men usually have a lower mortality, which then increases stronger at old age to get relatively close to the male mortality, Fig. 1. Thus, between 70 and 100 years a strong curvature is seen, while the Gompertz slope for 30 to 70 years is ignored. These lawless women are thus no good tests for the exponential Gompertz law. The male mortalities are more appropriate but are based on smaller statistics at old age.

ii) Of course, if q is discretized as the fraction of people dying within one year, this fraction cannot be larger than one and because of this triviality cannot obey the Gompertz law. It is better to discretize $q(a+1/2) = \ln[S(a)/S(a+1)]$ since monthly mortalities seem available only for babies, not for old people. In this form, q is called the hazard factor, force of mortality μ [45], or mortality function, and is used in Figs.1 and 2.)

iii) The probabilities of Fig. 3 give *average* numbers; the actual maximum age in a population of millions fluctuates by a few years [3], but not by a dozen years. Thus the French maximal age of 122 is an extreme outlier [45] and should not be compared with Fig. 3; the typical age of the oldest French is similar to Sweden, slightly below 110 [3], and this age should be compared with Fig. 3.

iv) Horiuchi and Wilmoth [48] define as "deceleration" a negative second derivative not of the mortality function but of its logarithm: $d^2 \ln q/da^2 < 0$. Then our Fig. 1, or the Perls suggestion, shows a slight deceleration for Swedish men, but the Gompertz law with $d^2 \ln q/da^2 = 0$ then means neither an acceleration nor a deceleration. Human evidence for a change-over from acceleration to deceleration seems missing.

The Penna model alone cannot answer these questions: It gives a maximum age which is so high that it is difficult to find by computer simulations, but with suitable parameters or modifications it also gives a reduction of mortalities at old age below the Gompertz extrapolation [1,25]. For example, assuming a probabilistic instead of deterministic killing through mutations [50], an ageing curve similar to Swedish men, Fig. 1, was obtained [49]. The Penna model could reproduce the heritability of longevity [51].

The medical progress in the short run has been responsible for the increase of life expectancy and could also be simulated on a computer [52]. However, it is possible that after many generations this progress is partially offset by changes in the human genome caused through selection and mutation as a result of modern medicine.

A more computational aspect is the question whether the number of bits in the bit-string is a crucial quantity, or cancels out if all ages are expressed in units of this length. The answer seems to depend on details like a fixed or a fluctuating number of mutations [53]. Bit lengths up to 10^3 have been used [29].

5 Summary

This review summarized some essential simulations of the asexual Penna model, emphasising material too recent to be included in [1]. Physics journals and the interdisciplinary Theory in Biosciences not only published papers of physicists in the field, but also constructive criticisms and alternative modifications of the Penna model by geneticists Cebrat and Pletcher, or by physician Klotz. Such interdisciplinary cooperation might help in the final aim of gene therapy [54] or other ways to help ageing people.

The mutation-accumulation theory was selected as the explanation of ageing since it is one of several possibilities, seems similar to statistical physics (mutations = entropy versus selection = energy) and is suitable for computer simulations. The alternative model presented in the appendix is too young to be judged. The other explanations summarized in the introduction may also be true, do not necessarily contradict the mutation-accumulation theory, but require very different type of simulatiuons, like quantum chemistry for oxygen radicals. More work on these alternatives would be nice; the anti-oxidant effects of red wine [55] justify the old German wisdom: "Rotwein ist für alte Knaben eine der besondren Gaben."

Within the methods discussed here, a practical application would be a simulation of the demographic transition from a growing population with lots of young people to a stable or shrinking population with lots of old people. How can we adjust birth rates (including immigration) and retirement age such that this transition becomes less painful for retirement funds. For Germany, such a study based on Eqs. (1,2) and refs. [2,3] may come too late, but it could help Brazil, for example.

I thank S. Moss de Oliveira and P.M.C. de Oliveira for hospitality in Brazil, and them, N. Jan, T.J.P. Penna, A.T. Bernardes, S. Cebrat and M.Ya. Azbel for helping my ageing over many years.

6 Appendix: A simpler alternative

This appendix presents a simpler evolutionary computer model of ageing, which does not rely on mutation-accumulation in the sense of the Penna model. It has some similarity with the idea that we die to make place for our children, with antagonistic pleiotropy [25], and with Ito's self-organization of a minimum reproduction age [56,36]. Neither a bit-string of mutations nor a set of real numbers for survival probabilities at various ages is used; only a minimum reproduction

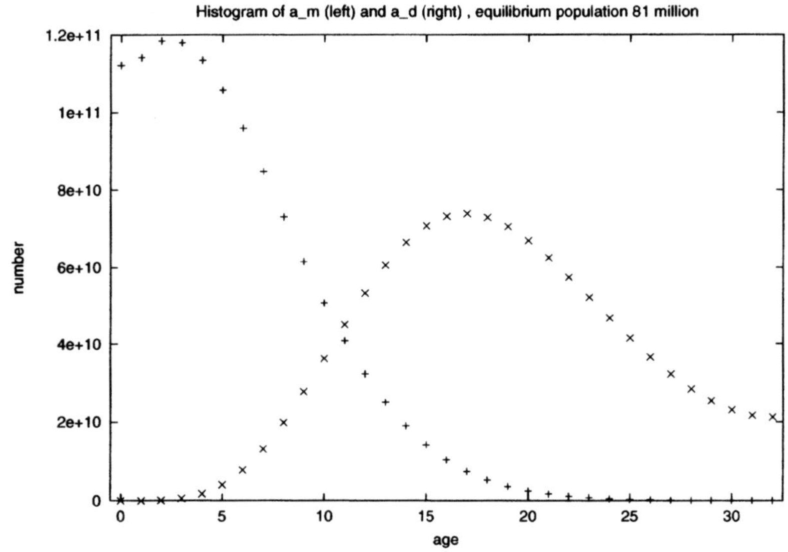

Fig. 4. Distribution of minimum age of reproduction (+) and of genetic death age (x) after a stationary state has been reached for the model in the appendix.

age a_m and a maximal genetic lifespan a_d is transmitted, with mutations, to the next generation by asexual reproduction.

Our individuals i ($i = 1, 2, \dots N$) live at most $a_d(i)$ time intervals and at each interval they give with probability b birth to one offspring, provided the parental age $a(i)$ is not below the minimum reproduction age $a_m(i)$. Due to space and food limitations, at each time interval every survivor dies with the Verhulst probability N/N_{max} given by the "carrying capacity" N_{max}. The values for a_m and a_d for the offspring are mutated, randomly by ± 1 for each child separately, away from the paternal values, within the constraints $0 \le a_m < a_d \le 32$. The birth rate b is roughly $1/(a_d(i) - a_m(i))$; more precisely, we add as a crucial free parameter a small constant $\epsilon = 0.08$ to avoid a divergence:

$$b = (1+\epsilon)/(a_d(i) - a_m(i) + \epsilon) \quad .$$

Thus late births and early genetic deaths increase the birth rate.

We start with $a_m = 1$, $a_d = 16$ for all $N_{max}/300 = 50$ million individuals and check for self-organization of a_m and a_d away from these initial values. Indeed, Fig. 4 shows asymmetric distributions for a_m and a_d with peaks away from 0 and 32, and averages $\langle a_m \rangle = 5.6$, $\langle a_d \rangle = 18.6$, from 10,000 to 20,000 iterations. The age distribution corresponds to a mortality increasing with age as required for senescence.

While this model shows self-organization of senescence and a typical age of death, it has not yet been applied to the numerous special situations in which the Penna model agreed with reality. A first test for catastrophic senescence of salmon looks promising [57]. The program death.f is available from stauffer@thp.uni-koeln.de

References

1. S. Moss de Oliveira, P.M.C. de Oliveira, D. Stauffer: *Evolution, Money, War and Computers*, Teubner, Stuttgart and Leipzig 1999.
2. M.Ya. Azbel: Phys. Repts. 288, 245 (1997); Proc. Natl. Acad. Sci. USA 96, 3303 and 15368 (1999); Physica A 269, 564 (1999) and preprint.
3. J.R. Wilmoth, L.J. Deegan, H.Lundström, S. Horiuchi: Science 289, 2366 (2000); J.R. Wilmoth, S. Horiuchi: Demography 36, 475 (1999); J.R. Wilmoth, H. Lindström: Eur. J. Population 12, 63 (1996). See also S.J. Olshansky, B.A. Carnes and A. Désesquelles: Science 291, 1491 (2001) versus S. Tuljapurkar, N. Li, C. Boe: Nature 405, 789 (2000).
4. D. Stauffer, P.M.C. de Oliveira, S. Moss de Oliveira, T.J.P. Penna, J.S. Sá Martins: An. Acad. Bras. Ci. 73, 15 (2001); see also D. Stauffer, J.S. Sá Martins, S. Moss de Oliveira: Int. J. Mod. Phys. C 11, 1305 (2000) and J.S. Sá Martins, D. Stauffer: Physica A, in press (2001).
5. K. W. Wachter, C. E. Finch, *Between Zeus and the Salmon. The Biodemography of Longevity*, National Academy Press, Washington DC 1997; La Recherche 322 (various authors), July/August 1999; Nature 408, No. 680 (various authors), November 9, 2000.
6. P.B. Medawar: *An Unsolved Problem of Biology*. H.K. Lewis, London 1952.
7. L. Hayflick: Exp. Gerontology 33, 639 (1998).
8. M. Wagner, P. Jansen-Dürr: Exp. Gerontology 35, 729 (2000).
9. J.W. Shay, W.E. Wright: Science 291, 839 (2001); M. Aragona et al., Int. J. Oncol. 17, 981 (2000).
10. H. Denis: La Recherche 322, 48 (1999)
11. C.A. Wolkow et al.: Science 290, 147 (2000).
12. M. Michikawa et al.: Science 286, 774 (1999); D.L. Ly et al.: Science 287, 2486 (1999); J. Vijg: Mutation Res. 447, 117 (2000).
13. C. Delcourt: La Recherche 322, 62 (1999).
14. M. Ya. Azbel: Proc. Natl. Acad. Sci. USA 91, 12453 (1994); but see W.A. Van Voorhies: Exp. Gerontology 36, 55 (2001).
15. J. Wanagel, D.B. Allison, R. Weindruch: Toxicological Sci. 52, Suppl., 35 (1999); see also D. Stauffer and T. Klotz: The Aging Male, in press; S. Rogina et al.: Science 290, 2137 (2000).
16. S.M. Jazwinski: Acta Biochim. Pol. 47, 269 (2000).
17. H.A. Tissenbaum and L. Guerente: Nature 410, 227 (2001).
18. E. Miglaccio et al.: Nature 402, 309 (1999).
19. M.R. Rose: Sci. Amer. 281, Dec. 1999, page 68 and Exp. Gerontology 34, 577 (1999).
20. T.J.P. Penna: J. Stat. Phys. 78, 1629 (1995). For reviews see A.T. Bernardes: page 359 in *Annual Reviews of Computational Physics*, vol. IV, World Scientific, Singapore 1996; T.J.P. Penna, M. Argollo de Menezes, A. Racco: Computer Physics Communications 122, 108 (1999); S. Moss de Oliveira, D. Alves, J.S. Sá Martins: Physica A 285, 77 (2000).

21. L. Partridge, N.H. Barton: Nature 362, 305 (1993); L.D. Mueller, M.R.Rose: Proc. Natl. Acad. Sci. USA 93, 15249 (1996); S.D. Pletcher, J.W. Curtsinger: Evolution 52, 454 (1998); S. Dasgupta: J. Physique (France) I 4, 1563 (1994); N.G.F. de Medeiros, R.N. Onody: "The Heumann-Hötzel Model revisited", preprint; Y. Cui, R.S. Chen, W.H. Wong: Proc. Natl. Acad. Sci. 97, 3330 (2000); D.P. Shanley, T.B.L. Kirkwood: Evolution 54, 740 (2000).
22. A.Z. Maksymowicz et al: Physica A 273, 150 (1999).
23. R.C. Desai, E. James, E. Lui: Theory in Biosciences 118, 98 (1999); M. Magdoń-Maksymowicz, A.Z. Maksymowicz, K.Kułakowski: Theory in Biosciences 119, 139 (2000).
24. J.S. Sá Martins and S. Cebrat: Theory in Biosciences 119, 156 (2000); see also J. Dąbowski, M. Groth, D. Makowiec: Acta Phys. Pol. B 31, 1027 (2000).
25. A.O. Sousa, S. Moss de Oliveira, Physica A, 294, 431 (2001); G. Medeiros, M.A. Idiart, and R. M. C. de Almeida, Int. J. Mod. Phys. C 11, 1283 (2000).
26. M. Eigen, J. McCaskill, P.Schuster: Adv. Chem. PHys. 75, 149 (1990).
27. S. Moss de Oliveira, P.M.C. de Oliveira, D. Stauffer: Braz. J. Phys. 26, 626 (1996).
28. L. Demetrius: J. Theor. Biol. 206, 1 (2000).
29. E. Niewczas, A. Kurdziel, S. Cebrat: Int. J. Mod. Phys. C 11, 775 (2000).
30. S. Moss de Oliveira, T.J.P. Penna, D. Stauffer: Physica A 215, 298 (1995).
31. T.J.P. Penna, S. Moss de Oliveira, D. Stauffer: Phys. Rev. E 52, 3309 (1995); see also remark by S. Tuljapurkar on page 70 in Wachter and Finch [5].
32. S.D. Pletcher, C. Neuhauser: Int. J. Mod. Phys. C 11, 525 (2000); Z.F. Huang, D. Stauffer: Theory in Biosciences 120, 20 (2001).
33. S.B. Hedges: Nature 408, 653 (2000).
34. G.J. Adcock et al.: Proc. Natl. Acad. Sci. USA 98, 537 (2001).
35. D. Makowiec, M. Groth, J. Dąbowski: Physica A 273, 169 (1999).
36. S. Moss de Oliveira, A.T. Bernardes, J.S. Sá Martins: Eur. Phys. J. B 7, 501 (1999).
37. H. Fisher, Sci. Amer. 281, Dec. 1999, page 98; similarly E.E. Baulieu: La Recherche 322, 72 (1999).
38. T.J.P. Penna, A. Racco, A.O. Sousa: talk IT-11 at FACS 2000, Maceió, Brazil, Oct. 2000. For trees see M. Argollo de Menezes, A. Racco, T.J.P. Penna, Physica A 233, 221 (1996).
39. S.J. Feingold: Physica A 231, 499 (1996) and Int. J. Mod. Phys. C 9, 295 (1998); S. Cebrat, J. Kąkol: Int. J. Mod. Phys. C 8, 417 (1997); D. Makowiec, Physica A 245, 99 (1997). For hunting see also A.Z. Maksymowicz: Comp. Phys. Comm. 122, 113 (1999); A.K. Altevolmer: Int. J. Mod. Phys. C 10, 717 (1999).
40. A.O. Sousa, S. Moss de Oliveira: Eur. Phys. J. B 9, 365 (1999); D. Makowiec: Physica A 289, 208 (2000); see also M.S. Magdoń, A.Z. Maksymowicz: Physica A 273, 182 (1999).
41. C. Chisholm, N. Jan, P. Gibbs, A. Erzan: Int. J. Mod. Phys. C 11, 1257 (2000).
42. P.M.C. de Oliveira, Theory in Biosciences 120, 1 (2001)
43. J.W. Vaupel et al.: Science 280, 855 (1998).
44. D. Stauffer: page 329 in *Annual Reviews of Computational Physics* , vol. VIII, World Scientific, Singapore 2001.
45. A. R. Thatcher, V. Kannisto and J. W. Vaupel, *The Force of Mortality at Ages 80 to 120*: Odense University Press, Odense 1998; A. R. Thatcher: J. Roy. Statist. Soc. A 162, 5 (1999) and priv.comm.
46. T.T. Perls, Sci. Amer. 272, Jan. 1995, page 50.
47. C. Finch, M.C. Pike: J. Gerontology A 51, B 183 (1996).
48. S. Horiuchi, J.R. Wilmoth: Demography 35, 391 (1998).

49. D. Stauffer: Int. J. Mod. Phys. C 10, 1363 (1999).
50. J. Thoms, P. Donahue, N. Jan: J. Physique I 5, 935 (1995).
51. P.M.C. de Oliveira, S. Moss de Oliveira, A.T. Bernardes, D. Stauffer: Lancet 352, 911 (1998).
52. P.M.C. de Oliveira, S. Moss de Oliveira, D. Stauffer, S. Cebrat: Physica A 273,145 (1999); E. Niewczas, S. Cebrat, D. Stauffer: Theory in Biosciences 119, 122 (2000).
53. K. Malarz: Int. J. Mod. Phys. C 11, 309 (2000); R.M.C. de Almeida and G.L. Thomas: Int. J. Mod. Phys. C 11, 1209 (2000); see also [29].
54. Editorial: Clinical Genetics 57, 13 (2000).
55. G.J. Troup, D.R. Hutton, D.G. Hewitt, C.R. Hunter: Free Radical Res. 20, 63 (1994); P. L. da Luz and S.R. Coimbra: An. Acad. Bras. Ci. 73, 51 (2001).
56. N. Ito: Int. J. Mod. Phys. C 7, 107 (1996).
57. H. Meyer-Ortmanns, Int. J. Mod. Phys. C 12, No. 3 (2001)

Spatio-temporal modes of speciation

Martin Rost and Michael Lässig

Institut für Theoretische Physik, Universität zu Köln,
50937 Köln, Germany

Abstract. The split of a population into two reproductively isolated subpopulations is studied within a model including spatial heterogeneity. We find three dynamical pathways of speciation resulting from a coupling of space, competition and mating behaviour: (i) sympatric at small habitat heterogeneity, (ii) sympatric with subsequent spatial differentiation at intermediate heterogeneity, and (iii) allopatric under strong heterogeneity.

1 Introduction

Speciation is the splitting of a species into two new ones. A population of individuals forms a species if they produce viable offspring among them. Essential steps at the beginning of speciation are differentiation in body characteristics and the formation of two reproductively isolated subpopulations. The dynamics of this division, which is the subject of this paper, can be rather fast. Eventually, if interbreeding of the two subpopulations is cut for long enough, the ability to mate successfully will be lost and the division gets irreversible. Speciation has been a long standing issue in evolutionary biology [1–8]. As we will show in this article, it is also a challenging problem of non-equilibrium statistical physics.

An obvious way of species division is *allopatric speciation* [2]. In this case a population gets separated by an external cause, e.g., a previously connected part of the sea may be divided after the formation of a natural barrier, and inhabitant organisms will be isolated from their relatives on the other side. Different characteristic forms may evolve independently and eventually the two groups will form two different species which cannot produce viable common offspring even if the barrier gets removed again.

In *sympatric speciation*, on the other hand, the two dividing groups continuously share a common habitat. Also in sympatry differentiation into subpopulations with different characteristics can be favoured, a process called *disruptive selection.* [he subpopulations fill different *ecological niches*, e.g., they feed from different resources. Niche populations can be stable if their gain in fitness through reduced competition outweighs the loss in fitness through specialisation [4,6,8].

An obstacle to sympatric speciation lies in sexual reproduction [2]. By the rules of genetics, mating between the two opposite subpopulations always produces offspring of intermediate phenotype and prevents their drifting apart. For a long time therefore speciation has been believed to take place predominantly in allopatry [2]. If speciation is to happen in sympatry, i.e., with both subpopulations sharing a common habitat, some mechanism has to prevent interbreeding.

Theoretical studies, e.g., [3,4], have addressed the possibility of and conditions for sympatric speciation, and tried to elaborate possible mechanisms to prevent interbreeding. One of the proposed mechanisms is assortative mating, where e.g. a female's mate choice is determined by a male's trait, how close or different it is from her own or her preferred one. Mating preference may also depend on a different male trait and both may mutually enhance their evolution towards extreme characteristics [5].

In more recent years, reconstruction of phylogenetic trees from molecular data obtained in new field studies have dramatically changed our understanding of the processes and brought up striking empirical evidence for sympatric speciation. The most prominent example is cichlid fishes in the great African lakes (Victoria, Malawi), where several hundred sister species of monophyletic origin, i.e., all descending from a single colonising species, have been found to coexist [9]. Even more surprisingly this evolution must have taken place in a surprisingly short time: Geological data from the bottom of Lake Victoria indicate that it was completely dry about 12 000 years ago [10]. All species inhabiting the lake today must have colonised it *after* that desiccation and subsequently have speciated. It seems unlikely that with different low water levels the lake would have been divided in many small lakes to give rise to the opportunity of allopatric speciation for so many species in such a short time.

Moreover, these observations also show spatial structure in the fish populations despite the absence of barriers and their sympatric origin. They prefer different parts of the lake due to different food supply and use. Some feed from organisms on the ground, others from floating particles throughout the lake [9]. Encounters between such different populations may well be reduced by this spatial separation. This has an important consequence on the picture of sympatric speciation: As the subpopulations start to differ in phenotypic characteristic they gain the possibility of a secondary spatial differentiation, also called patching.

Spatial separation plays an even larger role in organisms which are not constraint to a relatively small and spatially well mixed habitat such as a lake. It is more pronounced when, e.g., the splitting populations specialise as parasites on different host plants, or when the interaction range of individuals is small compared to the area inhabited by the entire species. This is e.g. the case for *salamadra* taxa in Europe who seem to consist of different clades whose colonisation steps after the last ice age can be tracked down [11].

Speciation is more than a single event, and only the *history* of the entire process yields an adequate understanding of spatially structured populations. Different evolutionary pathways describe primary spatial (and hence allopatric) division compared to secondary patching with originally sympatric differentiation.

Present models for sympatric speciation neglect space [6–8], in terms of statistical physics they are "mean field" models. In order to fully understand the history of a speciation process one needs more refined categories than merely the alternative of sym- and allopatry. In this article we combine the "internal" trait space of the populations with the "external" topographical space, i.e., to go from a mean field model to a spatio-temporal field theory. The resulting modes of speciation show indeed an intricate interplay of internal and external space.

This article is organised as follows: In the next Section we present the general setup of the spatio-temporal model. Characteristic *states* of the system and their biological interpretations are presented in Section 3. Section 4 then shows the possible *evolutionary pathways* and the transitions between states occurring along these pathways. A brief discussion ends the article.

2 Evolution of competing phenotypes

General form. The model in its most general form deals with *competitive* interaction of different groups of a population extending over a whole spectrum of phenotypes and living in a structured landscape. The dynamical variable is a density $N(x;\mathbf{r},t)$, depending on time t and the ("external") spatial coordinate $\mathbf{r}$, as well as some suitable variable x denoting a character in the ("internal") trait space. In the example of fish this is typically the body size [9], but one can imagine any other variable in which the split into subpopulations and finally speciation becomes first visible. A "minimal" model equation for speciation can be written in the form

$$\begin{aligned}\partial_t N(x;\mathbf{r},t) \;=\;& \nu\nabla_{\mathbf{r}}^2 N(x;\mathbf{r},t) - \nabla_{\mathbf{r}}\cdot(\lambda\mathbf{v}(x;\mathbf{r},t)N(x;\mathbf{r},t)) + \\ & f(x;\mathbf{r})R(x;\mathbf{r},t) - \left[\alpha(\mathbf{r}) + \int dy\;\beta(x-y)N(y;\mathbf{r},t)\right]N(x;\mathbf{r},t).\end{aligned} \tag{1}$$

It has the structure of a nonlinear reaction-diffusion equation and contains the following features:

- *Movement in space* is due to diffusion (first term on the right hand side) and to deterministic drift (second term), here expressed by a "velocity" field $\mathbf{v}$ which could, e.g., be due to a gradient in habitat quality.
- R is the *birth rate* of offspring. In a purely asexual model it would be proportional to N itself but here it reflects the genetic inheritance patterns, as will be specified below. R is multiplied by $f(x;\mathbf{r})$, which denotes the *habitat quality* at point $\mathbf{r}$ and plays the role of a fitness.
- The *competition* is quadratic, a simple general form, familiar from Lotka-Volterra equations, proportional to the number of encounters per time unit between two individuals. Some function $\beta(x-y)$ denotes the phenotype dependence of its strength. β should be maximal at $x=y$ and decay monotonically as $|x-y|$ increases.
- An overall *death rate* $\alpha(\mathbf{r})$ is used to tune global, e.g., climatic, changes in living conditions.

Equation (1) gives a rather general framework for spatio-temporal population dynamics. The results depend, in particular, on the spatial distribution of fitness values $f(x;\mathbf{r})$. Here we discuss the simplest topography relevant for speciation.

Two-habitat topography. Topographical space is assumed to consist of two homogeneous habitats, A and B. In our particular model choice a newborn individual with phenotype x has fitnesses

$$f^{\mathrm{A}}(x) = f^{\mathrm{B}}(-x) = \exp\left(-(x - x_0)^2/x_f^2\right) \tag{2}$$

in habitats A and B respectively, which is schematically represented in Figure 1 a. Habitat A favours positive values of x, B negative ones. The resource qualities in both habitats decay on a scale x_f and have relative difference $2x_0$. For simplicity

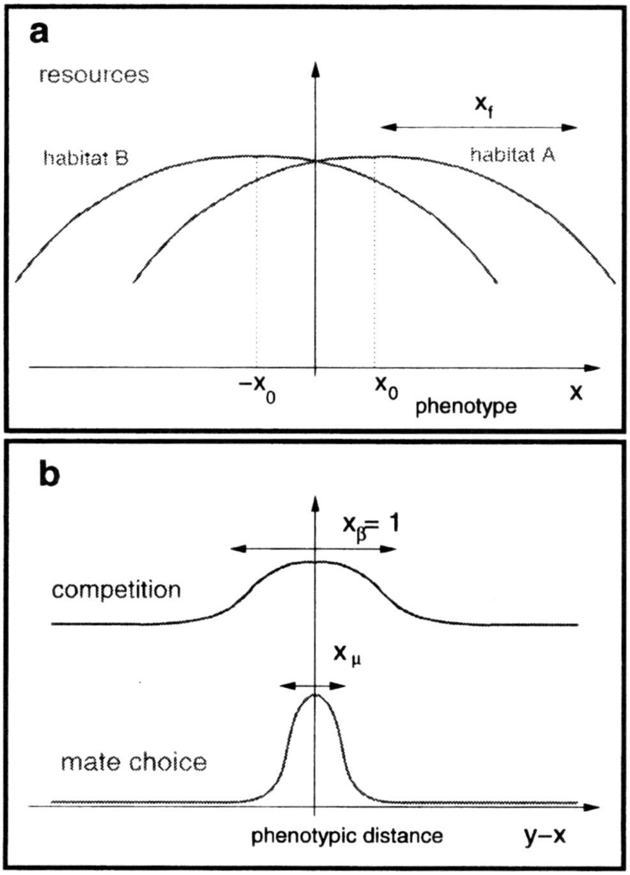

Fig. 1. Schematic representation of habitat quality (panel **a**) and phenotypic interactions (panel **b**). Habitat A favours phenotypes with larger values of x (maximal at x_0), habitat B offers maximal resource quality to phenotype $-x_0$. The resource curves decay on a lateral scale of x_f. Competition and mate preference are maximal between equal phenotypes and decay on scales of order $x_\beta = 1$ (top curve) and x_μ respectively (bottom curve).

we assume symmetry between the sign of x and interchange of A and B. Without loss of generality we take a Gaussian decay for the habitat quality. All properties of environmental quality are then expressed by these two parameters.

It is useful to define an order parameter $\rho \in [0, 1]$ describing the tendency of the population to settle into the more favourable habitats,

$$\rho = \frac{\int_{x>0} \left(N^{\mathrm{A}}(x) - N^{\mathrm{B}}(x)\right)}{\int_{x>0} \left(N^{\mathrm{A}}(x) + N^{\mathrm{B}}(x)\right)}, \tag{3}$$

where $N^{\mathrm{A}}(x)$ denotes the number density of individuals with trait x in habitat A, and $N^{\mathrm{B}}(x)$ the same for B. Without any habitat preference, both are equally populated and $\rho = 0$. Populations fully retreated into one habitat, i.e., those with positive x to A and negative x to B, give $\rho = 1$.

The distribution of subpopulations, and hence the value of ρ, follow from local fitness differences and the population flux from one habitat to another. These quantities are described by the first two term of Equation (1). The diffusion constant ν and the transport coefficient λ are subject to evolutionary changes. As a consequence also ρ changes under evolution. It turns out to be a simpler approach to take ρ itself as the primary evolving variable. This will be specified below in detail, once the other model components have been introduced.

Competition. Competitive interaction decays with phenotypic distance, see the top curve in Figure 1 b,

$$\beta(x-y) = \exp\left(-(x-y)^2/x_\beta^2\right). \tag{4}$$

Again the Gaussian shape is a particular choice without any qualitative difference to others. Moreover, we use competition to set the scale in phenotype space by choosing $x_\beta = 1$. This gives x_f in Equation (2) a simple interpretation as an estimate for the number of subpopulations that can coexist, or in other words for the number of possible ecological niches.

Reproduction rate. An expression for the reproduction rates R is straightforward to construct. As limiting factor we take the breeding capacity of the females, which should be a reasonable assumption in many cases. The number density of individuals of type x born per time unit is then given by

$$R(x) = \int_{y,z} C(x; y, z)\, m(z; y)\, N(y). \tag{5}$$

One has to sum up all possible couples of parents. $N(y)$ denotes the number of mothers, z is the phenotype of the males, and $m(z; y)$ the probability for a y-female to mate a z-male.

Inheritance and quantitative genetics. $C(x; y, z)$ is a genetic tensor giving the probability that a couple of parents y and z will have x-offspring. This needs the normalisation $\int_x C(x; y, z) = 1$. It reflects the underlying genetic representation of the trait with respect to which we study the possibility of speciation. Different explicit forms for C are possible, but as x is a quantitative trait depending on many loci on the genome, we can make use of the principles of quantitative genetics [7,12]. The distribution $C(x; y, z)$ of offspring trait values has a mean close to $(y + z)/2$ and a variance that depends on the number of loci involved.

Mate choice. Of course $m(z; y)$ depends on population sizes, the frequency of males to choose from, as well as the absolute number. In too sparse a population females may lack suitable mating partners or be forced to choose against their preference, which may cause interesting effects [13,14]. In this work we focus on what sometimes is called *saturated mating*, where each female finds a partner, so $\int_z m(z; y) = 1$ for all y and consequently $\int_x R(x) = \int_x N(x)$. Other than on the number of available males, $m(z; y)$ depends on the preference of females for certain types of males, which are expressed in terms of preference factors $\mu(z; y)$. Similar to competition, in our model female attraction towards a given male depends on their mutual phenotypic distance,

$$\mu(z; y) = \exp\left(-(z - y)^2/x_\mu^2\right), \tag{6}$$

relative to a distance x_μ which we define as the *mating range*. An example with relatively narrow mating range is shown in the bottom curve of Figure 1 b. The actual mating probabilities then are given by the numbers of available males weighted by the female preference factors,

$$m(z; y) = \frac{\mu(z; y)N(z)}{\int_{z'} \mu(z'; y)N(z')}. \tag{7}$$

Here it becomes clear how the mating range x_μ interpolates between indifferent or "random" mating with $x_\mu = \infty$ and strong mating preference or assortativity as $x_\mu \to 0$.

Discrete phenotypes. As one more simplifying step trait space is discretised into a finite number of "bins". Three bins, where the population at a point is described by the numbers N_1, N_2, and N_3, are needed in the simplest case to tell the difference between a population split into two independent subpopulations (when $N_2 = 0$) and a contiguous population profile extending over the whole range (when $N_2 > 0$).

The habitat preference order parameter ρ defined in Equation (3) gains a simple meaning: $(1 + \rho)/2$ is the fraction of 1-individuals in habitat A, and of 3 in B. Diffusion and migration between the habitats, still explicitly present in Equation (1) are fully included into the spatial order parameter ρ in this simple discrete case.

There is a simple way to mimic a continuous trait space even when only three bins explicitly appear in the model. Phenotype 2 is located in the middle at $x = 0$, but the positions of 1 and 3, $\pm x$, are adapted to an optimal value of phenotypic width. At a given value of x we ask, whether populations at $x \pm dx$ can invade and suppress the previous ones at x. If so, x is replaced by the new value, until a final stable value is reached.

Evolutionary adaptations and pathways. So far we have explained the *population dynamics* of competing phenotypes. For fixed external ecological parameters (resource quality and competition) and for fixed internal parameters (mating and habitat preference) the population dynamical equations lead to some stable fixed point describing a population dynamical equilibrium. This can be seen as a resident population subject to the appearance of mutants with different strategies in mating and habitat preference. Generally mutants will be able to invade and push out the previous residents until an evolutionary stable state is reached [6–8].

We assume that the rate at which mutants with new characteristics appear is much smaller than the relaxation rate of population dynamics. The same should be true for the rate of global changes in the environment. In Section 4 we present a slow decrease in the external death rate α from 1 to 0, in order to model a slow increase in habitat quality, e.g., as a result of climatic changes. The slowest changing variable parametrises time in the model.

For instance, α may change adiabatically and approach a transition point, where a small change in α causes a large jump in x_μ, an example of which is given in Section 4. During this change x_μ is the slowest variable and here a natural way to measure "time" is by the rate of mutations in x_μ. These evolutionary pathways define the history of evolutionary adaptations.

For a full description one still needs the initial conditions. It makes sense to assume *no* habitat and mating preference, $\rho = 0$ and $x_\mu = \infty$, before any diversification in phenotypic space appears, and to see if spatial structure and mating preference can evolve.

3 States of the system

In this Section we characterise the various states of stationary populations in the model and their meaning with respect to speciation.

If the range of resources is narrow, only the middle population is viable, $N_2 > 0$, but not the outer ones and therefore $N_1 = N_3 = 0$. Then we also find cases of coexistence where all three blocks are populated. The population profile may be structured and show some tendency of splitting into two independent blocks. However, there is gene flow across the population. Besides direct mating between the two opposite outer populations there will be indirect gene flow: If individuals of phenotype 2 mate with both 1 and 3, none of them is isolated from its counterpart. Differentiations can take place in three ways:

- **Spatial separation.** If the populations are fully retreated to their respective more favourable habitats, if 1 lives only in A and 3 in B (in the model parameters this is the case $\rho = 1$), individuals of opposite phenotype just don't meet each other and therefore don't mate. Gene flow then is suppressed by spatial separation.
- **Complete mating assortativity.** Under fully developed mating preference (or assortative mating), $0 \simeq x_\mu \ll x$, there will be no cross mating between 1 and 3 and no offspring of intermediate phenotype from such matings.
- **Trait differentiation.** Absence of intermediate traits separates the population profile. In the discrete case: $N_2 = 0$ and $N_1 = N_3 > 0$. Trait differentiation necessarily needs one of the first two types of separation. The inheritance rules give a parent couple of 1 and 3 offspring of type 2 (which is at the arithmetic mean). Absence of the middle phenotype therefore needs absence of mating between 1 and 3.

This shows that the types of differentiation are not independent. In the evolutionary model dynamics spatial separation or assortativity sets in as the *primary* separation. Both cases induce trait differentiation. Spatial separation may also come in as *secondary* differentiation. This case is of particular interest since trait differentiation is stabilised by a *double boundary* interrupting gene flow, spatially and by mate choice.

4 Evolutionary pathways

Let us now turn to the transitions between these states and present typical sequences of states in an evolutionary context. We focus on a setup inspired by the already mentioned field studies of recolonisation of Europe after the last glacial period [11]. Initially the colonising population finds very poor living conditions, in the model we have the external death rate $\alpha \simeq 1$. Under such conditions only the central population at $x = 0$ is viable, the system is in the state $N_1 = N_3 = 0$ described first in the previous Section. Now we model slowly warming climatic conditions with increasing habitat quality; α is gradually lowered and in wider regions of phenotype space populations become viable.

When this region has opened up wide enough, two neighbouring populations will become viable and make better use of the resources than a single one in the middle. This gives selective advantage to any properties in the population enhancing its outer parts. In the previous Section we have seen the different states of the system in which this is achieved. They are reached through evolutionary pathways during which the parameters for mating and habitat preference are adapted. We find three different kinds, which are also illustrated in Figure 2.

- The **sympatric mode** of speciation, shown in Figure 2 a, at small or no habitat heterogeneity. Mating preference slowly develops but remains only partial until at a critical value of habitat quality full assortativity develops with $x_\mu \to 0$. At this point, mating between opposite phenotypes is suppressed and so is its offspring as a source for the population of middle phenotype. On its own this phenotype is not viable, the competition load from

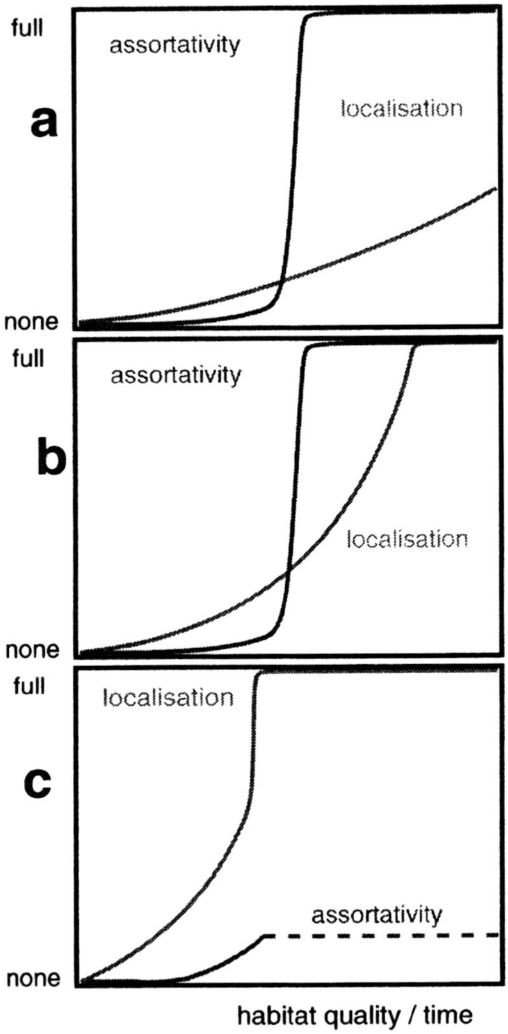

Fig. 2. Three examples of evolutionary pathways, panel **a** and **b** show two varieties of the *sympatric mode* of speciation, panel **c** the *allopatric mode*. **a** has lowest habitat heterogeneity (smallest value of x_0) and **c** highest. The slowest time scale with the gradual increase in habitat quality is shown on the horizontal axis. In panels **a** and **b** mating preference switches rapidly from partial to full assortativity when the habitat quality is good enough. At this point the middle phenotypes become suppressed, $N_2 \to 0$. In panel **b** full localisation develops after assortativity. Now the two subpopulations are *doubly isolated*, spatially and by mating preference. In panel **c** localisation develops more rapidly than assortativity and causes $N_2 \to 0$.

the two outer ones is too large; so it disappears together with x_μ. The entire population is now in a state where 1 and 3 coexist but are reproductively isolated from each other.

- The **sympatric mode with patching**, Figure 2 b, at moderate heterogeneity, where after establishing assortativity there is another transition in the model. If x_0 is large enough, the system will reach a point where $\rho = 1$ and both 1 and 3 have fully retreated into their respective more favourable habitats. Now there is a double boundary against mating and gene flow between the subpopulations, a sexual and a spatial one.
- The third type of pathway at strong habitat heterogeneity, the **allopatric mode** in Figure 2 c, is different. Here the tendency towards localisation is so strong that the two subpopulations become spatially isolated before assortativity can develop. Patching suppresses cross mating and causes the central phenotype to go extinct. Now mating preference becomes meaningless as different phenotypes do not share any pieces of habitat and x_μ remains at the value of partial assortativity reached last.

5 Discussion

The main goal of this work is to study the interplay of trait and real space, the internal and external degrees of freedom in a population at the onset of speciation. The model constructed for this purpose shows a complex interaction of its degrees of freedom despite its simplicity. *The pathways of speciation are the combined result of habitat topography, genetics, and ecology.* The traditional alternative of sympatry and allopatry is obsolete.

The evolutionary history is crucial for interpreting observed states of the system. Spatial separation can occur as primary differentiation in the allopatric mode or as a secondary step, after trait differentiation and mating assortativity have already evolved in sympatry.

The effect of spatial heterogeneity on (the possibility of) sympatric speciation is twofold: On one hand it may prevent the evolution of mating preference, as emerging extreme phenotypes retreat into their respective favourite habitat patches before differentiation in sympatry together with mating assortativity has fully developed. On the other hand, if they retreat *after* differentiation, patching forms an additional boundary against gene flow. *Sympatric speciation with subsequent patching is the most efficient way to cut the gene flow between two subpopulations.*

Clearly, the model can be generalised in several ways. For example a more detailed spatial model, closer to Equation (1), can be used to study the profile of the phenotypical population structure at the interface between habitats and the gene flow across the boundary. Moreover, mating can and will be unsaturated in certain cases, in contrast to the definition of Equation (7). If not all females are able to find a desired mating partner, in particular at small population sizes, the transitions between the different states of the system and hence the evolutionary pathways will be modified [13,14].

Acknowledgements

It is a pleasure to thank Alex Kondrashov, Arne Nolte, Sebastian Steinfartz, and Diethard Tautz for sharing their insight. In particular, Diethard Tautz has pointed out to us the ubiquity of spatial structure in speciating populations.

References

1. C. Darwin, *The Origin of Species*, John Murray, London (1859).
2. E. Mayr, *Animal species and evolution*, Belknap Press, Cambridge (1963).
3. J. Maynard Smith, American Naturalist **100**, 637–650 (1966).
4. M.L. Rosenzweig, Biological Journal of the Linnean Society (London) **10**, 275–289 (1978).
5. R. Lande, Proceedings of the National Academy of Sciences of the USA **78**, 3721–3725 (1981).
6. S.A.H. Geritz, É. Kisdi, G. Meszéna, and J.A.J. Metz, Evolutionary Ecology **12**, 35–57 (1998).
7. A.S. Kondrashov and F.A. Kondrashov, Nature **400**, 354–357 (1999).
8. U. Dieckmann and M. Doebeli, Nature **400**, 351–354 (1999). Clarendon Press, Oxford (1980).
9. A. Meyer, T.D. Kocher, P. Basasibwaki, and A.C. Wilson, Nature **347**, 550–553 (1990).
10. T.C. Johnson, C.A. Scholz, M.R. Talbot, K. Kelts, R.D. Ricketts, G. Ngobi, K. Beuning, I. Ssemmanda, and J.W. McGill, Science **273**, 1091–1093 (1996).
11. S. Steinfartz, M. Veith, and D. Tautz, Molecular Ecology **9**, 397–410 (2000).
12. M.G. Bulmer, *The mathematical theory of quantitative genetics*, Clarendon Press, Oxford (1980).
13. M. Rost and M. Lässig, in preparation.
14. A. Atik, Diploma Thesis, University of Cologne (2002).

Part IV

Large-scale evolution

Food web structure and the evolution of ecological communities

Christopher Quince[1], Paul G. Higgs[2], and Alan J. McKane[1]

[1] Department of Theoretical Physics and
[2] School of Biological Science,
University of Manchester, Manchester M13 9PL, UK

Abstract. Simulations of the coevolution of many interacting species are performed using the Webworld model. The model has a realistic set of predator–prey equations that describe the population dynamics of the species for any structure of the food web. The equations account for competition between species for the same resources, and for the diet choice of predators between alternative prey according to an evolutionarily stable strategy. The set of species present undergoes long-term evolution due to speciation and extinction events. We summarize results obtained on the macro-evolutionary dynamics of speciations and extinctions, and on the statistical properties of the food webs that are generated by the model. Simulations begin from small numbers of species and build up to larger webs with relatively constant species number on average. The rate of origination and extinction of species are relatively high, but remain roughly balanced throughout the simulations. When a 'parent' species undergoes speciation, the 'child' species usually adds to the same trophic level as the parent. The chance of the child species surviving is significantly higher if the parent is on the second or third trophic level than if it is on the first level, most likely due to a wider choice of possible prey for species on higher levels. Addition of a new species sometimes causes extinction of existing species. The parent species has a high probability of extinction because it has strong competition with the new species. Non-parental competitors of the new species also have a significantly higher extinction probability than average, as do prey of the new species. Predators of the new species are less likely than average to become extinct.

1 Introduction

In this article we discuss the Webworld model which, by describing the interaction of species over a wide range of timescales, allows us to start from a very few species and evolve food webs which represent the predator–prey relationships amongst a large community of diverse species. A typical food web generated by the model is shown in Figure 1.

The web shown in the figure is typical in the sense that the webs which are evolved by Webworld resemble real food webs by having very few predator–prey links between species on the same level, and also very few cycles; most of the links, and certainly the major ones, start on one level and end on a higher one. A more quantitiative comparison between the model and ecological data confirms this broad similarity.

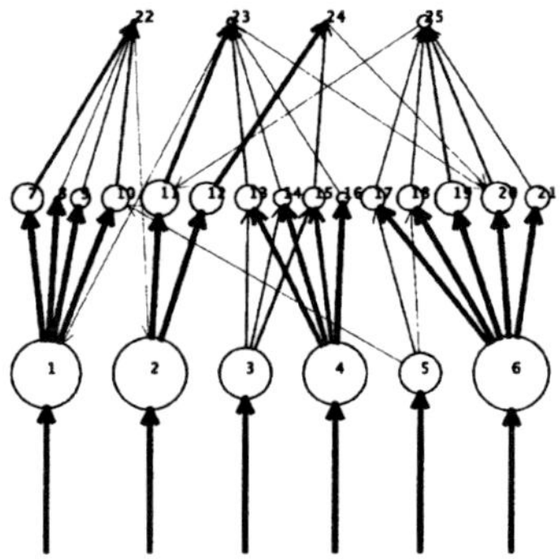

Fig. 1. Typical food web produced by the model

Webworld was designed as a model of an evolving community of many interacting species. It is intended to address questions of interest on both ecological and evolutionary timescales. Models of population dynamics accounting for predator–prey interactions between species work on the ecological scale. These date back to Lotka and Volterra and form a familiar part of the theoretical biology literature (Pielou, 1977; Roughgarden, 1979). Many studies have considered the dynamical behaviour and type of attractors that arise with systems of coupled equations representing a few (usually two or three) species (Emlen, 1984; Hallam, 1986; Hastings & Powell, 1991; McCann & Yodzis, 1994; Post *et al.* 2000). Some studies have addressed the problem of stability of these dynamical equations when many interacting species are present (May, 1974; Svirezhev & Logofet, 1983; Logofet, 1993; Hofbauer & Sigmund, 1998). When one considers many species, the structure of the food web that connects them becomes relevant. Which species prey on which other species? Which species compete with one another for the same resources? Recent studies of general food web structures including these features have been carried out by Bastolla *et al.* (2000) and Lässig *et al.* (2001).

There is also a body of literature in theoretical biology that considers the statistical properties of food webs (Pimm, 1982), both as observed in the field (e.g. Hall & Raffaelli, 1991; Goldwasser & Roughgarden, 1993; Martinez & Lawton, 1995) and as predicted by various types of random graph models (Cohen, 1990; Cohen *et al.* 1990). These studies consider how many trophic levels there are in food webs, how the species are distributed between the levels, and how many predators and prey are possessed by each species in the web. Patterns such as these are static rather than dynamic, but it is clear that the arrangement of the links in the food web will influence the population dynamics, and that the community structure we observe has been created by the dynamics of the system, rather than being simply thrown together. We therefore require theoretical models that can address both static and dynamic questions.

In addition to the relatively rapid ecological dynamics, it is important to consider long-term evolutionary dynamics. Speciation will create new species and extinction will remove some of the old ones. When a new species arises, this will have knock-on effects on other species in the community through predation and competition interactions. There is the potential for indirect effects to influence many other species. There are now a large number of evolutionary models, mostly in the physics literature, that investigate the dynamics of extinction events and the possibility of large scale avalanches of extinctions. Mass extinctions are an important feature observed in the fossil record. Biologists have tended to ask what causes these events – climate changes, meteorite strikes etc. An idea stemming from the theory of self-organized criticality is that the dynamics of the system itself may be inherently unstable and subject to large scale avalanches of extinctions (Bak & Sneppen, 1993; Solé *et al.* 1997). Theoretical models of macro-evolution and extinction have been reviewed recently by Drossel (2001). It is not surprising that changes in the non-living environment of an ecosystem can sometimes have catastrophic effects on the living species. The key question here is what would the macro-evolutionary dynamics be like in the absence of external changes. Does evolution lead to communities of stable interacting species that cannot be replaced by new ones? Does it create a continual turn-over of new species replacing old ones (a Red Queen scenario), or does it create critical food web structures prone to large scale fluctuations?

Webworld has been designed to fill a gap between the different types of models described above. Neither the models of food web structure or those of population dynamics consider evolution, whilst the macro-evolutionary models contain very little biological detail and often ignore population dynamics. The model we have arrived at involves lengthy computer simulations and is more complex than many of the other models referred to above. Whilst we recognize that simplicity is a virtue, we also feel that the level of detail included here has the payoff of allowing us to address a very wide range of biological questions. We also would like to resist the tendency to assume that effects observed in very simple models will always prove to be 'universal'. We will argue that it is important to have qualitatively correct population dynamics, and point out cases where omitting seemingly unimportant effects, leads to significant changes in the biological conclusions.

The Webworld model has already been described in detail in Drossel *et al.* (2001) and Caldarelli *et al.* (1998), therefore we only describe it briefly in the following section. This paper will try to summarize some of the effects observed in simulations with Webworld in our previous papers, and will then focus on the question of what happens when a new species is added to a food web. Diagrams will be shown of a few representative cases, and new statistical results will be given describing the response of the web to a single speciation event.

2 Population dynamics

Each species in the model is represented by a set of L features or phenotypic characters chosen from a set of K possible features. Typically $L = 10$ and $K = 500$ in the model, which means that the number of possible species is extremely large and evolution never runs out of scope for innovation. Each species has a "score" against any other species that is calculated as a function of the set of features possessed by each of the species. The score S_{ij} is positive if species i is adapted to prey on species j, and is zero if not. These scores are used as the numerical coefficients in the population dynamics equations discussed below.

Let the rate at which one individual of species i consumes individuals of species j be denoted by $g_{ij}(t)$. This is usually called the 'functional response', and it depends in general on the population sizes. We suppose that the population size N_i of each species satisfies an equation of the form:

$$\frac{dN_i(t)}{dt} = -N_i(t) + \lambda \sum_j N_i g_{ij}(t) - \sum_j N_j g_{ji}(t). \tag{1}$$

The first term on the right represents a constant rate of death of individuals in absence of interaction with other species. The final term is the sum of the rates of predation on species i by all other species, and the middle term is the rate of increase of species i due to predation on other species. Where there is no predator–prey relationship between the species the corresponding rate g_{ij} is zero. The factor λ is less than 1, and is known as the ecological efficiency. It represents the fraction of the resources of the prey that are converted into resources of the predator at each stage of the food chain. Throughout this paper, we have taken $\lambda = 0.1$, a value accepted by many ecologists (Pimm, 1982).

The external environment is treated as an additional 'species 0'. For primary producers, the middle term includes a non-zero rate g_{i0} of feeding on the external resources. External resources (e.g. sunlight) enter the ecosystem at a constant rate R. In the equations this is implemented by defining $N_0 = R/\lambda$, and keeping N_0 fixed. We have deliberately chosen the form of Eq. (1) to be the same for all species. We do not want to define different equations for primary producers, herbivores, and carnivores etc, because species can change their position in the ecosystem as it evolves, and most species are both predators and prey.

The most straightforward form for the functional response would be to have g_{ij} proportional to the prey population size N_j, as is the case in the Lotka–Volterra equations (Pielou, 1977; Roughgarden, 1979). A variety of other forms have been proposed that account for the fact that when prey are scarce or when many predators choose the same prey, competition between predators reduces the amount of prey available to each predator, whilst when prey are abundant, the consumption rate per predator must saturate rather than continue to increase indefinitely with the prey population size (Holling, 1959; Beddington, 1975; Huisman & De Boer, 1997). The form of the functional response used in the recent versions of Webworld is

$$g_{ij}(t) = \frac{S_{ij} f_{ij}(t) N_j(t)}{bN_j(t) + \sum_k \alpha_{ki} S_{kj} f_{kj}(t) N_k(t)}. \tag{2}$$

This is based on the ratio-dependent functional responses used by Arditi & Ginsburg (1989) and Arditi & Michalski (1995). We have described in detail in Drossel *et al.* (2001) how we generalized these studies to give equation (2). In the denominator, the sum runs over all the species that are predators of species j. The factor α_{ki} determines the strength of competition between species for the same resources. This depends on the degree of similarity between the species:

$$\alpha_{ij} = c + (1 - c)q_{ij}, \tag{3}$$

where c is a constant such that $0 \leq c < 1$, and where q_{ij} is the 'overlap', or fraction of features of species i that are also possessed by species j. This means that competition is strongest between similar species (or members of the same species), and is weaker for different species because they can use the resources in slightly different ways.

The factor f_{ij} is the fraction of its effort (or available searching time) that species i puts into preying on species j. These efforts must satisfy $\sum_j f_{ij} = 1$. We suppose that the efforts of any species i are chosen so that the gain per unit effort g_{ij}/f_{ij} is equal for all prey j. If this were not true, the predator could increase its energy intake by putting more effort into a prey with higher gain per unit effort. This choice of efforts leads to the condition

$$f_{ij}(t) = \frac{g_{ij}(t)}{\sum_k g_{ik}(t)}. \tag{4}$$

We showed (Drossel *et al.* 2001) that this choice is an evolutionarily stable strategy (ESS) (Parker & Maynard Smith, 1990; Reeve & Dugatkin, 1998). If the population has efforts chosen in this way, there is no other strategy with a different choice of efforts that can invade the population. Predator diet choice is often discussed using optimal foraging theory (Stephens and Krebs, 1986). The basic idea is that each predator maximises its individual rate of resource input. In our model, the success of the predators' strategies is a function of predator population size and the strategies of the other predators. The ESS solution is not equivalent to maximizing the resource input for a single predator. In most other models of optimal foraging, there is no competition between predators, hence there is no distinction between the ESS and the strategy with maximal resource input rate.

There are only four principal parameters that determine the behaviour of the model: R, the rate of input of external resources; λ, the ecological efficiency; b, which controls the saturation level of the functional response; and c, which controls the competition strength. All other quantities are generated automatically in the simulation. For example, the scores S_{ij} and the values α_{ij} are determined by the features of the species. The efforts are also determined at each time point in the dynamics by ensuring that they remain at their ESS value (see Drossel *et al.* 2001 for more details).

3 Evolutionary dynamics

For any set of species in the web the population dynamics can be described using the above equations. Evolution in Webworld occurs by speciation events.

An existing species is chosen at random to undergo speciation, and a new species is created that differs by one randomly chosen feature from the parent species. The population dynamics is then followed with the additional species until a new stable state is reached. Species whose population falls below a threshold of 1.0 are considered extinct and removed from the web. In this way, new species gradually replace older ones, and the number of species present rises and falls.

It is worth stressing that the features should not be thought of as having any genetic basis – they are purely phenotypic characteristics – nor should the random replacement of randomly chosen features be thought of as genetic mutations. We do not attempt to model the extremely complex process of speciation. Instead, we imagine observing the variation of the species in the system on such a coarse time scale that the process of speciation appears to be stochastic. This is the process that we are modelling when we randomly change features. Of course, there are several variants of the scheme which we have adopted which it could be argued might have done just as well. For instance, the parent could be chosen according to criteria dependent on population size or trophic level, and not purely at random, and the child might not always have just one modified feature. We hope in the future to check that changes such as these do not alter the general predictions of the model.

Results in this paper are obtained from 80 different runs of Webworld using the same set of parameters ($\lambda = 0.1$, $c = 0.5$, $b = 0.005$, and $R = 1.0 \times 10^5$). Each run lasted for 100000 speciation events. The mean number of species in these 80 runs is shown as a function of time in Figure 2. This shows a fairly rapid increase initially, followed by stabilization to a fairly constant state where speciations are balanced by extinctions. Our previous paper (Drossel *et al.* 2001) shows the variation in the number of species in several individual runs. There is considerable fluctuation in each individual run, even at the later times when the average number is relatively constant. This shows that there is a continual turnover of species. There is considerable fluctuation between different runs with the same parameters due to the random choice of the feature that is changed at each speciation event.

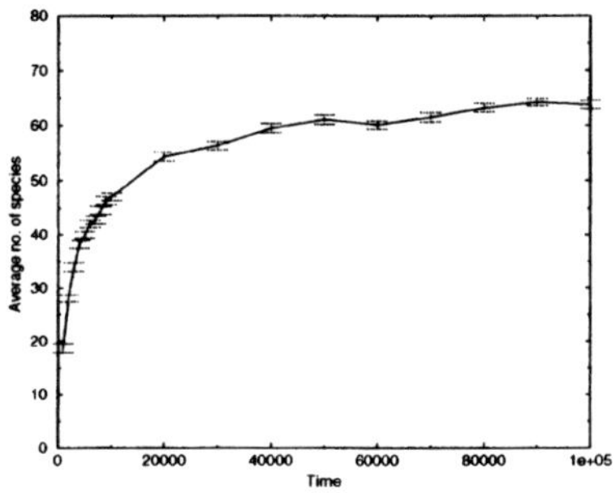

Fig. 2. Mean number of species vs. time

Figures 3–6 show some examples of particular food webs generated by these simulations. In order to indicate population sizes, circles are shown with radius proportional to the log of the population size. Typical population sizes decrease by an order of magnitude on each successive trophic level. The arrows represent the flow of resources from prey to predator. The thickness of the arrow represents the amount of effort the predator puts into each prey. Many of the second and third level species split their effort between more than one prey. The level assigned to each species is determined by the shortest path to the external resources. For example, species 20 in Figure 3 is on level 2 because it feeds on species 6, which feeds on external resources. There are also longer pathways from species 20, via species 23 and 24. We use the shortest path to define the level because this can always be calculated unambiguously (even in the presence of cycles) and because longer pathways provide relatively little resources. One observation from these webs is that the majority of interactions are simply between a prey species on one level and a predator on the level above (an 'upward' arrow in the diagrams). More complex situations, such as 'downward' or 'horizontal' arrows occur rarely.

In Figure 4, species 1–25 were in a state with stable populations. Species 1 has just undergone speciation to create the new species 26, which is also a level 1 species, like its parent. Figure 4 shows the situation after the population sizes have reached a new equilibrium. No extinction events have occurred. The population of species 26 has risen to be comparable to that of species 1. There is increased competition on level 1, hence some of the level 1 species populations have decreased (e.g. species 5). In Figure 3, species 1 had four predators feeding almost exclusively on it. When species 26 rises to a large population size, it pays these predators to put some of their effort into the new prey. This can be seen

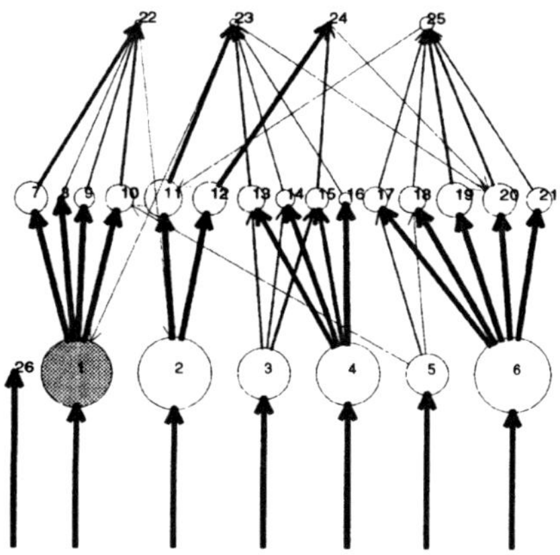

Fig. 3. Stable addition event: iteration 0

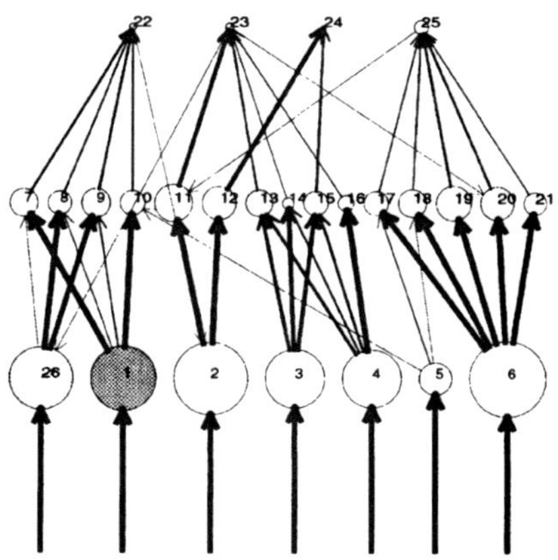

Fig. 4. Stable addition event: equilibrium

in Figure 4, where species 8 and 9 have switched predominantly to feeding on species 26, whilst species 7 has species 26 as a minor food source.

Another example is shown in Figures 5 and 6. Here we begin with the same 25 initial species, and species 26 arises on level 2, as a result of speciation of species 11. This new species causes the extinction of its parent species, 11, and three other species, 4, 16 and 24. The species that go extinct are all shaded in Figure 5. Figure 6 shows the situation after the new stable state is reached. This example shows that extinctions can occur on all levels, and that the nature of the interaction between the new species and the species that become extinct is not always straightforward. This point is pursued in the following section.

We referred to the continual turnover of species that occurs with this model. The main reason for this is that coevolutionary effects act both upwards and downwards in the web. New species on level one, for example, can out-compete older level 1 species, leading to their extinction, and possibly to the extinction of any level two species that fed on the old species (an upwards effect). However, a new level 2 species could arise that is a better adapted predator that causes its level 1 prey to go extinct (a downward effect). The way these effects operate depends on the form of the population dynamics equations. In our original paper (Caldarelli *et al.* 1998) we used a much simpler form of the population dynamics in which downward effects did not occur. We observed that the rate of turnover of species became slower and slower, and that the chance of a newly added species surviving decreased to virtually nil. Without downward effects, level one species can arise that are increasingly better adapted to the external environment, and it becomes increasingly more difficult to improve on them. With coevolutionary effects in both directions, species that are successful at one time do not remain successful for ever.

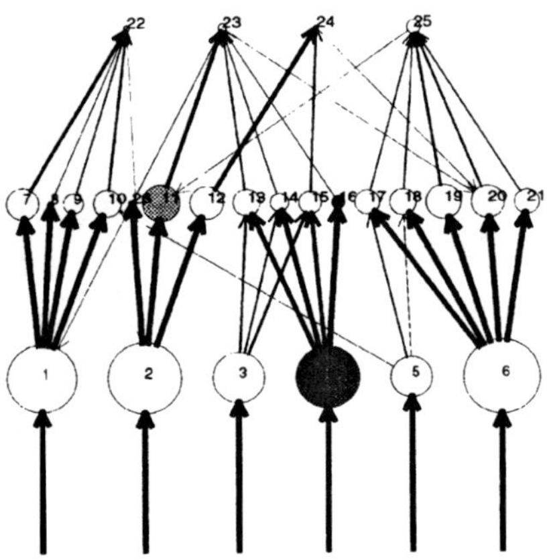

Fig. 5. Deletion event: iteration 0

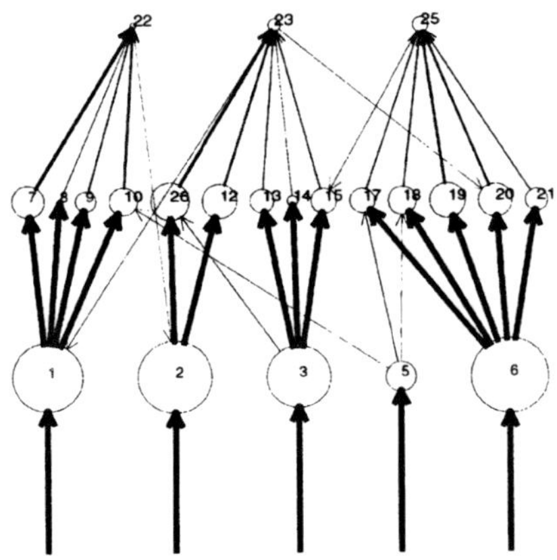

Fig. 6. Deletion event: equilibrium

The sensitivity that we observed of the evolutionary dynamics to the short time scale population dynamics makes us somewhat sceptical of the results of some of the simpler macro-evolutionary models in which there is essentially no population dynamics included at all. One question that has frequently been studied is the distribution of sizes of extinction events. Figure 7 shows the distribution of the number of species going extinct in each speciation event, given that at least one existing species goes extinct. The parameters are as for the runs in Figure 8. The majority of events are small, although the largest events (up to 14 species) represent a considerable fraction of the total number of species (average of 59 species per web). If we wished to look for evidence of critical phenomena it would be natural to ask how the size of the largest extinction events depends on the size of the web. However, in our model, the web size is not an independent parameter. We can increase the number of species by increasing the external resources, R, or decreasing the competition strength, c (see examples in Drossel *et al.* 2001). Computer time increases rapidly with the number of species, however, and it would be impractical to consider webs that were an order of magnitude larger than the present ones.

4 Food web properties and co-evolution

For the same 80 runs discussed above, details of the web were stored at intervals of 10000 speciation events from time 10000 to 100000, thus giving 800 largely independent webs which we investigated in detail. Table 1 shows average properties of these webs. In this analysis a link was defined as being present if the effort of the predator against the prey was greater than 1%. Since the efforts are proportional to the consumption rates (because of the ESS criterion), this means that a link is counted if greater than 1% of the resources consumed by the predator come from the prey. This is slightly different to the definition used in our previous papers, which required g_{ij} to be greater than 1.0 for a link to be counted. The figures in Table 1 can be compared with those in Drossel *et al.* (2001), which use a range of different parameters, and those in Caldarelli *et al.* (1998), which were generated using a simpler form for the population dynamics equations.

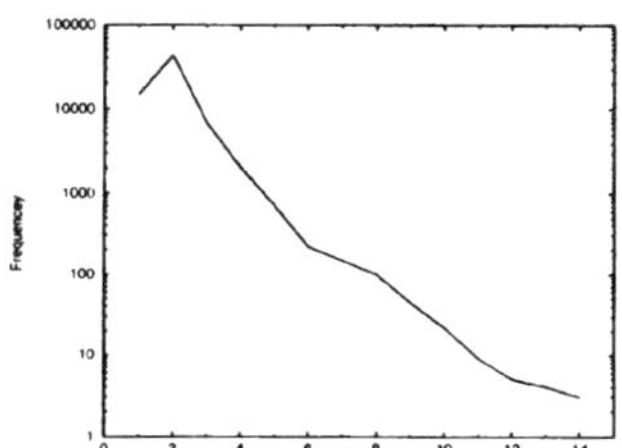

Fig. 7. Deletion event size distribution

Table 1. Statistics on food web structure

Result	Mean	Error in mean	Std. dev.
No. of species	59.1	0.3	8.6
Links per species	1.694	0.005	0.128
Av. level	2.313	0.002	0.059
Av. max. level	3.85	0.01	0.36
Basal species(%)	12.44	0.06	1.80
Intermediate species(%)	77.48	0.18	4.92
Top species(%)	10.06	0.16	4.45
Mean overlap level 1	0.260	0.002	0.047
Mean overlap level 2	0.105	0.001	0.019
Mean Overlap level 3	0.089	0.001	0.026

Table 2. Distribution of species between trophic levels

Level	Number of Species	Std. dev.	Proportion of Species	Std. dev.
1	7.30	1.14	0.125	0.018
2	27.43	5.21	0.462	0.042
3	22.97	4.29	0.389	0.046
4	1.64	0.70	0.029	0.014

In all the webs contributing to Table 1, the maximum trophic level present is either 3 or 4, with a mean of 3.85. The average level of all species is 2.31. The distribution of species between levels is given in Table 2. With these parameters, level 4 species tend to have very low populations because there are barely sufficient resources coming up through the web to maintain them. Species can also be classed as basal (having no prey), top (having no predators) or intermediate (having both predators and prey). A majority of species are intermediate for most reasonable parameter values. In fact the 10% of top species observed here is rather higher than the values in the examples in our previous papers. This is due to the rule used to define a link. Very weak interactions that would have been counted with our previous definition are discounted here using the criterion that the effort must be greater than 1%. We will consider in more detail elsewhere the way web statistics depend on the link definition. Also shown in Table 1 are the mean overlaps between species on each level (see equation (3)). This shows that considerable phenotypic diversity has accumulated in the species: only 2.6 out of 10 features are shared between species on level 1, and less than this on the higher levels.

The general patterns observed in real food webs are reproduced fairly well by the model. The values of quantities in real webs fluctuate greatly from web to web (see examples in Caldarelli *et al.* 1998). This can be put down to three factors. Firstly, the random nature of the evolutionary process leads to quite large fluctuations in web properties even if physical properties such as resource input and ecological efficiency are the same. In the model, this can be seen from the standard deviations in the web properties. Secondly, real webs are observed in different types of locations (lakes, deserts, estuaries etc.) that clearly do differ in terms of resource input and the nature of the limiting resource. Different groups of organisms exist in different locations that may differ in their ecological efficiency. In terms of the model, we began to consider the way the web properties changed with parameters such as R, λ and c in our previous papers. The third source of variation between the real webs is due to human observation. Different researchers use different definitions of what counts as a species, and what counts as a link. To what extent should similar species be lumped together? How often must a predation event be observed before a link is defined as being present? These questions are not straightforward, and they make detailed comparison between individual real webs, and between real and model webs difficult. Furthermore, while it is the case that the webs that we have generated are being compared to local communities, be they islands, lakes or forests, it is clear that the interaction between the community that we have evolved, and the species and individuals outside it, needs to be included in order to make the comparison with data valid. We hope to include the effects of immigration, and other such factors, in future work.

We have used the set of 800 stored webs to study in detail what happens when a single speciation event occurs. We performed 2000 independent speciation events on each of the starting webs and recorded the properties of each of the webs that arose when a new equilibrium situation was reached. Table 3 shows the statistics of these events. Events were defined as either addition (the child species survives and nothing goes extinct), substitution (the child species replaces the parent species but no other extinctions occur), no change (the child species goes extinct immediately, leaving the web the same as before the speciation event), and deletion (anything that is not addition, substitution or no change). In a deletion event, the new species must cause at least one species to become extinct that is not its parent. It can be seen approximately 89% of attempted

Table 3. Probabilities of event types arising from a single speciation

Event type	Mean	Error in mean	Std. dev.
Addition	0.0417	0.0005	0.0152
Substitution	0.0230	0.0003	0.0093
Deletion	0.0425	0.0035	0.1000
No change	0.8928	0.0033	0.0932
Origination rate	0.08853	0.0008	0.0231
Extinction rate	0.08949	0.0038	0.1075

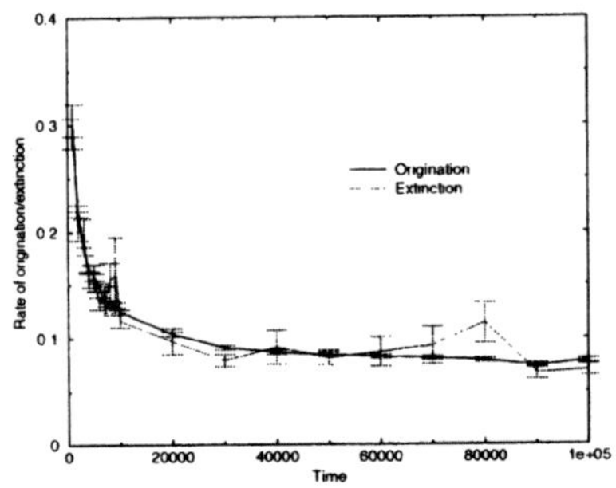

Fig. 8. Rate of origination/extinction vs. time

speciations lead to no change, and the other 11% lead to a change in the web of some description.

We also show in Table 3 the origination rate (i.e. the probability that the newly created species survives), and the extinction rate (the average number of species excluding the new species that go extinct per speciation event). These quantities are both approximately 8.9%, and do not differ significantly from each other. In other words, originations balance extinctions on the timescale of a single speciation event. However Figure 2 shows that, on a long timescale, there is a slight increase in the total number of species present over the period 10000 to 100000 speciation events during which the stored webs were collected. Thus there is a very slight excess of originations to extinctions. Figure 8 shows the way the extinction and origination rate vary with time. In the initial part of the simulation up to time 10000, both rates are high and are decreasing significantly with time. For the period 10000–100000 used for the statistics, the two rates are relatively constant. They seem to leveling off at roughly 8%, which represents a moderate non-zero rate of turnover of species.

The standard deviations shown in Table 3 measure variations between the 800 stored webs. The quantities were averaged over the 2000 speciation events sampled for each stored web, and the mean and standard deviations were then taken over the 800 webs. For most quantities the standard deviation is small compared to the mean, but this is not true for the deletion probability and the extinction rate. When the distribution of extinction rates is considered across webs, there is a significant tail of high extinction rate webs that causes the large standard deviation. In contrast, the origination rate varies much less between webs. We intend to consider in more detail in future what properties of the stored webs make them more or less prone to extinctions.

Table 4 considers the way the trophic level of the parent species undergoing speciation affects the process of addition of new species. We have defined the level-specific origination rate as the probability that the child species survives given that the parent was on a particular level. This varies substantially between levels. It can be seen that the figure for level 2 is roughly equal to the average rate for the whole web given in Table 3, whereas the origination rate is significantly

Table 4. The relationship of the trophic level of the child species to that of the parent, and the dependency of the origination rate on the parental level

Child level:	1	2	3	4	Origination rate	Error	Std. dev.
Parent level							
1	0.92	0.00	0.02	0.06	0.0200	0.0005	0.0139
2	0.00	0.90	0.09	0.02	0.0893	0.0009	0.0256
3	0.00	0.01	0.97	0.02	0.1117	0.0012	0.0334
4	0.00	0.03	0.29	0.68	0.0638	0.0025	0.0642

higher than average for level 3 and significantly lower for levels 1 and 4. Table 4 also gives the probability that the child ends up in each level given that it survives and given the level of the parent. Most of the child species occupy the same level as the parent (e.g. 92% of the children of level 1 species are also in level 1). There is a relatively small amount of movement between levels. The largest of the off-diagonal terms is the 29% probability of the child being in level 3 if the parent is in level 4. Given that most species enter the web on the same level as their parent, we can interpret the level-specific origination rates as measuring how easy it is to add species in a given level. The rate in level 1 is low because competition for external resources is strong, and because there is only one type of resource to compete for. For species on levels 2 and 3 there are many prey species on the level below from which to choose, and it becomes easier to find a niche in which to survive. The origination rate in level 4 is low, and we believe this is because level 4 species have a difficult time surviving due to lack of resources. Even a small amount of competition on this level is sufficient to drive the population below the minimum extinction threshold of 1. We have also measured the proportions of the species on each level as a function of time. These are roughly constant over the period 10000–100000, indicating that additions and deletions are roughly equal on each individual level.

Table 5 considers cases where a newly-created species survives, and asks what is the relationship between this species and any other species that go extinct as a consequence. The mean number of species affected is the mean number of species going extinct in each category as the result of the origination of one species. The mean probability of extinction is the probability that a species in a given category becomes extinct. The row 'all species' refers to all species previously in the web before the speciation event. 'Parent' refers to the parent of the new species. The 'Predators' and 'Prey' rows consider only species that are predators or prey of the new species. Note that the efforts of each species change during the population dynamics (see discussion in Drossel *et al.* 2001), so that predation links in the food web switch on and off as the population sizes change. For the purposes of this table, a species counts as a predator or prey if a link between the species is present at any time during the population dynamics (i.e. if the appropriate effort become greater than 1% for at least some of the time). Competitors are species that share at least one prey species with

Table 5. The relationship of species going extinct to the newly added species

Trophic relationship	Mean Number of Species Affected	Mean Probability of Extinction
All species	0.773	0.014
Parent	0.392	0.392
Predators	0.017	0.008
Prey	0.073	0.021
Competitors	0.499	0.147
Non-parental Competitors	0.108	0.034
Competitors' predators	0.082	0.014

the newly-added species for at least some of the time during the population dynamics. Non-parental competitors are competitors other than the parent of the new species. Competitors' predators are species that are predators of the competitors of the new species.

From the top row, we see that an average of 0.773 species of all types become extinct per origination. This corresponds to a probability of only 1.4% that any random species goes extinct. In comparison, there is a 39.2% probability that the parental species goes extinct. This is due to the strong competition between parent and child. Since they differ by only one feature out of 10, the overlap value will be 0.9, and hence the α value will be very high. Typical overlaps between species on a level are between 0.089 and 0.26 (from Table 1), hence competition between parent and child species is much stronger than for most species pairs. Also, since they share nearly all features, the two species will often be adapted to feed on the same prey, which again increases competition above that typical for two species on the same level. For non-parental competitors, the extinction probability is 3.4%, which is about two and a half times higher than the average. Note that the statistical errors in all the quantities in Table 5 are small and are therefore not given. Thus, competition with the new species is a significant factor in causing extinction of existing species.

The extinction probability for prey of the new species is also significantly higher than average, indicating that new, well-adapted predators can drive their prey extinct. However this value is less than the value for non-parental competitors. In contrast, the extinction probability for predators of the new species is significantly less than average. This is to be expected, because a predator of a successful new species is also likely to be successful. The final case of competitors' predators was considered because one might expect that if competitors of the new species are driven extinct then predators of those species would also be more likely to go extinct than average. The result shows that the competitors' predators extinction rate is equal to the average rate, hence the effect we looked for is not apparent. One reason for this is that there are a rather large number of competitors' predators, hence the figure is bound to be rather close to the average. Also, species on the upper levels tend to have several prey, so that extinction of any one of these is not particularly important.

5 Discussion

The most distinctive feature of Webworld, and the aspect which sets it apart from other models in this field, is the attempt to model phenomena which occur on very different time scales. On the shortest time scale in the model, the number and types of species are fixed, as are the number of individuals that belong to these species. Only the amount of effort individuals of species i put into preying on individuals of species j, f_{ij}, is allowed to vary. The choice of f_{ij} is a choice of foraging strategy, in our case given by the self-consistent solution of equations (2) and (4) (with the N_i fixed). On longer time scales the number and type of species is still fixed, but the number of individuals of a given species is now allowed to vary. This is the realm of traditional population biology, and in Webworld corresponds to determining the solutions of equation (1) in the long time limit. For simplicity, we have taken the death rate to be the same for all species. The choice of making the death rate equal to unity in equation (1) sets the timescale for the population dynamics. In all the simulations which we have carried out, we find that these solutions are steady states; no limit cycles or chaotic behaviour have been observed.

This point is worthy of further comment. A large fraction of the research which has been carried out in theoretical population dynamics has been concerned with models involving two or three species. There has been comparatively little work carried out on generic multispecies communities where a typical species will have several predators and several prey (whose number and identity may change with time) which is the situation of interest to us here. As a consequence, attention has focussed on the relatively simple equations found when only a few species are present, and especially on the phenomena of limit cycles and chaos frequently found in such equations. It is an open question as to whether these effects will be seen in food webs with a large number of species. One can argue that if one species is coupled to a large number of other species, these will act as a reservoir and blur out the details of the interactions that cause chaos, and so lead to a simpler dynamics. Alternatively, the adaptive nature of some aspects of the dynamics may lead to configurations where such behavior is less likely. This argument is similar to that put forward by Berryman & Millstein (1989a,b), who suggested that natural selection might favour parameter values which minimize the likelihood of chaotic dynamics. Of course, there may be other reasons why we have only seen steady states. Since the study of the long-time dynamics of equation (1) was not the ultimate goal of this investigation, we did not carry out a comprehensive investigation in the entire space of possible solutions, and it may be that some more complex behaviour does in fact exist. It might also be that the form of our equations has a particular structure which precludes more complicated long-time behaviour. Further work is required to decide which, if any, of these explanations is the correct one.

On the third, and longest, time scale the number and type of species is allowed to vary through a speciation mechanism. Effects of this kind have been modelled far less than have population dynamics or foraging mechanisms. One reason is that at the macroevolutionary level, such speciations will necessarily appear stochastic and modelling these will necessitate relatively long computer simulations. Another reason is that, at this stage in the development of the subject, there are few guidelines on how to model speciation. It is clear that we need to go beyond population dynamics, and give some internal characteristics to each species which are then allowed to vary with time according to an adaptive dynamics. We have chosen discrete features as these internal characteristics, but a set of continuous features might have also been a viable choice. It would be interesting to explore other ways of defining what is meant by a species in evolutionary models of this type. We always assume that the time for an ecosystem to reach a steady state is less than the time between speciation events. Thus it does not matter if the speciation rate varies with the number of species, or has other similar factors influencing it. In the Webworld model, ecological timescales are those which are longer than that defined by (1), but shorter than the time between speciation events. On the other hand, evolutionary timescales are those which are longer (typically, much longer) than the time between speciation events.

There has been a discussion in the literature between those opposed to and those favouring a functional response, g_{ij}, which is ratio-dependent (Berryman, 1992; Huisman & De Boer, 1997). Our main reason for choosing a ratio-dependent form was firstly, that it has many positive attributes from an ecological point of view (Berryman, 1992) and secondly, that it has fewer parameters associated with it. Indeed, our generalization of the ratio-dependent functional response to a multispecies food web (2), involves only two parameters: b and c. Nevertheless, it is not clear whether the precise form of the functional response matters much in the context of Webworld, where the ultimate state of interest is the one obtained after tens of thousands of speciation events. It may be that only general aspects such as the existence of strong competition, particularly between similar species, and the inclusion of a downward effect, as discussed earlier, are important. There may be a large degree of flexibility here, in the sense that the structure of the food web produced is insensitive to the precise form of g_{ij}, as long as the right qualitative structure is present. Further work is needed to clarify these points and to identify what the important ingredients are.

The food webs generated by Webworld, some of which are shown in Figs. 3–6, seem to evolve in a "natural" way when they are examined time-step by time-step. That is, the rules built into the model lead to consequences which can be interpreted according to rational criteria. Similarly, the consequences of a simple speciation event, which we have investigated in detail in this paper, seem very reasonable. Origination and extinction rates roughly balance on short time scales with slightly larger origination rates having an effect on longer time scales, child species tend to appear on the same level as parent species and there is strong competition between parent and child. All this is in line with expectations.

In summary, Webworld is a model which covers time scales varying from the very short, typified by changing foraging strategies, to the very long, required for evolutionary dynamics to reach a state where the number of originations and the number of extinctions balance on average. It gives results which are intuitively appealing and in broad agreement with food web data from real ecosystems. Many aspects of the model remain to be investigated, but we believe that it provides a realistic picture of the evolution of ecological communities which throws light on the nature of the basic mechanisms present in all such communities.

Acknowledgment

CQ wishes to thank EPSRC for the award of a postgraduate grant.

References

1. Arditi, R. & Ginzburg, L.R. (1989). Coupling in predator–prey dynamics: ratio-dependence. *J. Theor. Biol.*$\mathbf{139}$, 311–326.
2. Arditi, R. & Michalski, J. (1995). Nonlinear food web models and their responses to increased basal productivity. In: *Food webs: integration of patterns and dynamics* (Polis, G.A. & Winemiller, K.O., eds), pp. 122–133, Chapman & Hall, London.
3. Bak, P. & Sneppen, K. (1993). Punctuated equilibrium and criticality in a simple model of evolution. *Phys. Rev. Lett.* **71**, 4083–4086.
4. Bastolla, U., Lässig, M., Manrubia, S.C. & Valleriani, A.(2000). Diversity patterns from ecological models at dynamical equilibrium. eprint: arXiv: nlin.AO/0009025.
5. Beddington, J. (1975). Mutual interference between parasites or predators and its effect on searching efficiency. *Anim. Ecol.* **51**, 597–624.
6. Berryman, A.A. & Millstein, J.A. (1989a). Are ecological systems chaotic – and if not, why not? *Trends Ecol. Evol.* **4**, 26–28.
7. Berryman, A.A. & Millstein, J.A. (1989b). Avoiding chaos – reply. *Trends Ecol. Evol.* **4**, 240–240.
8. Berryman, A.A. (1992). The origins and evolution of predator–prey theory. *Ecology* **73**, 1530–1535.
9. Caldarelli, G., Higgs, P.G. & McKane, A.J. (1998). Modelling coevolution in multispecies communities. *J. Theor. Biol.* **193**, 345–358.
10. Cohen, J.E. (1990). A stochastic theory of community food webs VI – Heterogeneous alternatives to the cascade model. *Theor. Pop. Biol.* **37**, 55–90.
11. Cohen, J.E., Briand, F. & Newman, C.M. (1990). *Biomathematics* Vol. 20. Community food webs, data and theory. Springer Verlag, Berlin.
12. Drossel, B. (2001). Biological evolution and statistical physics. eprint: arXiv:cond-mat/0101409.
13. Drossel, B., Higgs, P.G. & McKane, A.J. (2001). The influence of predator–prey population dynamics on the long-term evolution of food web structure. *J. Theor. Biol.* **208**, 91–107.
14. Emlen, J.M. (1984). *Population biology.* Macmillan, New York.

15. Goldwasser, L. & Roughgarden, J. (1993). Construction and analysis of a large Caribbean food web. *Ecology* **74**, 1216–1233.
16. Hall, S.J. & Raffaelli, D. (1991). Food web patterns: lessons from a species-rich web. *J. Anim. Ecol.* **60**, 823–842.
17. Hallam, T.G. (1986). Community dynamics in a homogeneous environmemt. In: *Mathematical ecology.* Biomathematics Vol. 17. (Hallam, T.G. and Levin, S.A., eds), pp. 241–285, Springer-Verlag, Berlin.
18. Hastings, A. & Powell, T. (1991). Chaos in a three-species food chain, *Ecology* **72**, 896–903.
19. Hofbauer, J. & Sigmund, K. (1998). *Evolutionary games and population dynamics.* Cambridge University Press, Cambridge.
20. Holling, C.S. (1959). Some characteristics of simple types of predation and parasitism. *Can. Entomol.* **91**, 385–398.
21. Huisman, G. & De Boer, R.J. (1997). A formal derivation of the Beddington functional response, *J. Theor. Biol.* **185**, 389–400, and references therein.
22. Lässig, M., Bastolla, U., Manrubia, S.C. & Valleriani, A. (2001). The shape of ecological networks. eprint: arXiv:nlin.AO/0101026.
23. Logofet, D.O. (1993). *Matrices and graphs: stability problems in mathematical ecology.* CRC Press, London.
24. Martinez, N.D. & Lawton, J.H. (1995). Scale and food web structure – from local to global. *Oikos* **73**, 148–154.
25. May, R.M. (1974). *Stability and complexity in model ecosystems.* Monographs in population biology, Vol. 6. Princeton University Press, Princeton. Second edition.
26. McCann, K. & Yodzis, P. (1994). Biological conditions for chaos in three-species food chain. *Ecology* **75**, 561–564.
27. Parker, G.A. & Maynard Smith, J. (1990). Optimality theory in evolutionary biology. *Nature.* **348**, 27–33.
28. Pielou, E.C. (1977). *Mathematical ecology.* Wiley, New York, Second edition.
29. Pimm, S.L. (1982). *Food webs.* Chapman & Hall, London.
30. Post, D.M., Conners, M.E. & Goldberg, D.S. (2000). Prey preference by a top predator and the stability of linked food chains. *Ecology* **81**, 8–14, and references therein.
31. Reeve, H.K & Dugatkin, L.A. (1998). Why we need evolutionary game theory. In: *Game theory and animal behaviour* (Dugatkin, L.A & Reeve, H.K., eds), pp. 304–311. Oxford University Press, Oxford.
32. Roughgarden, J. (1979). *Theory of population genetics and evolutionary ecology.* Macmillan, New York.
33. Solé, R.V., Manrubia, S.C., Benton, M. & Bak, P. (1997). Self-similarity of extinction statistics in the fossil record. *Nature* **388**, 764–767.
34. Stephens, D. W. & Krebs, J. R. (1986). *Foraging theory.* Princeton University Press, NJ.
35. Svirezhev, Yu.M. & Logofet, D.O. (1983). *Stability of biological communities.* Mir Publishers, Moscow.

Dynamics and topology of species networks

Ugo Bastolla[1], Michael Lässig[2], Susanna C. Manrubia[1], and Angelo Valleriani[1]

[1] Max Planck Institute of Colloids and Interfaces, 14424 Potsdam (Germany)
[2] Institut für theoretische Physik, Universität zu Köln, Zülpicher Strasse 77, 50937 Köln (Germany)

Abstract. We study communities formed by a large number of species, which are an example of dynamical *networks* in biology. Interactions between species, such as prey-predator relationships and mutual competition, define the links of these networks. They also govern the dynamics of their population sizes. This dynamics acts as a selection mechanism, which can lead to the extinction of species. Adaptive changes of the interactions or the generation of new species involve random mutations as well as selection. We show how this dynamics determines key topological characteristics of species networks. The results are in agreement with observations.

1 Introduction

Complex networks are ubiquitous in biology. Recent attention has focused on examples at the intra-cellular or inter-cellular level, such as transcription regulatory networks, gene networks, protein networks, or cell signalling. These are regulatory elements transforming genetically encoded information into structure and function or coordinating the actions of several cells. They are often quite complicated, and we are only beginning to understand their important structural elements, let alone their evolutionary genesis.

Two complementary properties of biological networks are often discussed: their *robustness* under short-term environmental fluctuations or deleterious mutations and their *evolvability* in response to longer term selective forces. These response characteristics reflect a fundamental dynamical property: A biological network can have multiple and widely differing characteristic time scales. It remains a challenge for experimentalists and for theorists to quantify the dynamical properties and relate them to observed topological features.

In this article, we discuss a macroscopic type of biological networks, namely communities of many species that interact via predation and competition. A number of field data are available, describing their large-scale dynamics as well as their topological structure. We discuss a simple theoretical model that can be compared to these observations on a quantitative basis. The main conceptual result is the intimate statistical connection between the long-term dynamics and typical network structures: Evolution shapes the network topology.

Due to the relationship between dynamics and structure, these networks are an interesting subject for theoretical physicists. Clearly, a similar link is expected also for their molecular counterparts, and in this aspect the example of species networks may be useful for model building in a more general context. However, topological characteristics are seen to depend quite sensitively on, for example,

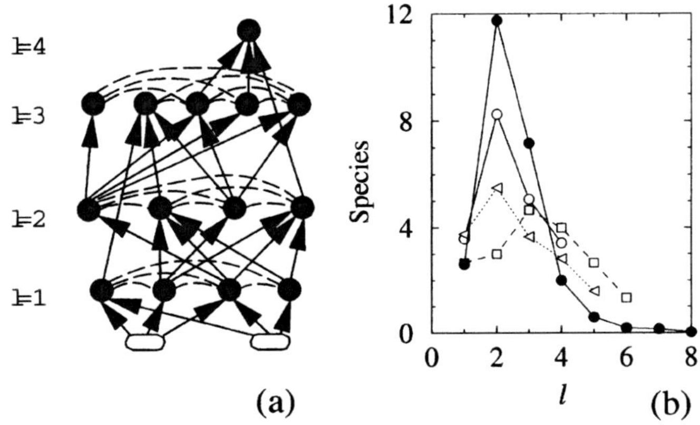

Fig. 1. (a) The Pamlico estuary foodweb in North Carolina, consisting of 14 species (filled circles) at four trophic levels. *Detritus*, *dinolagellates* and *diatoms* are at the bottom level ($l = 1$) and feed from external resources (empty symbols). There is a single trophic group at the highest level ($l = 4$), formed by the predatory fishes *Roccus* and *Cynoscion*. Arrows point from prey to predator; dashed lines connect species pairs with a nonzero link overlap (see text). Data from [1]; the level is defined here by the longest relevant food chain. (b) Average species numbers per level for a set of natural ecosystems, taken from Ref. [1] (empty symbols) and [9] (filled circles). This last case corresponds to an average over 61 food webs, most of which are empty at high levels.

the level of competition between species. Thus, the reader should be cautioned not to expect one universal theory of biological networks.

Species networks in nature have been studied on quite different scales of space and time. Ecologists' attention has focused on *food webs*, i.e., communities of animal species in a closed environment where food chains can be observed. Fig. 1(a) shows the graph of such a network, each arrow representing a prey-predator relationship. Despite large variations in size and environmental conditions, large ecosystems share a few important topological characteristics: (i) Every species lives at a certain *trophic level*. The level number can be defined as the minimum, the maximum or a suitable average length of its relevant 'downward' food chains; the differences between these definitions turn out not to be significant for the statistical properties of food webs we discuss here. Species at level one feed from external resources. (ii) The number of trophic levels is small, typically between three and seven. (iii) Most species have a small number of relevant prey species (typically around three), mainly from the next lower level. (iv) The number of species at level l increases with l for lower values of l and decreases again sharply for higher l [1,2], see Fig. 1(b). Networks of co-evolving species thus have a characteristic *shape*.

Evolutionary biologists and paleontologists have a different point of view on species communities. They record adaptive changes and extinctions of species and the arrival of new species. More than 99% of the species that have ever existed are extinct. It is only the slight excess of speciations over extinctions in the last 600 million years that has produced our diverse biosphere [3]. The dynamics of species numbers can be quite intermittent. Periods of relative stasis ($\sim 10^{4-6}$ years) alternate with bursts of extinctions and or speciations, leading to large fluctuations in the number of species [4,5]. It has been argued that these temporal patterns are generated by the complex interactions between species [6,7] and the resulting correlations, rather than by external variations alone.

These complementary descriptions refer to different time scales. Over short intervals, the relevant variations in a species network are in the population numbers (this is referred to as population dynamics), while the network structure (i.e., the species and their interaction links) remains fairly robust. The average time between network changes (extinctions, speciations, or adaptive changes) is much longer. However, as shown by recent field observations and theoretical results on sympatric speciation, also structural changes can be rather rapid (see the article by Rost and Lässig in this volume). So a strict separation of the relevant time scales should not always be assumed. Moreover, the network interactions induce large correlations and fluctuations. For example, an extinction may trigger rapid other extinctions and subsequent speciations. A statistical ensemble emerges from averaging over yet longer time intervals containing many speciations and extinctions, and this statistics is the subject of the article.

Classical work in mathematical biology has established stability criteria for networks with random interactions g_{ij} [8]. They are, however, of limited use for real ecosystems where the interactions are not random, but are themselves subject to selection. Only recently, a model for species networks with rather detailed interactions has been formulated and analyzed by numerical simulations [10] (see the article by Quince et. al. in this volume). Another class of recent models focuses directly on the dynamics of extinctions and speciations. These models have no explicit population dynamics and mostly random topology, with the important exception of Ref. [11].

Here we discuss the simplest models of species networks that can be compared to observations on a quantitative basis. These models are introduced in Section 2. The *global* shape of the model networks can be obtained from an approximate analytical calculation (see Section 3). The underlying dynamics and the resulting *local* fluctuations are discussed in Section 4.

2 Modelling species networks

A model for species communities has to specify the type of interactions and the resulting population dynamics, as well as the slower dynamics of the network itself, that is, of its nodes and links.

2.1 Species interactions and population dynamics

Lotka-Volterra equations. The simplest population dynamics for a community of species with population numbers $N_i(t)$ is

$$\frac{1}{N_i}\frac{\mathrm{d}N_i}{\mathrm{d}t} = \sum_{j=1}^{s} g_{ij} N_j + h_i \qquad (i = 1, \ldots, s) , \tag{1}$$

a set of coupled differential equations for the relative growth rates. The coefficients g_{ij} represent interactions between species, the most important of which are predation and competition. Accordingly, we decompose the interaction matrix into a predation part and a competition part,

$$g_{ij} = \gamma_{ij} - \beta_{ij} , \tag{2}$$

which are defined below. The terms h_i denote intrinsic production or decay rates. This type of equations, as well as many generalizations thereof, has been used to model coexistence, invasions, and adaptive change of populations. Of great importance is the conceptual connection to mathematical *game theory* [12]. A set of populations $(N_1, \ldots, N_S)$ represents a mixed strategy in a game with payoff matrix g_{ij}. An optimal strategy of this game – called Nash equilibrium – can often be realized as a stable fixed point $(N_1^*, \ldots, N_S^*)$ of an associated Lotka-Volterra dynamics. This explains how strategic optimization is reached in biological systems through reproductive success, with no need for rational thinking.

Predation denotes here any interaction between two species i and j that is advantageous for i and disadvantageous for j. It is described by matrix elements $\gamma_{ij} > 0$ and $\gamma_{ji} < 0$, taken for simplicity to be proportional, $\gamma_{ij} = \Delta|\gamma_{ji}|$. The matrix γ_{ij} is sparse in natural systems. Its nonzero matrix elements define the *predation network*, which is represented by solid lines in Fig. 1(a). The *productivity* of a species i is defined as the net contribution of predation to its growth rate,

$$P_i \equiv \sum_{j \in \pi(i)} \gamma_{ij} N_j - \sum_{j \in \Pi(i)} \gamma_{ji} N_j , \tag{3}$$

where $\pi(i)$ is the set of its prey species and $\Pi(i)$ the set of its predators.

Competition is the mutual interference of two species i and j in each other's livelihood. Again for simplicity, it is described by a symmetric matrix, $\beta_{ij} = \beta_{ji} > 0$. Competition takes place for nesting places, mating opportunities, and other resources not explicitly represented in the model. It is strongest between individuals of the same species, but also occurs between different species [13]. This interaction turns out to be the main limiting factor for the coexistence of species in a common network.We set the intra-species competition $\beta_{ii} = 1$; this normalization amounts to an appropriate choice of the time scale in (1). It is then a natural choice to quantify the inter-species competition in terms of the *predation overlap*

$$\rho_{ij} \equiv \sum_{k \in \pi(i) \cap \pi(j)} \gamma_{ik}\gamma_{jk} \Bigg/ \sqrt{\sum_{k \in \pi(i)} \gamma_{ik}^2 \sum_{k \in \pi(j)} \gamma_{jk}^2} \; . \tag{4}$$

We set $\beta_{ij} = \beta\rho_{ij}$ for $i \neq j$ with a coupling constant $\beta < 1$. The nonzero matrix elements ρ_{ij} define the *overlap network*, which is represented by dashed lines in Fig. 1(a). It is typically sparse as well in natural systems. The inter-species competition load of a species i is defined as

$$Q_i \equiv \beta \sum_{j \neq i} \rho_{ij} N_j \,. \tag{5}$$

Fixed points, viability threshold. The population dynamics (1) can now be written as

$$\frac{1}{N_i}\frac{\mathrm{d}N_i}{\mathrm{d}t} = (P_i - Q_i - \alpha_i)N_i - N_i^2 \,. \tag{6}$$

In general, the population numbers will converge to a stable fixed point of the form

$$N_i^* = P_i(N_1^*, \dots, N_s^*) - Q_i(N_1^*, \dots, N_s^*) - \alpha_i \,. \tag{7}$$

Furthermore, we require a minimum population size N_c for viable species, and count all species with $N_i^* < N_c$ as extinct. A uniform death rate $\alpha_i = \alpha$ is an equivalent cutoff for small population sizes.

External resources. The species community is maintained by a number of external resources, which are represented as extra 'populations' N_i with $h_i = \gamma_{i,0} R$ and predators only (i.e., $\gamma_{ij} \leq 0$ and $\beta_{ij} = 0$ for all j). The external resources and the viability threshold play the role of boundary conditions for the population dynamics. The dimensionless parameter $R/N_c \gg 1$ turns out to control the vertical size of the network, i.e., the length of food chains.

2.2 Network dynamics

As explained above, the population dynamics can lead to the extinction of one or more species, i.e., to the loss of nodes in the species network. The other changes in the network structure are adaptive mutations of the links γ_{ij} and speciations, which generate additional nodes.

At the molecular level, most mutations are neutral or deleterious. In this effective model, however, neutral mutations are not taken into account and the effect of deleterious mutations (the so-called mutation load) enters only through the viability threshold N_c and the death rate α. Only rare advantageous mutations leading to viable mutants are represented explicitly.

These mutations are modeled as follows. Consider the bare predation matrix with integer coefficients $x_{ij} = 0, 1, \dots, x_0$. These define the coefficients $\gamma_{ij} > 0$ by

$$\gamma_{ij} = \gamma_0 \frac{x_0 x_{ij}}{x_0 + \sum_{k \in \pi(i)} x_{ik}} \qquad \text{for } j \in \pi(i). \tag{8}$$

Here γ_0 is the overall predation strength, and $x_0 > 0$ is a saturation scale. For well-adapted species (i.e., for $\sum_{k \in \pi(i)} x_{ik} \gg x_0$), increasing predation on one resource implies an equal decrease of predation on the other resources.

We assume the population dynamics has settled at a fixed point with s viable species, $(N_1^*, \dots, N_s^*)$. One species i is chosen randomly as parent species. We denote by c_i the number of its prey species. Now we introduce a mutant of the species i, which is labelled as a new 'species' i' with initial population size $N_{i'}$ of order N_c. It differs from its parent species by a single bare predation coefficient. Either one of the c_i existing links $x_{ij} > 0$ is modified, $x_{i'j} = x_{ij} \pm 1$, or a new link $x_{i'j} = 1$ is created randomly; each of these $c_i + 1$ different cases is chosen with equal probability $1/(c_i + 1)$. The mutant has a different productivity $P_{i'}$ and a different competition load $Q_{i'}$, which includes a contribution $\rho_{ii'} N_i^*$ from the competition with the parent. The viability condition for the mutant reads $P_{i'} - Q_{i'} > \alpha + N_c$; the sign of the link modification is chosen at random.

A viable mutation generates an unstable perturbation of the fixed point $(N_1^*, \ldots, N_s^*)$. The mutant population $N_{i'}$ grows, leading to the temporary coexistence of $(s + 1)$ populations. We assume mutations are so rare that the population dynamics reaches a fixed point $(N_1^{*\prime}, \ldots, N_{s'}^{*\prime})$ before a new mutation takes place[1]. This population dynamics acts as selection. The new fixed point has $s' \leq s + 1$ species. In most cases, the parent species i is replaced by the mutant i', which is counted as an adaptive mutation of i. In some cases, if the overlap $\rho_{ii'}$ is small enough, the species i and i' coexist at the new fixed point; this is a speciation.

3 The shape of a species network

Here we derive the global shape of a food web in a 'mean-field' approximation, following ref. [14]. Details of the dynamics are not needed at this stage. It is sufficient to assume that the mutation-selection process maintains a broad distribution of productivitities and, hence, of biomasses. This asssumption is well supported by field observations; it reflects the diversity of habitats and ecological niches. In the framework of the network model, it can be verified by the numerical simulations discussed in section 4.

Throughout this section, we disregard fluctuations of the predation coefficients and set $\gamma_{ij} = \gamma_+$ if j is prey of i and $\gamma_{ij} = -\gamma_- = -\Delta\gamma_+$ if i is prey of j, with $0 < \Delta < 1$.

To illustrate of the relative roles played by predation and competition, we analyse first two types of simple networks, before we turn to species networks of general topology.

3.1 A single food chain

A food chain is a community of L species on L trophic levels where species at level l feed from that at level $l-1$. All the competition loads vanish, and thus the

---

[1] Of course, any mathematical fixed point of Eq. (1) is reached only after infinite time. However, the time-dependent population numbers $N_i(t)$ get exponentially close to the corresponding fixed point values for large times. In practice, the numerical integration of (1) is carried out until the number of species with $N_i(t) < N_c$ no longer varies.

population numbers at each level are given simply by $N_l^* = P_l - \alpha$. Hence, the entire chain is viable if $P_l > P_c$ for all species l, with the productivity threshold

$$P_c = \alpha + N_c \, . \tag{9}$$

The productivity of a species at level l is

$$P_l = \gamma_+ N_{l-1}^* - \gamma_- N_{l+1}^* \qquad (l = 1, \ldots, L) \tag{10}$$

and $P_0 = -\gamma_1 N_1^*$, with the boundary condition $N_{L+1}^* = 0$. These equations can be solved exactly by recursion starting from the top level $l = L$. Asymptotically, we obtain an exponential decrease in the biomasses for increasing level number. In the biologically relevant case $\gamma_+ \ll \gamma_-$, we get

$$N_l^* = \gamma_+ N_{l-1}^* - \frac{\alpha}{1 + \gamma_-} \, ; \tag{11}$$

the resulting length of the food chain scales as

$$L \sim \log \left(\frac{R}{N_c + f(\gamma_-)\alpha} \right) \tag{12}$$

with some function $f(\gamma_-)$. The important qualitative conclusion is that food chains are always short, as observed in real systems. The parameters α and N_c are seen to be equivalent viability cutoffs for the length of the chain.

3.2 A single trophic layer

A trophic layer is a group of S species at the same level. These species have a significant overlap in their predation links and a resulting competition load. We assume there is no predation within the layer, that is, the productivities P_i depend only on the interactions with 'external' species. Here we consider the P_i as fixed and concentrate on the effects of the direct competition terms Q_i. In the mean field approximation, we replace the individual link overlaps ρ_{ij} between different species by their expectation value $\bar{\rho}$, which has to be determined self-consistently. Consider, for example, a trophic layer feeding from a set of S' prey species with an average number of $\bar{c}$ prey species per predator species. We use a simple approximation for the link overlap which (i) takes into account that a configuration with zero overlap exists for $S \leq \max(S'/\bar{c}, 1)$ and (ii) assumes random predation for larger values of S. This approximation reads

$$\bar{\rho}(S, S') = \begin{cases} 0 & \text{if } S \leq \max(S'/\bar{c}, 1) \\ \min(\bar{c}/S', 1) & \text{if } S > \max(S'/\bar{c}, 1) \end{cases} \, . \tag{13}$$

The competition load in eq. (7) can now be evaluated approximately, yielding the fixed point populations

$$N_i^* = \frac{P_i - \beta\bar{\rho}S\bar{N} - \alpha}{1 - \beta\bar{\rho}} \, . \tag{14}$$

The average $\bar{N} \equiv S^{-1} \sum_{i=1}^{S} N_i^*$ is given by

$$\bar{N} = \frac{\bar{P} - \alpha}{1 + \beta\bar{\rho}(S-1)} . \tag{15}$$

Inserting the viability condition $N_i^* > N_c$ into (14) determines the productivity threshold

$$P_c = \alpha + (1 - \beta\bar{\rho})N_c + \beta\bar{\rho}S\bar{N} . \tag{16}$$

For small values of S, we have $\bar{\rho} = 0$ by (13), and (16) reduces to (9). With increasing S, the threshold increases as well.

The actual productivites P_i are constantly changing as a consequence of the mutations of species i as well as those of the other species. The statistical assumption used here (and verified in the next section) is that the productivities P_i are drawn from a broad probability distribution given by $\Phi(q) \equiv \mathrm{Prob}(P_i/\bar{P} < q)$. This distribution is assumed to be independent of S; that is, the number of species enters only via the average $\bar{P}$. The qualitative results do not depend strongly on the form of $\Phi(q)$; we use the simple approximation

$$\Phi(q) = \frac{q - q_0}{2(1 - q_0)} . \tag{17}$$

The expectation value of the smallest productivity $P_{\min}$ in a community of S species can then be estimated from the relation $S\Phi(P_{\min}/\bar{P}) = O(1)$, giving

$$P_{\min} = \left(q_0 + \frac{1 - q_0}{S}\right)\bar{P} . \tag{18}$$

The species community becomes unstable if the least productive species falls below the viability threshold. Equating (16) and (18) therefore gives an implicit relation for S as a function of $\bar{P}/N_c$, α/N_c, the relative productivity spread q_0, and the average pairwise competition load $\beta\bar{\rho}$ given by (13). That is, competition determines the number of 'ecological niches' in a trophic level as a function of the prey diversity and the competition strength β. For sufficiently large β, only non-overlapping species can coexist, i.e., $S = \max(S'/\bar{c}, 1)$. The number of ecological niches increases with decreasing β and increasing productivity spread q_0. This result generalizes the well known theorem of competitive exclusion [15], which states the condition for coexistence of two competing species. Note that this limiting effect on the number of species exists independently of the population numbers. It is indeed crucial for the buildup of high population numbers at the lower trophic levels. For example, a trophic level feeding from resources of size $R \gg \max(\alpha, N_c)$ acquires an extensive population number per species, $\bar{N} \sim R/S$ with S asymptotically independent of R. Without competitive exclusion ($\beta = 0$), speciations would further increase S and eventually lead to an extensive number of marginally viable species, $S \sim R/\bar{N}$ with $\bar{N}$ of order N_c. Such a level could not support sizeable predation from above.

3.3 The full network

We now turn to a full ecological network with L trophic levels. In the mean field approximation, we treat all species at the same level on an equal footing and derive self-consistent equations for the level averages of population and species number, $\bar{N}_l$ and S_l ($l = 1, \ldots, L$). The average productivities $\bar{P}_l$ satisfy the recursion relations

$$\bar{P}_l = \gamma_+ \bar{c} \bar{N}_{l-1} - \gamma_- \bar{c}(S_{l+1}/S_l)\bar{N}_{l+1} , \qquad (19)$$

where we assume that the species at every level predate only on the species at the next lower level. The average number $\bar{c}$ of predation links per predator is taken to be independent of l; this is indeed suggested by field data. The average number of predators per prey is then simply $\bar{c}S_{l+1}/S_l$. The productivity $\bar{P}_l$ is linked to $\bar{N}_l$ and S_l as in (15), using for $\bar{\rho}(S_l, S_{l-1})$ the approximation (13). Hence, the relations (19) determine the population numbers given the species numbers. The latter are again limited by the stability criteria $S_l \Phi(P_{c,l}/\bar{P}_l) = O(1)$ with the minimum productivities $P_{c,l}$ given as in (16); these relations determine the S_l given the $\bar{N}_l$. The coupled set of equations can be solved iteratively. Finally, the number of levels L follows from the condition $\bar{N}_L \approx N_c$, which is equivalent to $S_L \approx 1$.

Over a wide range of relevant parameters, these networks have the characteristic shape shown in the example of Fig. 2: The species numbers S_l increase with l at low levels due to the increasing prey diversity, which opens up more and more niches. They reach a maximum at an intermediate level and decrease again at higher levels, because more and more species have population numbers

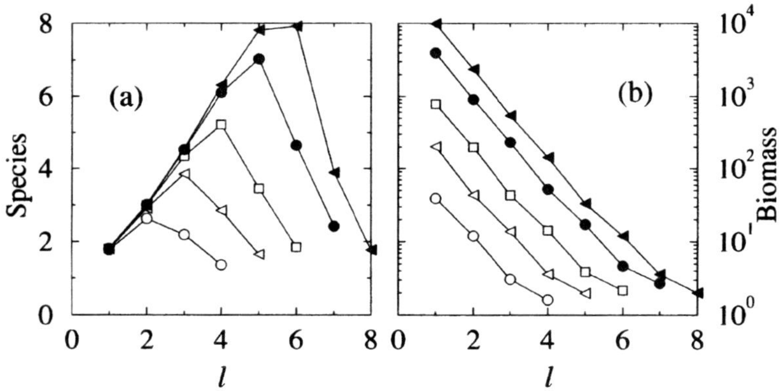

Fig. 2. The shape of ecosystems. (a) The species numbers S_l ($1 \le l \le L$) for networks with L trophic levels. The parameters are $\bar{c} = 3$, $\gamma_+ = 0.3$, $\gamma_- = 2.0$, $\lambda = 0.2$, $q_0 = 0.35$, $\alpha/N_c = 1$, and $R/N_c = 2 \times 10^3$, 10^4, 4×10^4, 2×10^5, and 5×10^5 for the cases $L = 4$, 5, 6, 7, and 8, respectively. (b) The average population numbers $\bar{N}_l$ for the same cases as in (a).

too low to support further predation. Hence, these two regimes reflect the two kinds of species interactions. The population numbers show an approximately exponential decrease in both regimes, just like for a single vertical chain. Hence, L is always small, in agreement with observations and with the results of [10,16].

4 The local structure and fluctuations

We now discuss the long-term dynamics of species networks, following ref. [17]. In the present model, a single step of this dynamics consists of a mutation and the subsequent selection by population dynamics; see section 2.1 above. The cumulative effect of many such steps is large fluctuations in the size of the networks and in many of its local characteristics. Here we illustrate this for the particularly simple case of a trophic *bilayer* in a stationary environment, as shown in Fig. 3. The number of species at the lower level and their predation gain are kept constant in time as appropriate for external resources. We focus on the dynamics of the species at the upper level (the predators) and of the prey-predation links between the two levels. Links to higher levels are neglected.

Fig. 4 shows the stationary fluctuations in the number of viable species at the upper level. Here the 'time' coordinate T is just the number of mutation-selection steps.

This should be compared to the time series for the population number $N(T)$ of a randomly chosen 'tracer' species i shown in Fig. 5(a). The fluctuations originate from the 'noise' of its own adaptive mutations and speciations and those of the other species. We can think of them as a random walk in N with variable step size. Since large values of $N(T)$ are suppressed, the random walk is essentially constrained to the range $N_c < N(t) < O(\bar{N})$ with an absorbing boundary at $N = N_c$. Hence, every species faces a continuous threat of extinction, which leads to an exponentially decaying survival probability. It will eventually get extinct when the pace of its own adaptations cannot keep up with the changes in its environment. This is the well-known *Red Queen effect* of co-evolutionary

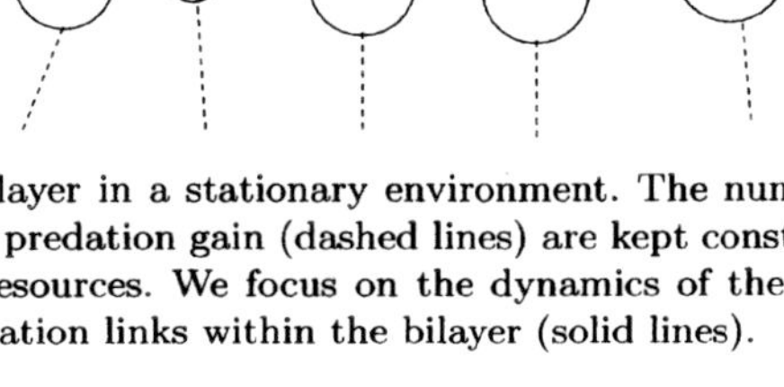

Fig. 3. A trophic bilayer in a stationary environment. The number of species at the lower level and their predation gain (dashed lines) are kept constant in time as appropriate for external resources. We focus on the dynamics of the species at the upper level and of the predation links within the bilayer (solid lines).

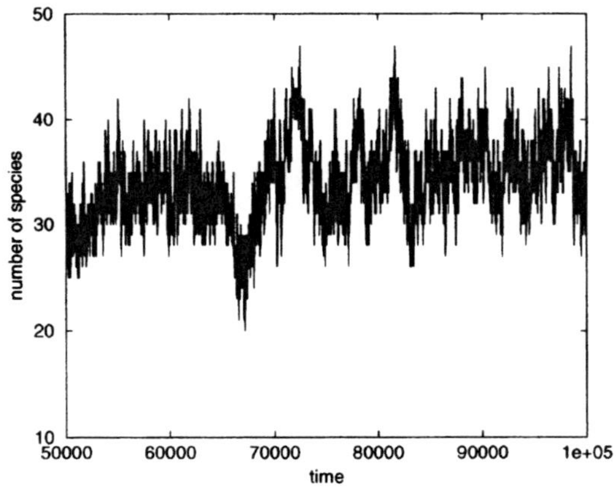

Fig. 4. Fluctuations in the number $s(T)$ of viable species. Each time-step corresponds to an attempted mutation or speciation.

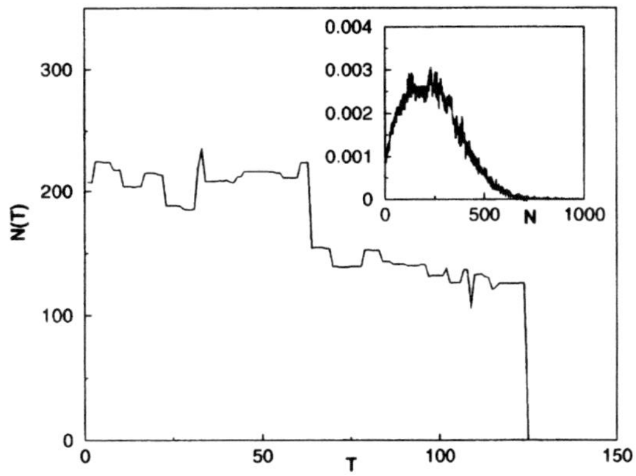

Fig. 5. The fate of a given species. Large fluctuations in the population number or biomass $N(T)$ can be given by the extinction or arrival of a strong competitor. Typically, the sudden decline of a population is caused by the arrival of a daughter that takes its place in the network. Inset: The corresponding stationary probability distribution $\mathcal{P}(N)$.

systems. Extinct species are replaced by speciations, leading to a long-term balance. The resulting stationary probability distribution of population numbers is shown in Fig. 5(b). It is indeed a broad distribution as anticipated in the previous section.

We now return to the effect of interactions on the network structure. Fig. 6(a) shows a typical snapshot of the overlap network at $\beta = 1$. The network is seen to be rather sparse: the species are forced into different ecological niches with little mutual overlap. In this example, there are 10 different resources and 9 predators.

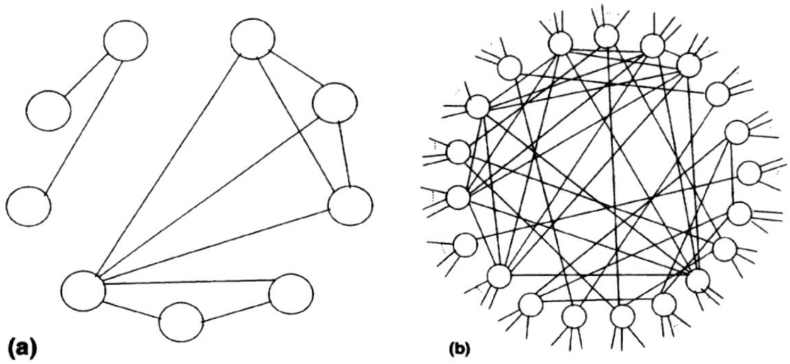

Fig. 6. (a) The overlap network at the upper level of a trophic bilayer for large competition ($\beta = 1$). The network is sparse, the species are organized into ecological niches with little mutual overlap. (b) At smaller competition (here $\beta = 0.5$), the size of the network strongly increases. Only a fraction of the nodes is shown. A species is connected to many other species, the ecological niches have disappeared.

A typical predator species feeds on about 3.5 resources. It has nonzero overlap with 2 other predators on average; a random pair of predator species has an average overlap of 0.04 (clearly, these network characteristics depend on the overall size of the network).

If competition is weak (e.g. $\beta = 0.5$), typical overlap networks are much denser, see fig. 6(b). The number of predators strongly increases, while their population numbers decrease. In this example, there are now 300 predators, each having a nonzero overlap to about 40 others. Hence, the system is no longer organized in ecological niches.

5 Conclusion

We have seen that the structure of species networks is shaped by evolutionary forces over long periods of time. Some of the resulting features are far from random: these networks are graded into trophic levels with a characteristic 'shape' defined by the level dependence of the number of species.

Darwinian evolution is a coupled process of mutations and selection. At the level of species communities, selection takes place through the dynamics of population numbers, and this dynamics depends on the interactions between species. We have identified two universal selective interactions, which are remarkably simple. Predation is the basic transport of energy in the system, competition forces the species into states with little overlap. In physics, mutual avoidance is a well known property of fermions. Competitive exclusion may thus be regarded as the Pauli principle of co-evolution: It generates the complexity of species networks just as its quantum-mechanical counterpart does for atoms and molecules.

Fig. 6 shows an example of the dependence of the local network structure on species interactions. In a more general context of biological networks, such relationships pose an interesting set of what physicists call *inverse problems.* Can we deduce from observed structures the evolutionary forces that have produced them?

Biologists have another important way of looking into the past. Phylogenetic trees can be constructed from phenotypical characteristics or from molecular sequences of today's species; molecular trees are becoming more and more accurate and complex with the rapidly increasing amount of available data. They are a partial record of speciations for those species that have survived to date. Uncovering the connections of their statistical properties with the underlying co-evolutionary dynamics is a challenging problem for the future.

Acknowledgment

M.L. is grateful to the Max-Planck-Institute for Colloids and Interfaces for its hospitality throughout the duration of this work.

References

1. J.E. Cohen, F. Briand, and C.M. Newman: *Community Food Webs,* Biomathematics Vol. 20 (Springer-Verlag, Berlin Heidelberg 1990)
2. P.H. Warren: Trends Ecol. Evol. **9**, 136 (1994); L.-F. Bersier and G. Sugihara: Proc. Natl. Acad. Sci. USA **94**, 1247 (1997)
3. J.J. Sepkoski Jr.: Phil. Trans. R. Soc. Lond. B **353**, 315 (1998)
4. D.M. Raup: Phyl. Trans. R. Soc. London B **325**, 421 (1989)
5. M.J. Benton: Science **268**, 52 (1995)
6. R.V. Solé, S.C. Manrubia, M.J. Benton, and P. Bak: Nature **388**, 764 (1997); J.W. Kirchner and A. Weil: Nature **395**, 337 (1998); R.V. Solé, S.C. Manrubia, J. Pérez-Mercader, M.J. Benton, and P. Bak: Adv. Complex Systems **1**, 255 (1998)
7. J.W. Kirchner and A. Weil: Proc. R. Soc. Lond. B **267**, 1301 (2000)
8. R.M. May: *Stability and complexity in model ecosystems* (Princeton University Press, 1973)
9. M.L. Rosenzweig: *Species diversity in space and time* (Cambridge University Press, 1995)
10. G. Caldarelli, P.G. Higgs, and A.J. McKane: J. theor. Biol. **193**, 345 (1998); B. Drossel, P.G. Higgs, and A.J. McKane: J. theor. Biol. **208**, 91 (2001).
11. P. Bak and K. Sneppen: Phys. Rev. Lett. **71**, 4083 (1993); R.V. Solé and S.C. Manrubia: Phys. Rev. E **54**, R42 (1996); L.A.N. Amaral and M. Meyer: Phys. Rev. Lett. **82**, 652 (1999).
12. J. Maynard Smith: *Evolution and the theory of games* (Cambridge University Press, 1982)
13. For a discussion of the 'scale' dependence of competition, see S.J. Gould: Phil. Trans. Roy. Soc. B **353**, 307 (1998)
14. M. Lässig, U. Bastolla, S.C. Manrubia, and A. Valleriani: Phys. Rev. Lett. **86**, 4418 (2001)
15. J. Maynard Smith: *Models in ecology,* (Cambridge University Press, 1974)
16. U. Bastolla, M. Lässig, S.C. Manrubia, and A. Valleriani: J. Theor. Biol. **212**, 11 (2001)
17. M. Lässig and A. Valleriani, *to appear.*

Modelling macroevolutionary patterns: An ecological perspective

Ricard V. Solé

[1] Complex Systems research Group, Department of Physics, FEN-UPC, Campus Nord B4, 08034 Barcelona, Spain
[2] Santa Fe Institute, 1399 Hyde Park Road, Santa Fe, NM 87501, USA

Abstract. Complex ecosystems display well-defined macroscopic regularities suggesting that some generic dynamical rules operate at the ecosystem level where the relevance of the single-species features is rather weak. Most evolutionary theory deals with genes/species as the units of selection operating on populations. However, the role of ecological networks and external perturbations seems to be at least as important as microevolutionary events based on natural selection operating at the smallest levels. Here we review some of the recent theoretical approximations to ecosystem evolution based on network dynamics. It is suggested that the evolutionary dynamics of ecological networks underlie fundamental laws of ecology-level dynamics which naturally decouple micro from macroevolutionary dynamics. Using simple models of macroevolution, most of the available statistical information obtained from the fossil record is remarkably well reproduced and explained within a new theoretical framework.

1 Macroevolution and extinction

Looking at today's biosphere, it is hard to realize how much it has changed through millions of years of evolution. Some groups of organisms, once successful and ecologically dominant, went extinct. Extinction is the eventual fate of all species. Even for some of the most succesful groups that flourished over very long periods of time became extinct and vanished. Their remains are provided by the fossil record, an incomplete but rather informative data set [7]. As David Jablonski points out: "it is hard to resist the fossil record as a source of spectacular evolutionary triumphs, grotesqueries and catastrophes" [25]. From the Cambrian explosion of Metazoan life (about 550 million years ago) complex forms have evolved on land and sea and a pattern of increasing diversity (figure 1) is matched by a pattern of extinction punctuated by large-scale events of devastating consequences (figure 2).

Extinction has been seldom considered as a relevant ingredient in neodarwinian theories. The classical view suggested by Darwin involved a slow process of decline: " species and groups of species gradually disappear, one after another, first from one spot, then from another, and finally from the world". The rapid, sometimes massive extinction of entire groups was assumed to be due to the incompleteness of the record. But we certainly know that this is not the case: extinctions happen to occur at different intensities in different moments of life's history. The record shows many small events together with some few, mass

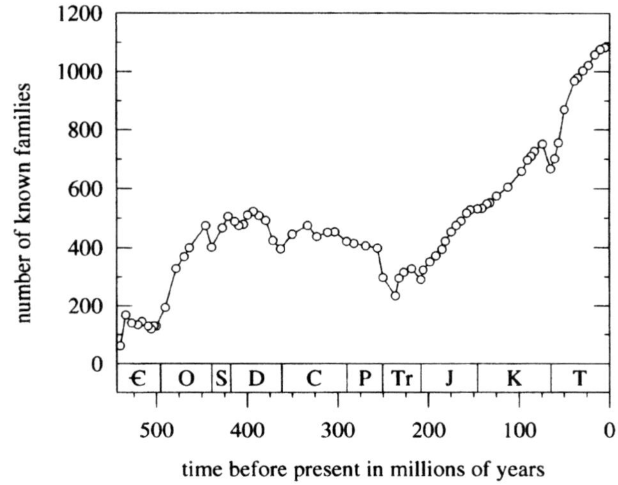

Fig. 1. Number of known marine families alive over the time interval from the Cambrian to the present. Data compiled by J. Sepkoski.

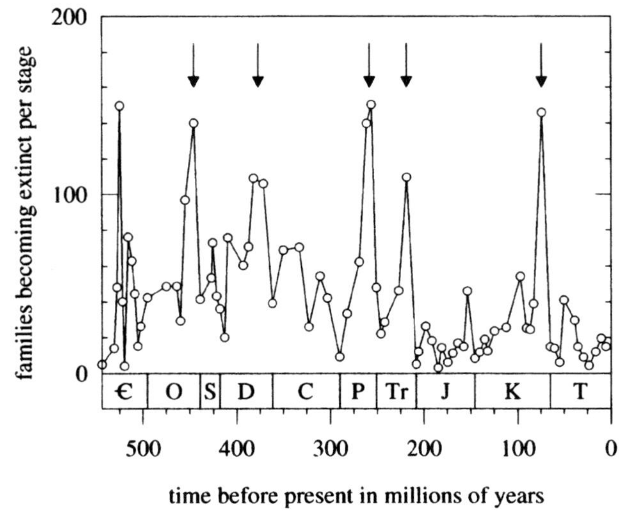

Fig. 2. Estimated extinction of marine animals in families per stratigraphic stage since the Cambrian. The arrows indicate the positions of the "big five" mass extinctions.

extinctions that wiped out a great part of Earth's diversity (see table 1). Of particular note are the five large peaks in extinction marked with arrows (figure 2). These are the "big five" mass extinction events which marked the ends of the Ordovician, Devonian, Permian, Triassic, and Cretaceous periods.

Table 1. Extinction intensities at the genus and species level for the big five mass extinctions of the Phanerozoic. Estimates of genus extinction are obtained from directed analysis of the fossil record while species loss is inferred using a special statistical technique.

	Genus loss (observed)	Species loss (estimated)
End Ordovician	60%	85%
Late Devonian	57%	83%
Late Permian	82%	95%
End Triassic	53%	80%
End Cretaceous	47%	76%

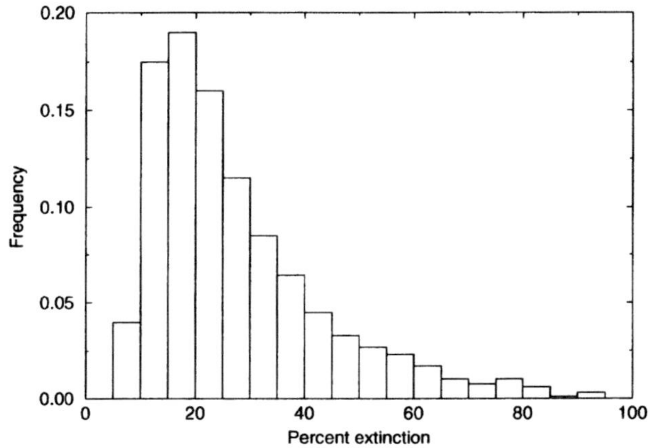

Fig. 3. Frequency distribution of extinction events. It shows a continuous range of values (instead of a bimodal one, as would be expected from a two-regime process, see text). We can see a maximum indicating a possible characteristic scale.

Although most neodarwinian theory almost ignores extinction, the fact is that the number of species extinctions in the history of life is almost the same as the number of originations [52]. Early analyses suggested that two basic regimes were involved in the overall pattern of extinction. The first would be "background extinctions" (possibly due to biological competition) and a second one, the "mass extinction" regime (perhaps associated to external stress). The observation of a continuous distribution (figure 3) does not support this view. Instead, it suggests a common causal origin for both large and small events. The problem is, of course, the nature of such an explanation.

Two major types of explanation have been suggested over the last decades. In one of them extinctions (particularly the large ones) result from external (non-biotic) events such as meteorite impacts, volcanic eruptions or changes in the magnetic field of Earth [52] [15] [50] [51]. This view has an obvious interest and relevance. There is clear evidence for external perturbations of the biosphere throughout the Phanerozoic and any theory of macroevolution should incorporate them. The end-Cretaceous event (K/T) is particularly well known and is consistent with a high-energy asteroid impact which generated severe darkening with a temporal cesation of photosynthetic activity on a very large scale and a rapid decrease in primary productivity (see below).

However, one should ask if these external events explain the previously mentioned features or are instead the trigger points of a cooperative biotic response. In this sense, it has been shown that the response of the biosphere to the size of the perturbation is far from linear [20] and the evidence does not suggest mass extinctions generally caused by such impacts. Actually a rather extensive search for extraterrestrial signatures at other stratigraphic intervals recording mass extinctions has been essentially negative [16]. In fact some impact structures have no link with known extinction events. This is the case of the Montagnais impact structure, with a size of 45-Km wide and an estimated age of ≈ 51 million years, with no associated extinction event. And a potentially gigantic impact crater found in the Kalahari desert (with a diameter of ≈ 350 Km) has been dated around the Jurassic-Cretaceous boundary, were no evidence for severe extinctions is known.

Most published studies on externally-driven extinctions involve the analysis of available data together with a number of (usually qualitative) hypothesis concerning the correlations between physical and biotic patterns. Few theoretical models in the paleobiological literature have developped quantitative predictions of statistical patterns and in this sense their conclusions are mainly based in a priori assumptions of what mechanisms are at work. There are a few relevant properties of the fossil record that should be explained by any plausible theory of macroevolution [64]:

1. The Extinction pattern of species (or families or other taxonomic units) is clearly "punctuated" (strictly speaking this term is not properly used in the same context as it was first introduced in evolutionary biology). This means that rapid changes can be seen in the system in terms of large extinction (or diversification) events. This pattern has been shown to display long-range correlations [63] [65] [24].
2. The distribution of extinctions $N(m)$ of size m follows a power-law decay with $N(m) \propto m^{-\alpha}$ with $\alpha \approx 2$ [62] [41]
3. The lifetime distribution of family durations $N(t)$ follows a power-law decay $N(t) \propto t^{-\kappa}$ being $\kappa \approx 2$ [58].
4. The statistic structure of taxonomic systems also shows fractal properties. For example, the number of genus formed by S species, $N_g(S)$, follows a power-law distribution with $N_g(S) \propto S^{-\alpha_b}$ with $S^{-\alpha_b} \approx 2$ [11] [12].
5. A study of the rates at which different groups of organisms go extinct through time shows that a species might disappear at any time, irrespective of how long it has already existed. This result, first reported by Leigh Van Valen strongly modified the ecological view of macroevolution [68] [8]. The fine-scale structure of these patterns is, however, episodic (see figure 4) and re-

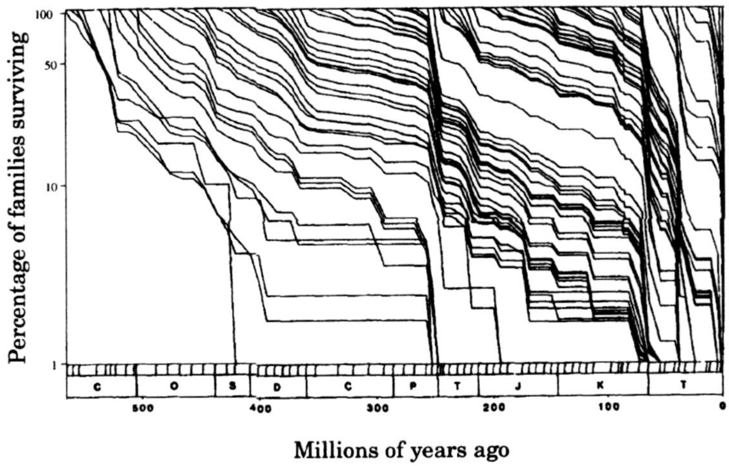

Fig. 4. Survivorship of 2316 families of marine animals over the past 600 Million years. Each line is a so called *pseudocohort* which starts (upper left) with th efamilies present in the fossil record at a point in time. Mass extinctions appear as sharp drops in survivorship (adapted from Raup, 1986).

flect the interplay between slow dynamical processes and rapid changes that appear as drops in the diversity curves [54].

The presence of long-range correlations in the fossil record time series has been a source of controversy [43] [32]. One clear conclusion from the disagreements between different studies is that the direct application of spectral techniques to the fossil record data has been far from appropriate, and the only clear conclusion is that there are long-range correlations [1], although their range and origins are debatable. Some authors have recently explored the problem of properly analysing the FR data sets [18] by means of the Lomb's method, which allows to obtain appropriate characterizations of uneven time series with nonstationary behavior. In this context, V. Dimri and M. Prakash have found evidence of long-range correlations together with a periodic component, thus confirming the presence of at least to types of structures in the large-scale dynamics of the biosphere. These results have been confirmed by means of wavelet analysis (Solé and Valverde, unpublished). The wavelet transform [49] replaces the Fourier transform's sinusoidal waves by a family generated by translations and dilations of a window called a wavelet. In this sense, a big disadvantage of a Fourier expansion is that it has only frequency resolution and no time resolution. This means that although we might be able to determine all the frequencies present in a signal, we do not know when they are present. The wavelet analysis takes advantage of nonstationary behavior and allows to see that the fossil record shows a fractal pattern over long time scales.

More recently, Plotnick and Sepkposki have presented a re-analysis of the available data suggesting that it is better understood in terms of a conceptual model, based on a hierarchy of levels that interact in a multiplicative fashion [48]. The authors compare their model outcomes with improved extinction and origination data and find a good agreement, thus concluding that it provides a better understanding of macroevolutionary patterns than the ones presented by previous models (to be discussed below). In this sense, it is important to establish appropriate criteria allowing to evaluate the value of a given model when compared to the fossil record data. Most of the published literature (both in paleontology and physics) present models or analysis that concentrate in one or two basic traits, completely ignoring the whole picture that emerges when all the available data is taken into account. Although there are many open questions emerging from the new theoretical approaches [39] [27] one clear test for any sensible theoretical explanation of macroevolutionary patterns of extinction and diversification is to be able to reproduce as many quantitative traits as possible.

A number of mathematical models of long-term evolutionary patterns have been proposed by several authors. The earliest and most appealing of them is Sepkoski's model of competition, which assumed that the Cambrian, Paleozoic and Modern evolutionary faunas each diversified logistically as a consequence of early (exponential) growth followed by a slowing down as ecosystems became filled [56] [26]. By tuning several parameters and a set of external perturbations similar to those suggested by the fossil record, a very similar pattern of diversification is obtained. Although this model is simple and appealing, the lack of a unique parameter determination and the number of assumptions implicit in the competition model makes it essentially descriptive. A similar criticism can be applied to other models, although their value as a theoretical framework is undeniable.

More recently, a new generation of models give support to a scenario where externally-driven ecological responses might play a relevant role [5] [64] [47]. These models have shown that it is possible to reproduce many quantitative traits displayed by the fossil record and even a new theoretical interpretation of the macroevolutionary process. The implications for macroevolution are significant. They suggest that multispecies interactions are a key ingredient in shaping the structure of evolving ecosystems and that the fate of individual species would be the result of collective phenomena, not reducible to a list of independent fitnesses. In this context, it has been suggested that long-term, ecological-level network dynamics provides the natural decoupling between micro- and macroevolutionary patterns [59] [64].

Evolution does not take place in an ecological vacuum. Even the first ecologies emerging from the Cambrian boundary have been shown to display some of the characteristic features of modern ecologies [17]. Besides, the aftermath of mass extinctions show that ecological-based responses underlie the extraordinarily protracted lag-times for recovery before similar diversity levels are reached again [33] [21] [22] [23]. Besides, in many well-documented cases, changes in the pattern of extinction and diversification are directly associated with ecological responses (such as the emergence and evolution of mineralized exoskeletons triggered by predation).

This review has been writen with a partisan view of macroevolution based on an ecological representation of evolving biological structures. In that sense, I am not considering other types of models where such a network of interactions is essentially ignored. This is certainly a limitation, since I am sure that other approaches, such as Newman's stress model [41] [42] or Sibani's reset model [57] have a very important value and are close to reality in many ways (not to mention the fact that in many ways the Phanerozoic involves different sources of innovation and thus of nonstationarity). In truth, the final answers to the problems arising from the patterns displayed by the fossil record are likely to be understood by using appropriate ingredients provided by these different approximations.

2 Coevolution on a rugged fitness landscape

The first attempt to understand large-scale evolution in terms of a complex adaptive system with interactions among different species was introduced by Kauffman and Johnsen [29], who used previous theoretical work on fitness landscapes [30]. The model is inspired in previous theoretical work by Per Bak and co-workers on self-organized criticality [3], [5], [6]. The basic idea of the fitness landscape metaphor is that single species can be characterized in terms of a string of genes or traits, $S_1S_2...S_N$ which constitute the "genotype" and have an associated real number $\Phi(S_1S_2...S_N) \geq 0$ usually normalized to one. This quantity is the *fitness* of the string and the distribution of fitness values over the space of genotypes defines the *fitness landscape* (figure 5). Depending upon the distribution of the fitness values, the fitness landscape can be more or less *rugged*. The rugeddness of the landscape is a crucial property, strongly constraining the dynamics. If we consider a population of strings, then the way this population evolves depends on how mountainous the landscape, how large is the population size and on mutation rates [30].

Rugged landscapes (RL) are a common feature of many different complex systems, from RNA viruses to glasses. They have been studied from various viewpoints in disparate areas such as biophysics of macromolecules to combinatorial optimization problems.

If we want to model macroevolutionary dynamics, then in principle many different, coupled species have to be taken into account. Each species is characterized by the number of traits N and by another parameter K which is in fact a measure of the degree of ruggedness (it gives the number of epistatic interactions among genes/traits). The fitness of a given string is obtained by means of a table of values, as the one shown in figure 6. Here a $N = 3, K = 2$ system is shown, together with the corresponding landscape, here just a simple three-dimensional cube. Adaptive walks only occur in the direction of increasing W, and so the system is finally frozen at one of the two local maxima (here indicated by means of circles).

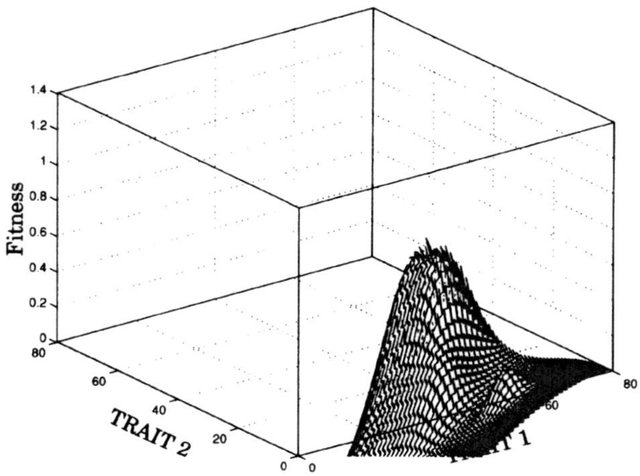

Fig. 5. Simple, two-dimensional continuous rugged landscape. This corresponds to the early metaphor suggested by Sewall Wright. Here a given species is defined as a two-trait pair (x_i, y_i). For each pair a fitness $\phi(x_i, y_i)$ can be defined. The ruggeddness of the landscape controls the population flow through trait space. If the landscape is very rugged, historic effects play a dominant role.

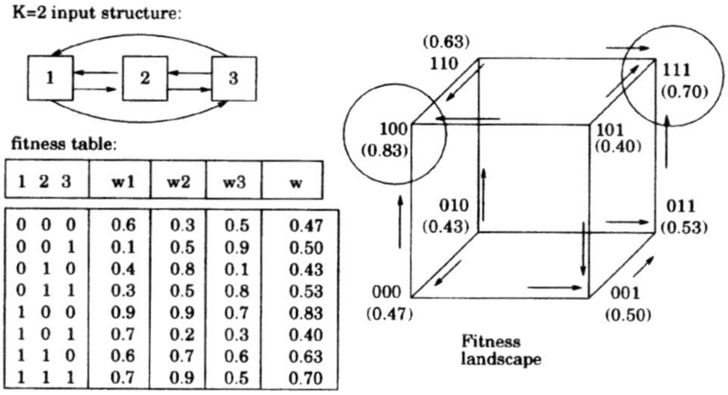

1 2 3	w1	w2	w3	w
0 0 0	0.6	0.3	0.5	0.47
0 0 1	0.1	0.5	0.9	0.50
0 1 0	0.4	0.8	0.1	0.43
0 1 1	0.3	0.5	0.8	0.53
1 0 0	0.9	0.9	0.7	0.83
1 0 1	0.7	0.2	0.3	0.40
1 1 0	0.6	0.7	0.6	0.63
1 1 1	0.7	0.9	0.5	0.70

Fig. 6. How to built a fitness landscape. We have a $N = 3$ string with $K = 2$ interactions among each trait (so called epistatic interactions) and a table providing the fitness $W(i)$ of each individual trait given a particular string sequence.

This NK model has been widely explored and many relevant results concerning its statistical properties have been derived. This is not surprising, given its close similarity with spin glass (SG) models. As in SG, frustration takes place and allows to understand the distribution of peaks when K is tuned. The basic dynamics in this model involves adaptive walks. Here for a single species we choose a given trait and flip a coin (i. e. mutate) the bit. Then we look at the fitness table and if the average fitness of the new configuration is larger than the last one, an adaptive walk occurs and so a movement in the fitness landscape.

If not, no walk is allowed to occur. This simple procedure leads to a hill climbing in the landscape until a local peak is reached. Afterwards, nothing happens. For $K = 0$ (the Fujiyama landscape) no interactions among different traits are present and a very smooth landscape is obtained, with a single global maximum and an expected number of walks $L_w = N/2$ to reach the optimum. This is a highly correlated, simple landscape.

At the other extreme, when $K = N - 1$, the landscape is fully random. Several interesting properties have been reported, among others: (a) the number of local fitness optima is maximum; (b) the expected number of fitter one-mutant variants drops by 1/2 at each improvement step; (c) the length of adaptive walks to optima are short, with a characteristic value $L_w \approx \log(N)$.

For an uncorrelated landscape, each bit string is assigned a fitness at random, so even single-bit changes may have very different fitness values. Although this is not a biologically realistic model, it allows to obtain some basic analytic results and will help to understand the coevolutionary patterns arising from the Kauffman-Johnsen model.

The number of maxima is easily calculated. If only one-bit mutations per string are allowed to occur, each string has N one-bit neighbors. The probability that any one string has higher fitness than any of its neighbors is:

$$P_1 = \frac{1}{N+1} \tag{1}$$

and thus the average number of local maxima is $\langle M_1 \rangle = P_1 2^N$. The length of the walks can also be estimated [55]. Let us assume that we start with the least fit string, and that the range of fitness values is constrained to the unit interval, with a uniform distribution over fitness space. Any mutation will give a fitness increase with an average value, for the first walk, of $\langle F_1 \rangle = 1/2$. The second walk will increase the fitness to an average $\langle F_2 \rangle = 3/4$ and after k walks we will have:

$$\langle F_k \rangle = 1 - \left(\frac{1}{2}\right)^k \tag{2}$$

It is easy to show that the probability of a string not reaching a local maximum after k walks is:

$$P_k = \left(1 - \frac{1}{2^k}\right)^{N-1} \tag{3}$$

If we define $\omega_k = 1 - P_k$, the probability $\mathcal{L}_k$ that the walk will last through the k-th step is:

$$\mathcal{L}_k = \prod_{r=1}^{k} \omega_r = \prod_{r=1}^{k} \left[\left(1 - \frac{1}{2^r}\right)^{N-1}\right] \tag{4}$$

Most walks will proceed until $\mathcal{L}_k < 0.5$, i. e. until $\approx \log(N-1)$ steps.

Now the problem is how to obtain a more complete picture of an evolving system formed by many species in interaction. This can be done by using the so called NKC model [29]. The parameter C introduces the number of couplings between species. Again each species is represented by just a string (instead of a population of individuals) which somehow defines the average characteristics (the phenotype) for that particular species. Now each trait receives "inputs" from C other traits belonging to different species. These traits are chosen at random between the S species.

The NKC model shows two well defined dynamical regimes (phases). These regimes are the high-K, chaotic phase, where changes in the ecosystem are always taking place (i. e. the system does not settle down in a number of local optima) and the low-K, ordered (frozen) phase where local optima are reached by all species (the so called Nash equilibria in economic theory). At the boundary separating the two phases, complex dynamics takes place. For a given N, if C is small (below a given threshold) then the whole population evolves into a state where no further changes take place and all species are at Nash equilibria. However, after a critical point is reached, the dynamics becomes chaotic and no final steady state is obtained. Just at the boundary, species in a finite system reach local peaks but any small perturbation generates a *coevolutionary avalanche* of changes through the system. The distribution of these avalanches is a power law, as expected for a critical state. Kauffman and Johnsen mapped these avalanches into extinction events, suggesting that *the number of changes* in species is proportional to the extinction of less-fit variants. If this analogy is used then the obtained scaling relation for avalanches of S *changes* is $N(s) \propto S^{-1}$, which does not agree with the value reported from the fossil record. However, a further version of this model (allowing evolution in the parameters) has been shown to self-organize to the critical state [31] with avalanches following the correct $\tau = -2$ exponent (figure 7). The final picture that emerges from this last model is that as species tune their own landscapes (by readjusting the ruggedness) they poise the entire ecosystem close to the critical boundary.

Some analytic work on the NKC model has been done by Per Bak and coworkers. If we restrict ourselves to the $K = N - 1$ case and assume a large number of species, the existence of two well-defined phases in the (N, C)-plane can be derived [4]. Let us assume that the fitness values are uniformly distributed in the interval $U = [0, 1]$. Additionally, let us assume that instead of keeping the C randomly chosen foreign genes that any species depends on, we exchange this "quenched" randomness for "annealed". This just means that the C connections are randomly assigned at each time step. Finally, let us consider a very large ecosystem so that a probability density can be defined. Here $\rho_M(F, t)$ will be the fraction of species with fitness F and M less-fit 1-mutan neighbors at time t.

Bak et al. define a quantitative measure of the evolutionary activity in the system [4]. This quantity, $A(t)$, gives the probability that a change in a random gene leads to higher fitness (and therefore is accepted):

$$A(t) = \sum_{M=0}^{N} \left(1 - \frac{M}{N}\right) \int_0^1 dF \rho_M(F; t) \tag{5}$$

where $1 - M/N$ is the probability that the change of a single random unit leads to higher fitness.

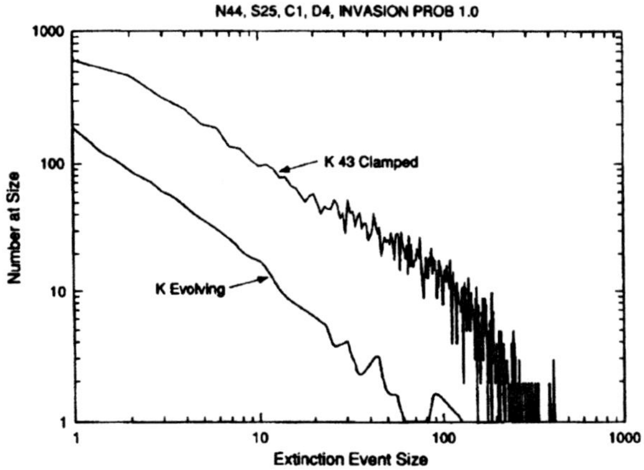

Fig. 7. Power law distribution of coevolutionary avalanches in the modified Kauffman-Johnsen model. Here two different results are shown. One is for a NKC network where the K-parameter has been fixed to a high value (in the chaotic regime) and the lower one is obtained in a system where the landscape ruggedness is allowed to evolve. The parameters are $C = 1, S = 25, N = 44$ and K evolves towards an intermediate value ($K \approx 22$).

Now the probability that such a mutation is accepted and leads to a fitness F (for the changed species) is:

$$\Phi(F;t) = \int_0^F dF' \phi(F';t) \tag{6}$$

where

$$\phi(F',t) = \frac{1}{1-F'} \sum_{M=0}^{N} \left(1 - \frac{M}{N}\right) \rho_M(F;t) \tag{7}$$

Using these quantities, a master equation can be derived, leading to:

$$\frac{\partial}{\partial t}\rho_M(F,t) = -\left(1 - \frac{M}{N}\right)\rho_M(F;t) + B_{M,N}(F)\Phi(F,t)$$
$$- \frac{C}{N}A(t)\rho_{M,N}(F,t) + \frac{C}{N}A(t)B_{M,N}(F,t) \tag{8}$$

This equation gives the time evolution of the (relative) number of species with fitness F and M less-fit neighbors. Here $B_{M,N}(F)$ is a binomial distribution with mean F, i. e.:

$$B_{M,N} = \frac{N!}{M!(M-N)!} F^M (1-F)^{N-M} \tag{9}$$

standing for the probability thet M out of N one-mutant neighbors to a genome with fitness F are less fit than F.

Although a detailed analysis of the parameter space can be derived, here only an estimation of this phase space will be obtained. A trivial solution of the master equation is given by all species placed in local fitness maxima:

$$\rho_M^*(F,t) = \delta_{M,N}\rho(F) \tag{10}$$

As usual, the relevance of this solution depends on the connectivity C. Here $\rho_M^*(F,t)$ will be attractive if $C = 0$ and at the other extreme, when $C/N \ll 1$ we get

$$\rho_M^*(F,t) = B_{M,N}(F) \tag{11}$$

which corresponds to maximum activity $A = 1/2$.

The interesting properties are observed at intermediate $(0 < C < N)$ values of the connectivity. If some stationary activity is present for a given C value, we could ask whether this activity stops or not. We can obtain an approximate relation between C and N that will give us the critical line separating frozen from chaotic phases.

We have previously mentioned that in NK landscapes with $K = N - 1$, an average number of adaptive walks (until a local maximum is reached) follows a logarithmic dependence

$$\mu_1 \approx \ln(N) \tag{12}$$

For a NKC landscape this is a lowest bound to the number of changes per species by which the NKC model can evolve to the fixed point $\rho_M^*(F,t)$.

Now, suppose that, for a given C-value, the species have been arranged in order to satisfy (10). A small perturbation is then introduced: the fitness of one species is changed to a random value, being the others in the same state. Such a perturbed species will need an average of μ_1 steps before to reach a local maximum. But in fact the fitness of other species depends, through C, on the values taken by other genes/traits. If any of these genes/traits are among the μ_1 ones that changed through the walk, the affected species will set back in evolution. Our question of course is whether or not the initial change can trigger a "chain reaction" able to percolate through the system. The critical condition is easily obtained:

$$\mu_1 \frac{C_{crit}}{N} = 1 \tag{13}$$

i. e. when, on the average, one out of C randomly chosen genes is among the μ_1 changed genes. This gives the critical line in the (N, C) space:

$$C = \frac{N}{\ln(N)} \tag{14}$$

This line separates the two phases.

3 Network model of macroevolution

One of the criticisms to the previous model (and other early models based on self-organized critical behavior, such as the Bak-Sneppen model [5]) is that they lack true extinction and diversification [35]. Although the evolutionary activity in these models has something to do with the underlying extinction dynamics, it is not obvious how to map the first into the second. On the other hand, one of the obvious rules to be considered by any reasonable model is replacement of empty niches by surviving species. And these species will interact through a new, evolving network of connections.

The standard mathematical approach to population dynamics is the Lotka-Volterra (LV) n-species model,

$$\frac{dN_i}{dt} = N_i\left(\epsilon_i - \sum_{j=1}^{n} \gamma_{ij}N_j(t)\right) \tag{15}$$

where $\{N_i\}$, $i = 1, ..., n$ are the populations of each species. These models have been explored in deep. Two main qualitative problems have been considered: (i) small-n problems, involving two or three species and (ii) large-n models, involving a full network of interacting species.

The Lotka-Volterra equations used in most models of multispecies ecosystems are too difficult to manage if the matrix of interactions Γ is formed by time-dependent terms (as one would expect in an evolving ecosystem). We want to retain the basic qualitative approach, but our interest is shifted from population sizes to the appearance and extinction of species. Here species are defined as a binary variable: $S_i = 0$ (extinct) or $S_i = 1$ (alive). The state of such species evolves in time (now assumed discrete) according to

$$S_i(t+1) = \Phi(\phi_i(t)) = \Phi\left(\sum_{j=1}^{n} \gamma_{ij}(t)S_j(t)\right) \tag{16}$$

with $i = 1, ..., N$. Here $\Phi(z) = 1$ if $z > 0$ and zero otherwise. Equation (2) can be understood as the discrete counterpart of (1), but involving a much larger

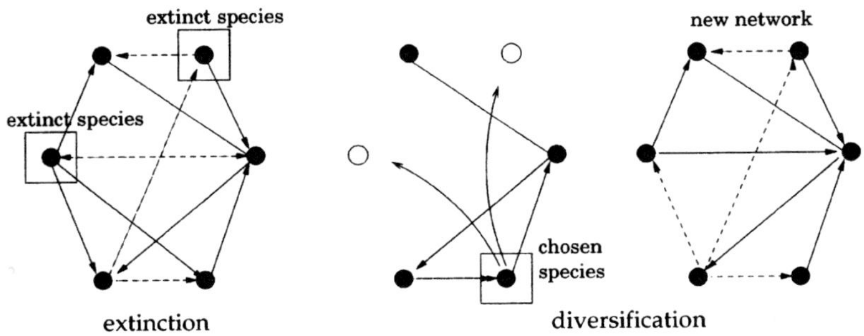

Fig. 8. Basic rules of the network model of evolution. After a number of species are extinct (here two of them), one of the survivors is chosen (center) to repopulate the network. The diversification is performed by copying the connections of the chosen species. These rules are completed with a random change of connections involving both internal and external changes.

time scale. In this model, first introduced by Solé [59] [60] [61], the i–th species is in fact represented by the set of connections $\{\gamma_{ij}, \gamma_{ji}\}$, $\forall j$. The elements γ_{ij} are the *inputs* and define how the the other species influence it. The symmetric elements γ_{ji} are the *outputs* and represent the influence of this species over the remaining ones in the system.

The dynamics of this model is defined in three steps:

1. Changes in connectivity. Each time step we change one connection γ_{ij} which takes a new, random value $\gamma_{ij}(t+1) \in [-1, 1]$, for each $i = 1, ...N$, with $j \in \{1, ..., N\}$ chosen at random. This rule is linked to changes in species interactions. They could be associated with external causes or simply be the result of small changes as a consequence of coevolution. This rule introduces random, small changes into the network.
2. Extinction. The local fields $\phi_i(t) = \sum_j \gamma_{ij}(t) S_j(t)$ are computed, and all species are synchronously updated following (2). If the k–th species goes extinct, then all the connections that define it are set to zero, that is $\gamma_{kj} = \gamma_{jk} \equiv 0$, $\forall j$. This updating introduces extinction and selection of species. Those sets of connections which make a species stable will remain. But in removing a given species, some positive connections, with a stabilizing effect on other species can also disappear, and the system can become more unstable.
3. Replacement. Some species are now extinct (i. e. $S_k = 0$), empty sites are available and diversification is introduced. A living species is picked up at random and "copied" in the vacant niches. The new species are basically identical to the one randomly choosen, except for a small random change in all their connections. Specifically, let S_c the copied species. For each extinct species S_j (vacant spaces), the old connections are set to zero, and the new connections γ_{ij} and γ_{ji} are given by $\gamma_{kj} = \gamma_{cj} + \eta_{kj}$ and $\gamma_{jk} = \gamma_{jc} + \eta_{jk}$. Here η is a small random variation (typically $\eta_{kj} = O(10^{-2})$. In this way, the new species are the result of the diversification of one of the survivors.

Fig. 9. Power law distribution of extinction events in the network model (left). The corresponding time series is shown (right).

The random changes in the network connections make the trophic links between species more and more complex. We can quantify such complexity by means of an adequate statistical measure. Let us first consider the time evolution of connections. Let $P(\gamma^+)$ and $P(\gamma^-) = 1 - P(\gamma^+)$ be the probability of positive and negative connections, respectively. The time evolution of $P(\gamma^+, t)$ is defined by the master equation

$$\frac{\partial P(\gamma^+,t)}{\partial t} = P(\gamma^-,t)P(\gamma^- \to \gamma^+) - P(\gamma^+,t)P(\gamma^+ \to \gamma^-) \quad (17)$$

From the definition of the model, we have a transition rate per unit time given by $P(\gamma^+ \to \gamma^-) = P(\gamma^+ \to \gamma^-) = 1/(2N)$ and so we have an exponential relaxation

$$P(\gamma^+,t) = (1 + (2P_0 - 1)\exp(-t/N))/2$$

where $P_0 = P(\gamma^+, 0)$. This result leads to an exponential decay in the local fields, $\phi_i(t) \propto \exp(-t/N)$. As a result, the system evolves towards a (critical) state where the sum of the inputs introduced by the coevolving partners are small and so small changes involving single connections can generate extinctions.

We can use the entropy of connections per species, i. e. the Boltzmann entropy

$$H(P(\gamma^+,t)) = -P(\gamma^+,t)\log(P(\gamma^+,t)) - (1 - P(\gamma^+,t))\log(1 - P(\gamma^+,t)) \quad (18)$$

as a quantitative characterization of our dynamics. The Boltzmann entropy (also known as the Shannon entropy) gives us a measure of disorder but also a measure of uncertainty. It is bounded by the following limits: $0 \leq H(P(J^+,t)) \leq \log(2)$. These limits correspond to a completely uniform distribution of connections (i. e. $P(\gamma^+,t) = 1$ and $P(\gamma^-,t) = 0$) with zero entropy and to a random distribution with $P(\gamma^\pm,t) = 1/2$ which has the maximum entropy. Our rules make possible the evolution to the maximum network complexity, here characterized by the upper limit of the entropy.

As $H(P(\gamma^+,t))$ grows, after a large extinction event, towards its maximum value $H^* = \log(2)$, sudden drops take place near large extinctions. So our system slowly evolves towards an "attractor" characterized by a randomly connected network. At such state, small changes of strength $1/N$ can modify the sign of ϕ_i and extinction may take place. At this point, one clearly sees what is the role that external perturbations play: for them to trigger a large extinction, it is necessary that they act on a system located close to the critical state (here, the network close to the maximum entropy). A large extinction will never be found in a system with a low entropy of connections even with a reasonably large external perturbation. We can see that a wide distribution of extinctions is obtained: it is a power-law distribution, $N(s) \approx s^{-\tau}$ with $\tau = 2.05 \pm 0.06$, consistent with the information available from the fossil record. Actually, the previous rules are able to reproduce the most relevant features of the observed dynamics in the fossil record, including the presence of well-defined transient trends (see table 2 and figure 10).

A mathematical model can be derived for this mean-field approximation [36]. The starting point is again a set of N species, now characterized by a single

Table 2. some basic trends of macroevolutionary patterns. Observed and predicted by the SOC model (see text). All the quantitative reported exponents from the FR are reproduced by the SOC model as well as the qualitative features like the diversification curves. ((1): recent studies seem to suggest a lower exponent, of about $\gamma \approx 1.6$ [44]).

Property	Observed	SOC Model
Dynamics	Punctuated	Punctuated
Mass extinctions	Few events	Expected
Diversity	Increasing	Transiently increasing
Species decay	Exponential	Exponential
Extinction pattern, $N(E)$	Power law ($\alpha \approx 2$)	Power-law ($\alpha \approx 2$)
Hurst exponent, H	persistence, $H > 1/2$	persistence $H > 1/2$
Genera lifetimes $N(T)$	Power law$^{(1)}$ ($\gamma \approx 2$)	Power law ($\gamma \approx 2$)
Genera-species $N_g(S)$	Self-similar, ($\tau \approx 2$)	Self-similar,($\tau \approx 2$)

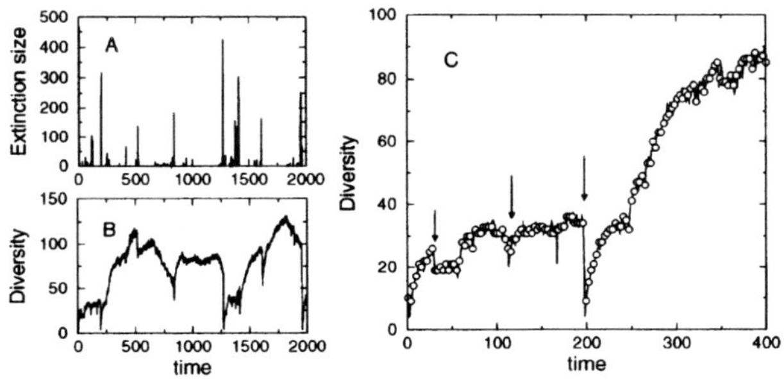

Fig. 10. Patterns of extinction and diversification for the mean field approximation to the network model. (A) Extinction dynamics, (B) diversity dynamics, here defined in terms of the number of genera in the system. In (C) we show the transient increase in diversity. It fits quite well the observed trends displayed by the fossil record. We also show the occurrence of large extinctions, as indicated by the arrows.

integer quantity ϕ_i ($i = 1, 2, ..., N$). This quantity will play the role of the internal field, as before. Each species is now represented by this single (now assumed integer) number $\phi_i \in \{-N, -N+1, ..., -1, 0, 1, ..., N-1, N\}$.

The dynamics consists of three steps: (a) with probability $P = 1/2$, $\phi_i \to \phi_i - 1$, otherwise no change occurs (this is equivalent to the randomization rule in the network model); (b) all species with $\phi_i < \phi_c$ (below a given threshold) are extinct. Here we use $\phi_c = 0$ but other choices give the same results. The number of extinct species, $0 < E < N$, defines the extinction size. All E extinct species are replaced by survivors. Specifically, for each extinct site (i. e. when $\phi_j < \phi_c$) we choose one of the $N - E$ survivors ϕ_k and $\phi_j = \phi_k$; (c) after an extinction event, a wide reorganization of the web structure occurs. In this simplified model this is introduced as a coherent shock. Each of the survivors are updated as $\phi_k = \phi_k + q(E)$, where $q(E)$ is a random integer between $-E$ and $+E$.

The master equation for the dynamics involves the following three-step process:

$$N(\phi, t+1/3) = \frac{1}{2}N(\phi, t) + \frac{1}{2}N(\phi+1, t) \tag{19}$$

$$N(\phi, t+2/3) = N(\phi, t+1/3) + N(\phi, t+1/3)\sum_m \frac{m}{N-m}P(m) \tag{20}$$

if $\phi > 0$ and zero otherwise. Finally:

$$N(\phi, t+1) = N(\phi, t+2/3) - N(\phi, t+1/3) + \sum_{q>-\phi} N(\phi+q, t+1/3)P(q) \tag{21}$$

equations (19–21) lead to the full master equation for the dynamics:

$$N(\phi, t+1) = \frac{1}{2}\sum_{q=-\infty}^{+\infty}\sum_m \frac{P(m)}{2m+1}\theta(m-|q|) \times \Big[N(\phi+q, t) - N(\phi+q+1, t)\Big] + \frac{1}{2}[N(\phi, t) + N(\phi+1, t)]\sum_m \frac{mP(m)}{N-m} \tag{22}$$

Where two basic statistical distributions, which are self-consistently related, have been used. These are:

$$P^*(q) = \sum_m \frac{P_e(m)}{2m+1}\theta(m-|q|) \tag{23}$$

which is an exact equation giving the probability of having a shock of size q. Here $P_e(m)$ is the extinction probability for an event of size m, and now we have a mean-field approximation relating both distributions:

$$P_e(m) = \sum_q P^*(q)\delta\Big[\sum_{\phi=1}^{q-1} N(\phi) - m\Big] \tag{24}$$

The last equation uses the so called average profile $N(\phi)$. This function is the time-averaged distribution of ϕ-values. Here we follow the Manrubia-Paczuski argument [36] for the mesoscopic regime $1 \gg q \gg N$. By Taylor-expanding the

the master equation, i. e.

$$N(\phi) = \frac{1}{2}\sum_{q=-\infty}^{+\infty}\sum_{m}\frac{P(m)}{2m+1}\left\{2N(\phi+q) + \frac{\partial N}{\partial \phi}\bigg|_{\phi+q} + \frac{1}{2}\frac{\partial^2 N}{\partial \phi^2}\bigg|_{\phi+q} + \dots\right\}$$

$$+\frac{1}{2}\left\{2N(\phi) + \frac{\partial N}{\partial \phi}\bigg|_{\phi} + \frac{1}{2}\frac{\partial^2 N}{\partial \phi^2}\bigg|_{\phi} + \dots\right\}\sum_{m}\frac{mP(m)}{N-m} \tag{25}$$

and using a continuous approximation, it is easy to see that the previous equation reads:

$$\frac{1}{2}\int dm \int_{-m}^{m}\frac{P(m)}{2m}\left\{2\Big[\frac{N(\phi+q)}{N(\phi)} - 1\Big] + \frac{\partial LnN}{\partial \phi}\bigg|_{\phi+q} + \dots\right\}$$

$$+\frac{1}{2}\left\{2 + \frac{\partial LnN}{\partial \phi}\bigg|_{\phi} + \dots\right\}\int\frac{mP(m)}{N-m}dm = 0 \tag{26}$$

Using an exponential *ansatz* for the average profile, i. e. $N(\phi) = \exp(-c\phi/N)$, we can integrate each part of the last equation, assuming $N(\phi+q)/N(\phi) = \exp(-cq/N) \approx 1 - cq/N$. It is easy to check that the first term cancels exactly, the second gives $-2c/N$ and the third scales as $(1 - O(1/N))N^{1-\tau}$. So the previous equation leads to:

$$-\frac{2c}{N} + N^{1-\tau}G\Big(1 - O(\frac{1}{N})\Big) = 0 \tag{27}$$

in order to satisfy this equality, we need $\tau \approx 2$, which gives us the scaling exponent for the extinction distribution. This is confirmed by numerical simulations, which give $\tau = 2$.

This model (actually both of them) also display the observed exponential decay in species survival displayed by fossil record data. In figure 11 we show an example of our runs. This corresponds to the law of constant mean extinction rate, also known as Van Valen's law [68]. As mentioned in the introduction, this law maintains that the probability of extinction within any group remains essentially constant through time. This is a consequence of the Red Queen theory and an observational result. This is, however, an average: on average, extinction rates are constant but a close inspection of the decay curves shows both continuous and episodic decays. The sudden, episodic drops are often associated with mass extinctions and are usually assumed to be the result of external perturbations.

The episodic nature of the species decay is easily explained by our model. Though long periods of stasis and low extinction rates give a constant decay, *the same* intrinsic dynamics generates the episodes of extinction involving several (some times many) species. The survivorship curves shown in figure 11 are generated by starting at a given (arbitrary) time step in the simulation and following all the species present at this time step. The exponential decay in the number of survivors is closely related to the monotonous drift that the system experiences towards the extinction threshold, due to the constant change of connections to

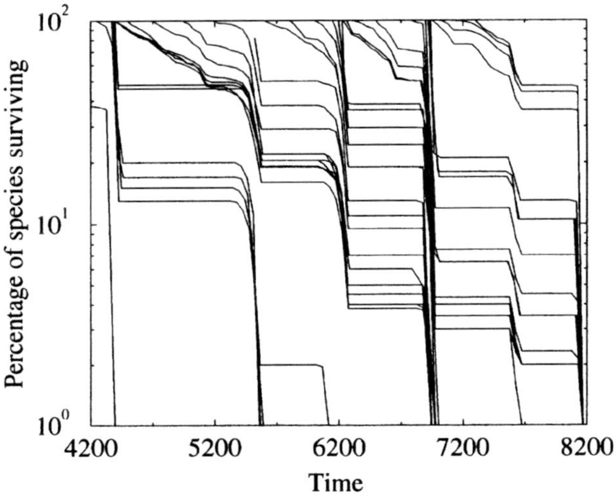

Fig. 11. Pattern of species extinction in the network model of macroevolution: on average, species get extinct in an exponential way, but this average pattern is punctuated by coextinction associated with mass extinction events.

random values. As we can see (and this is rather typical) both constant and episodic decays are observed. We do not need to seek for a special external explanation for the episodic decay. Obviously, an external cause can trigger a large extinction event by altering the network dynamics at the critical state. Extinctions are an unavoidable outcome of network dynamics. Though some selection of connections is present after each extinction event, unpredictability is always present. A given species with a high "fitness" (defined in terms of the total input field it receives) can get extinct in a few steps due to an extinction avalanche propagating throughout the network.

4 Extinction in layered networks

The third class of models to be analysed here involves a very important aspect of real ecologies: the presence of different trophic levels. This is important not only because it adds an ingredient of realism to the topology of species interactions, but also because it has deep consequences on the dynamics. Real ecologies have some amount of hierarchical organization that can be described (on a first approximation) as a set of layers of interacting species. Such layers go from primary producers at the bottom to top predators in the upper floor. The number of layers is limited through both energetic and dynamic constraints [46]. This is, again, an oversimplification, since real ecosystems are not fully hierarchical. Actually, they show small world topology [40].

It is well known that the ffects of perturbations on primary producers can have important (if not devastating) consequences on other parts of the food chain. Such effects can be direct: an example is provided by the K/T event. Here the extinction pattern in marine habitats is entirely compatible with the effects of decreased food supply for higher trophic levels due to the collapse of phytoplankton productivity [2]. At the end-Cretaceous, primary productivity declined suddenly and considerably. The further decay of many species at higher trophic levels in oceanic plankton communities was mainly due to this decline.

But cascade effects are also triggered by the removal of keystone species at any level in the web. Keystone species are specially relevant because of their large impact on the community dynamics, stability and composition. Their loss can cause extinctions to cascade throughout the system. An example provided by the fossil record is the extinction of megaherbivores (such as mastodon and mammoth) at the end-Pleistocene [45]. These species, mainly large herbivores, were essentially invulnerable to non-human predation on adults. As a consequence, they attained high (saturating) densities. Their effects on vegetation patterns were huge (as it occurs today with elephants) and their loss had a catastrophic impact. As a very attractive prey to humans, their extermination brought extensive vegetational changes, eventually generating the concomitant disappearance of so many other vertebrates.

Such cascading effects have been reported in different field studies [9] [10], and their ecological effects have long been suspected for the major mass extinctions. They have been incorporated in a simple model ecosystem with layered structure by Amaral and Meyer [1].

The model considers L levels ($l = 0, 1, ..., L-1$) with N niches per level (figure 12). We can indicate the presence or absence of species at the i-th niche in the l-th level by a binary variable $S_i(l)$. The bottom species, belonging to the $l = 0$ layer, define the group of organisms which do not feed on others. From $l = 1$ to $l = L - 1$ the species of these layers feed on k or less species on the lower $(l-1)$-level.

The rules are very simple: (a) a new species is created at each niche at a rate μ. If $l > 0$, then k prey species are randomly chosen from the $(l-1)$-layer; (b) at a given rate p, species at the bottom layer are extinct. Then for species at $l > 0$ layers, extinction takes place if no input links are present. Thus if $W(i, j; l-1 \to l)$ indicates the connection between species $i \in l-1$ and species $j \in l$ (here this connection is either one or zero) then:

$$S_j(l) = \Phi\left[\sum_{i=1}^{N} W(i, j; l-1 \to l)\right] \tag{28}$$

where now $\Phi(z) = 1$ for positive z and zero for $z = 0$. Clearly, the link among connected species will be able to generate avalanches of coextinction through different layers. From numerical simulations, Amaral and Nunez showed that in fact the distribution of avalanches is a power law with an exponent -2, consistent with the fossil record data.

An elegant analytic derivation of this result was obtained by Barbara Drossel [19]. Let us consider the $k = 1$ case. In this specific situation, each species feeds only on one prey species at the lower layer. Several species on different

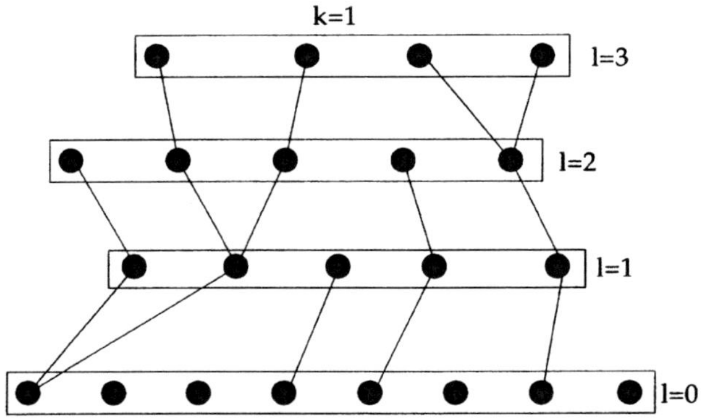

Fig. 12. Network structure of the simplest ($k = 1$) Amaral-Nunez layered ecosystem. The basal layer is formed by those species not feeding on others.

layers will eventually feed on the same bottom species, and the trophic structure looks like a set of trees starting at a single species in the lower level. From rule (b) each time a bottom species is removed (with probability p) a whole tree is gone and thus the size distribution of extinction events must be identical to the tree size distribution.

Assuming stationarity (and so a given mean number of species per level) a mean-field equation for the density of species at each layer ρ_l can be derived:

$$\frac{d\rho}{dt} = \mu(1-\rho) - p\rho_l \tag{29}$$

This equation has a fixed point $\rho_l^* = \mu/(p+\mu)$. Let $\Lambda_l^i(t)$ the number of species at the l-th layer connected to the i-th bottom species. The growth of $\Lambda_l^i(t)$ follows the dynamical equation:

$$\frac{d\Lambda_l^i}{dt} = \left[\frac{\mu(1-\rho_l)}{\rho_{l-1}}\right]\Lambda_{l-1}^i = p\Lambda_{l-1}^i \tag{30}$$

And as a consequence the size of a whole tree $\mathcal{T}^i = \sum_l \Lambda_l^i$ will follow a linear equation

$$\frac{d\mathcal{T}^i}{dt} = p\mathcal{T}^i \tag{31}$$

i. e. we have

$$\mathcal{T}^i(t) = \mathcal{T}^i(0)\exp(pt) \tag{32}$$

with $\mathcal{T}^i(0) = 1$. The size distribution of trees, $P(\Lambda)$ is linked to the age distribution $P(t)$ through $P(\Lambda) = P(t)dt/d\Lambda$. Here $P(t)$ follows the simple decay equation

$$\frac{dP(t)}{dt} = -pP(t) \tag{33}$$

and thus $P(t) \propto \exp(-pt)$. Using the previous results, we finally get the scaling relation:

$$P(S) \propto S^{-2} \tag{34}$$

in agreement with simulations. By analysing different sources of finite-size effects, Drossel shows that the previous results are robust for a broad range of parameter values. The previous analysis can be extended to the general $k > 1$ problem.

The previous proof involves the distribution of extinctions *per dead species in the lowest level*, let us call it $P_{ds}(s)$. One can also compute the distribution of extinction sizes *per time step*, $P(s)$, which is what the fossil record actually supplies, and adds all the extinctions of the first one along a time step [14]. In this sense, it is worth to stress that the exponent -2 derived theoretically by Drossel is related but strictly is not the exponent of the extinction size distribution per time step, $P(s)$. In order to calculate this second distribution, one has to combine $P_{ds}(s)$ with the Poisson statistics, $P_o(l)$, for the number of extinctions in this level. Even if one neglects the correlations between one extinction and the next ones, the resulting expression, namely

$$P(s) = \sum_{l=1}^{s} P_o(l) \sum_{\sum_{j=1}^{l} s_j = s} P_{ds}(s_1)...P_{ds}(s_l),$$

is very cumbersome to deal with analytically, so that, in practice, only numerical simulations can solve it. If one simulates a Poisson process with $P_{ds} \sim s^{-2}$ for the parameters used in our simulations, one finds the same scaling relation. Therefore, there is no contradiction between both distributions. In fact, as long as the average number of dead species per unit time in the lowest level is not high, if one removal gives rise to a big extinction, it is likely that this is the only big extinction taking place in that time step, since $P_{ds} \sim s^{-2}$ decreases rapidly with s. Therefore $P(s) \simeq P_{ds}(s) \sim s^{-2}$.

One can also estimate the location of the maximum in the extinction distribution, $P(s)$. We know that when one species in the lowest level dies, the probability for an extinction of size s scales as s^{-2}. This means that 60% of the extinctions due to an extinction in the lowest level are of size 1 (since $\zeta^{-1}(2) \simeq 0.6$). Then, most of the time a species in $l = 1$ is gone with no further cascade. Now, if N_1 denotes the average number of species in the first level at the stationary state, this implies that most of the time the number of extinctions in a time step will be $\simeq N_1 p$. It is easy to see that $N_1 \approx N(1-p/\mu)$, so that for $N = 1000, \mu = 0.02$ and $p = 0.01$, $N_1 p \approx \%5$, meanwhile for the case $\mu = 0.004$ and $p = 0.002$, one has $N_1 p \approx \%1$. These predictions for the location of the maxima agree well with numerical simulations, as shown in the last: the maximum is found at $s \sim 5$ in the first case, but it is absent in the second one (figure 13).

5 Discussion

It is generally agreed that species may go extinct because they are unable to evolve rapidly enough to meet changing circumstances or because their niches

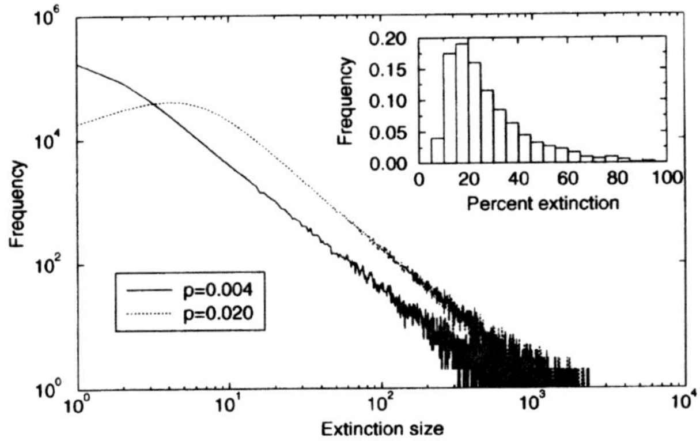

Fig. 13. Scaling in the Amaral-Meyer model. Here we show the distribution of extinction events for the AM model. The values of the parameters are: $\mu = 0.02$, $p = 0.01$, and $\mu = 0.004$, $p = 0.002$. A maximum is found in the first case, in agreement with the fossil record (see inset).

disappear. In the second case, no capacity for rapid evolution could save them from extinction. Theoretical models have been traditionally based on assumptions invoking either individual-based selection/adaptation mechanisms or externalities such as meteorite impacts. But in both cases the underlying ecological organization is essentially ignored. Species are effectively isolated entities whose extinction has little influence on ecosystem functioning (and thus in promoting further extinctions). But these assumptions are far from true: the trophic nature of ecological interactions makes a big difference. When a species is gone, its effect can be very small or very large.

Biotic responses have been of great importance in the past [37] [38] as they are today in our biosphere. Keystone species are probably an inevitable result of ecological complexity [28] and in that sense their removal from an ecosystem can have highly nonlinear effects [66]. Such nonlinearity and the high-level patterns emerging from network dynamics are likely to decouple micro- from macroevolutionary processes. It is the global features of the food web what matters, not the specific properties of the species constituting the web. In Bak's sandpile metaphor, we cannot understand the sandpile behavior by looking at gravitation and friction acting on individual grains. Instead, we recognized that the sandpile at the critical state must be analysed as a collective phenomenon, since new properties (interactions among grains) are at work. Similarly, natural selection and adaptation are operating on species in complex ecologies, but the complete picture requires the consideration of interactions among species. In the long run, the effects of selection pressures on single species are like gravity and friction operating on grains of sand: we need to take them into account, but they cannot explain the avalanches.

The future of this area, in my view, will need to incorporate a detailed understanding of the information provided by the fossil record at different scales in space and time. Instead of considering a few sets of large-scale patterns, available information dealing with evolutionary responses by well-defined groups whould be taken into account. This is particularly important when looking at recovery patterns [21] [22] [23]. Recovery patterns provide a unique window to explore the structure and evolution of paleoecosystems. Ecological links do not fossilize, but the underlying structure of ancient food webs can be inferred from the fossil record. In this context, new models of evolution [67] exploring the responses of different trophic levels to mass extinctions events will be very useful not only in order to understand how the biosphere reacted in the past to external challenges, but also to provide useful insight into the current human-driven mass extinction event.

The previous models and others not reviewed here (see in particular the work by Caldarelli et al. [13]) are a first step towards a complete theory of large-scale evolution. These models are typically non-historic, and in that sense they ignore some essential ingredients of the true dynamics of biological entities. But they are able to capture some large-scale trends that other, more detailed approximations cannot provide without an important loss of real understanding. In spite of their simplicity, they clearly indicate that such a theory is feasible, if we are able to go beyond the infinite details provided by the fossil record. As Niles Eldredge said: "...we should not despair. There is real order in all this apparent chaos. Life has had a long and complex, but ultimately comprehensible, history. There are patterns repeated over and over again as new species come and go, and as ecosystems form and fall apart. These organizing principles of life's history are the processes of evolution".

Acknowledgements

The auhor thanks S. Kauffman, G. Eble, S. Manrubia and J. M. Montoya for many discussions on evolutionary ecology. This work has been supported by a grant CICYT PB97-0693 and The Santa Fe Institute (RVS).

References

1. Amaral, L. A. N. and Meyer, M. Phys. Rev. Lett. **82**, 652 (1999)
2. Arthur, M. A., Zachos, J. C. and Jones, D. S. Cretaceous Research **8**, 43 (1987)
3. Bak, P., Tang, C. and Wiesenfeld, K. Phys. Rev. Lett. **59**, 381 (1987)
4. Bak, P., Flyvbjerg, H. and Lautrup, B. Phys. Rev. A **46**, 6724 (1992)
5. Bak, P. and Sneppen, K. Phys. Rev. Lett. **71** 4083 (1993)
6. Bak, P.*How Nature Works*, Springer (1996)
7. Benton, M. J. Science **268**, 52 (1995)
8. Benton, M.J. In: *Palaeobiology* (ed. D.E.G. Briggs & P. R. Crowther). Blackwells. Oxford (1995)
9. Brown, J. H. and Heske, E. J. *Science*, **250**, 1705 (1991)

10. Brown, J. H., in *Complexity: Metaphors, Models and Reality* Cowan, G., Pines, D., and Meltzer, D., eds., pp. 419-449, Addison Wesley (1994)
11. Burlando, B. J. theor. Biol. **146**, 99 (1990)
12. Burlando, B. J. theor. Biol. **163**, 161 (1993)
13. Caldarelli, G., Higgs, P. G. and McKane, A. J. J. Theor. Biol. **193**, 345 (1998)
14. Camacho, J. and Solé, R. V. Phys. Rev. E 62, 1119-1123 (2000)
15. Chaloner, W. and Hallam, A. (eds) Phil. Trans. R. Soc. London B325, 239-488 (1989)
16. Conway Morris, S. Phil Trans. Roy. Soc. Lond. B **353**, 327 (1998)
17. Conway Morris, S. *The Crucible of Creation.* Oxford U. Press (1998)
18. Dimri, V. P. and Prakash, M. R. Earth and Planetary Science Letters (to appear)
19. Drossel, B. Phys. Rev Lett. **81**, 5011 (1998)
20. Droser, M. L., Bottjer, D. J. and Sheehan, P. M. Geology **25**, 167 (1997)
21. Erwin, D. H. *The Great Paleozoic Crisis: Life and Death in the Permian.* Columbia U. Press (1993)
22. Erwin, D. H. Nature **367**, 231 (1994)
23. Erwin, D. H. Trends Ecol. Evol. **13**, 344 (1998)
24. Hewzulla, D., Boulter, M. C., Benton, M. J. and Halley, J. M. Proc. Roy. Soc. London B **354**, 463 (1999)
25. Jablonski, D. Science **284**, 2114 (1999)
26. Jablonski, D., Erwin, D. H. and Lipps, J. H. 1996. Evolutionary paleobiology. Chicago U. Press
27. Jablonski, D. Paleobiology Suppl. **26**, 15 (2000)
28. Jain, S. and Krishna, S. preprint arXiv:nlin.AO/0005039 (2000)
29. Kauffman, S. and Johnsen, J. J. Theor. Biol. **149**, 467 (1991)
30. Kauffman, S. *The Origins of Order*, Oxford U. Press, Oxford (1993)
31. Kauffman, S. *At Home in the Universe*, Oxford U. Press, Oxford (1997)
32. Kirchner, J. W. and Weil, A. Nature **395**, 337 (1998)
33. Kirchner, J. W. and Weil, A. Nature **404**, 177 (2000)
34. Korvin, G. 1992. *Fractal Models in the Earth Sciences*, Elsevier
35. Maddox, J. Nature **371**, 197 (1994)
36. Manrubia, S. C. and Paczuski, M. Int. J. Mod. Phys. C **9**, 1025 (1998)
37. Maynard Smith, J. Phil. Trans. R. Soc. Lond. B **325**, 241 (1989)
38. McKinney, M. L. and Drake, J. A. (eds.) *Biodiversity Dynamics*, Columbia U. Press (1998)
39. Miller, A. I. Paleobiology Suppl. **26**, 53 (2000)
40. Montoya, J. M. and Solé, R. V. Santa Fe Institute Working Paper 00-10-059 (2000)
41. Newman, M. E. J. Proc. Roy. Soc. London B **263**, 1605 (1996)
42. Newman, M. E. J. J. Theor. Biol. **189**, 235 (1997)
43. Newman, M. E. J. and Eble, G. Paleobiology **25**, 434 (1999)
44. Newman, M. E. J. and Palmer, R. Santa Fe Institute Working Paper 99-08-061 (1999)
45. Owen-Smith, N. Paleobiology **13**, 351 (1987)
46. Pimm. S. *The Balance of Nature*, Chicago Press (1991)
47. Plotnick, R. E. and McKinney, M. Palaios **8**, 202 (1993)
48. Plotnick, R. E. and Sepkoski. J. J. Paleobiology **27**, 126 (2001)
49. Rao, R. M. and Bopardikar, A. S. *Wavelet Transforms : Introduction to Theory and Applications*, Addison-Wesley (1998)
50. Rampino, M. R. and Sothers, R. B. Nature **308**, 709 (1984)
51. Rampino, M. R. and Sothers, R. B. Earth, Moon and Planets **72**, 441 (1995)

52. Raup, D. *Extinctions: Bad Genes or Bad Luck?*. Oxford U. Press (1993)
53. Raup, D. M. and Sepkoski, J. J., Jr. Proc. Natl. Acad. Sci. USA **81**, 801 (1984)
54. Raup, D. M. Science **231**, 1528 (1986)
55. Rowe, G. W. *Theoretical Models in Biology*, Oxford U. Press (1994)
56. Sepkoski, J. J., Jr. Paleobiology **10**, 246 (1984)
57. Sibani, P. Phys. Rev. Lett. **79**, 1413 (1997)
58. Sneppen, K. et al. Procs. Natl. Acad. Sci. USA **92**, 5209 (1995)
59. Solé, R. V. Complexity **1**, 40 (1996)
60. Solé, R. V., Bascompte, J. and Manrubia, S. C. Proc. Roy. Soc. London B **263**, 1407 (1996)
61. Solé, R. V. and Manrubia, S. C. Phys. Rev. E **54** R42 (1996)
62. Solé, R. V. and Bascompte, J. Proc. R. Soc. London **263**, 161 (1996)
63. Solé, R. V., Manrubia, S. C., Benton, M. and Bak, P. Nature **388**, 764 (1997)
64. Solé, R. V., Manrubia, S. C., Kauffman, S. A., Benton, M. and Bak, P. Trends in Ecol. **14**, 156 (1999)
65. Solé, R. V., Manrubia, S. C., Pérez-Mercader, J., Benton, M. and Bak, P. Adv. Complex Syst. **1**, 255 (1998)
66. Solé, R. V and Montoya, J. M. Santa Fe Institute Working Paper 00-10-060 (2000)
67. Solé, R. V., Montoya, J. M. and Erwin, D. H. Phil. Trans. Roy. Soc. London B. (to appear)
68. Van Valen, L. Evol. Theory **1**, 1 (1973)

Lecture Notes in Physics

For information about Vols. 1–549
please contact your bookseller or Springer-Verlag

Vol. 550: M. Planat (Ed.), Noise, Oscillators and Algebraic Randomness. Proceedings, 1999. VIII, 417 pages. 2000.

Vol. 551: B. Brogliato (Ed.), Impacts in Mechanical Systems. Analysis and Modelling. Lectures, 1999. IX, 273 pages. 2000.

Vol. 552: Z. Chen, R. E. Ewing, Z.-C. Shi (Eds.), Numerical Treatment of Multiphase Flows in Porous Media. Proceedings, 1999. XXI, 445 pages. 2000.

Vol. 553: J.-P. Rozelot, L. Klein, J.-C. Vial Eds.), Transport of Energy Conversion in the Heliosphere. Proceedings, 1998. IX, 214 pages. 2000.

Vol. 554: K. R. Mecke, D. Stoyan (Eds.), Statistical Physics and Spatial Statistics. The Art of Analyzing and Modeling Spatial Structures and Pattern Formation. Proceedings, 1999. XII, 415 pages. 2000.

Vol. 555: A. Maurel, P. Petitjeans (Eds.), Vortex Structure and Dynamics. Proceedings, 1999. XII, 319 pages. 2000.

Vol. 556: D. Page, J. G. Hirsch (Eds.), From the Sun to the Great Attractor. X, 330 pages. 2000.

Vol. 557: J. A. Freund, T. Pöschel (Eds.), Stochastic Processes in Physics, Chemistry, and Biology. X, 330 pages. 2000.

Vol. 558: P. Breitenlohner, D. Maison (Eds.), Quantum Field Theory. Proceedings, 1998. VIII, 323 pages. 2000

Vol. 559: H.-P. Breuer, F. Petruccione (Eds.), Relativistic Quantum Measurement and Decoherence. Proceedings, 1999. X, 140 pages. 2000.

Vol. 560: S. Abe, Y. Okamoto (Eds.), Nonextensive Statistical Mechanics and Its Applications. IX, 272 pages. 2001.

Vol. 561: H. J. Carmichael, R. J. Glauber, M. O. Scully (Eds.), Directions in Quantum Optics. XVII, 369 pages. 2001.

Vol. 562: C. Lämmerzahl, C. W. F. Everitt, F. W. Hehl (Eds.), Gyros, Clocks, Interferometers...: Testing Relativistic Gravity in Space. XVII,507 pages. 2001.

Vol. 563: F. C. Lázaro, M. J. Arévalo (Eds.), Binary Stars. Selected Topics on Observations and Physical Processes. 1999.IX, 327 pages. 2001.

Vol. 564: T. Pöschel, S. Luding (Eds.), Granular Gases. VIII, 457 pages. 2001.

Vol. 565: E. Beaurepaire, F. Scheurer, G. Krill, J.-P. Kappler (Eds.), Magnetism and Synchrotron Radiation. XIV, 388 pages. 2001.

Vol. 566: J. L. Lumley (Ed.), Fluid Mechanics and the Environment: Dynamical Approaches. VIII, 412 pages. 2001.

Vol. 567: D. Reguera, L. L. Bonilla, J. M. Rubí (Eds.), Coherent Structures in Complex Systems. IX, 465 pages. 2001.

Vol. 568: P. A. Vermeer, S. Diebels, W. Ehlers, H. J. Herrmann, S. Luding, E. Ramm (Eds.), Continuous and Discontinuous Modelling of Cohesive-Frictional Materials. XIV, 307 pages. 2001.

Vol. 569: M. Ziese, M. J. Thornton (Eds.), Spin Electronics. XVII, 493 pages. 2001.

Vol. 570: S. G. Karshenboim, F. S. Pavone, F. Bassani, M. Inguscio, T. W. Hänsch (Eds.), The Hydrogen Atom: Precision Physics of Simple Atomic Systems. XXIII, 293 pages. 2001.

Vol. 571: C. F. Barenghi, R. J. Donnelly, W. F. Vinen (Eds.), Quantized Vortex Dynamics and Superfluid Turbulence. XXII, 455 pages. 2001.

Vol. 572: H. Latal, W. Schweiger (Eds.), Methods of Quantization. XI, 224 pages. 2001.

Vol. 573: H. M. J. Boffin, D. Steeghs, J. Cuypers (Eds.), Astrotomography. XX, 434 pages. 2001.

Vol. 574: J. Bricmont, D. Dürr, M. C. Galavotti, G. Ghirardi, F. Petruccione, N. Zanghi (Eds.), Chance in Physics. XI, 288 pages. 2001.

Vol. 575: M. Orszag, J. C. Retamal (Eds.), Modern Challenges in Quantum Optics. XXIII, 405 pages. 2001.

Vol. 576: M. Lemoine, G. Sigl (Eds.), Physics and Astrophysics of Ultra-High-Energy Cosmic Rays. X, 327 pages. 2001.

Vol. 577: I. P. Williams, N. Thomas (Eds.), Solar and Extra-Solar Planetary Systems. XVIII, 255 pages. 2001.

Vol. 578: D. Blaschke, N. K. Glendenning, A. Sedrakian (Eds.), Physics of Neutron Star Interiors. XI, 509 pages. 2001.

Vol. 579: R. Haug, H. Schoeller (Eds.), Interacting Electrons in Nanostructures. X, 227 pages. 2001.

Vol. 580: K. Baberschke, M. Donath, W. Nolting (Eds.), Band-Ferromagnetism: Ground-State and Finite-Temperature Phenomena. IX, 394 pages. 2001.

Vol.581: J. M. Arias, M. Lozano (Eds.), An Advanced Course in Modern Nuclear Physics. XI, 346 pages. 2001.

Vol.582: N. J. Balmforth, A. Provenzale (Eds.), Geomorphological Fluid Mechanics. X, 579 pages. 2001.

Vol.583: W. Plessas, L. Mathelitsch (Eds.), Lectures on Quark Matter, XIII, 334 pages. 2002.

Vol.584: W. Köhler, S. Wiegand (Eds.), Thermal Nonequilibrium Phenomena in Fluid Mixtures. XVII, 470 pages. 2002.

Vol.585: M. Lässig, A. Valleriani (Eds.), Biological Evolution and Statistical Physics. XI, 337 pages. 2002.

Vol.586: Y. Auregan, A. Maurel, V. Pagneux, J.-F. Pinton (Eds.), Sound-Flow Interactions. XIV, 280 pages. 2002

Vol.588: Y. Watanabe, S. Heun, G. Salviati, N. Yamamoto (Eds.), Nanoscale Spectroscopy and Its Applications to Semiconductor Research. XV, 328 pages. 2002.

Monographs
For information about Vols. 1–30 please contact your bookseller or Springer-Verlag

Vol. m 31 (Corr. Second Printing): P. Busch, M. Grabowski, P.J. Lahti, Operational Quantum Physics. XII, 230 pages. 1997.

Vol. m 32: L. de Broglie, Diverses questions de mécanique et de thermodynamique classiques et relativistes. XII, 198 pages. 1995.

Vol. m 33: R. Alkofer, H. Reinhardt, Chiral Quark Dynamics. VIII, 115 pages. 1995.

Vol. m 34: R. Jost, Das Märchen vom Elfenbeinernen Turm. VIII, 286 pages. 1995.

Vol. m 35: E. Elizalde, Ten Physical Applications of Spectral Zeta Functions. XIV, 224 pages. 1995.

Vol. m 36: G. Dunne, Self-Dual Chern-Simons Theories. X, 217 pages. 1995.

Vol. m 37: S. Childress, A.D. Gilbert, Stretch, Twist, Fold: The Fast Dynamo. XI, 406 pages. 1995.

Vol. m 38: J. González, M. A. Martín-Delgado, G. Sierra, A. H. Vozmediano, Quantum Electron Liquids and High-Tc Superconductivity. X, 299 pages. 1995.

Vol. m 39: L. Pittner, Algebraic Foundations of Non-Com-mutative Differential Geometry and Quantum Groups. XII, 469 pages. 1996.

Vol. m 40: H.-J. Borchers, Translation Group and Particle Representations in Quantum Field Theory. VII, 131 pages. 1996.

Vol. m 41: B. K. Chakrabarti, A. Dutta, P. Sen, Quantum Ising Phases and Transitions in Transverse Ising Models. X, 204 pages. 1996.

Vol. m 42: P. Bouwknegt, J. McCarthy, K. Pilch, The W3 Algebra. Modules, Semi-infinite Cohomology and BV Algebras. XI, 204 pages. 1996.

Vol. m 43: M. Schottenloher, A Mathematical Introduction to Conformal Field Theory. VIII, 142 pages. 1997.

Vol. m 44: A. Bach, Indistinguishable Classical Particles. VIII, 157 pages. 1997.

Vol. m 45: M. Ferrari, V. T. Granik, A. Imam, J. C. Nadeau (Eds.), Advances in Doublet Mechanics. XVI, 214 pages. 1997.

Vol. m 46: M. Camenzind, Les noyaux actifs de galaxies. XVIII, 218 pages. 1997.

Vol. m 47: L. M. Zubov, Nonlinear Theory of Dislocations and Disclinations in Elastic Body. VI, 205 pages. 1997.

Vol. m 48: P. Kopietz, Bosonization of Interacting Fermions in Arbitrary Dimensions. XII, 259 pages. 1997.

Vol. m 49: M. Zak, J. B. Zbilut, R. E. Meyers, From Instability to Intelligence. Complexity and Predictability in Nonlinear Dynamics. XIV, 552 pages. 1997.

Vol. m 50: J. Ambjørn, M. Carfora, A. Marzuoli, The Geometry of Dynamical Triangulations. VI, 197 pages. 1997.

Vol. m 51: G. Landi, An Introduction to Noncommutative Spaces and Their Geometries. XI, 200 pages. 1997.

Vol. m 52: M. Hénon, Generating Families in the Restricted Three-Body Problem. XI, 278 pages. 1997.

Vol. m 53: M. Gad-el-Hak, A. Pollard, J.-P. Bonnet (Eds.), Flow Control. Fundamentals and Practices. XII, 527 pages. 1998.

Vol. m 54: Y. Suzuki, K. Varga, Stochastic Variational Approach to Quantum-Mechanical Few-Body Problems. XIV, 324 pages. 1998.

Vol. m 55: F. Busse, S. C. Müller, Evolution of Spontaneous Structures in Dissipative Continuous Systems. X, 559 pages. 1998.

Vol. m 56: R. Haussmann, Self-consistent Quantum Field Theory and Bosonization for Strongly Correlated Electron Systems. VIII, 173 pages. 1999.

Vol. m 57: G. Cicogna, G. Gaeta, Symmetry and Perturbation Theory in Nonlinear Dynamics. XI, 208 pages. 1999.

Vol. m 58: J. Daillant, A. Gibaud (Eds.), X-Ray and Neutron Reflectivity: Principles and Applications. XVIII, 331 pages. 1999.

Vol. m 59: M. Kriele, Spacetime. Foundations of General Relativity and Differential Geometry. XV, 432 pages. 1999.

Vol. m 60: J. T. Londergan, J. P. Carini, D. P. Murdock, Binding and Scattering in Two-Dimensional Systems. Applications to Quantum Wires, Waveguides and Photonic Crystals. X, 222 pages. 1999.

Vol. m 61: V. Perlick, Ray Optics, Fermat's Principle, and Applications to General Relativity. X, 220 pages. 2000.

Vol. m 62: J. Berger, J. Rubinstein, Connectivity and Superconductivity. XI, 246 pages. 2000.

Vol. m 63: R. J. Szabo, Ray Optics, Equivariant Cohomology and Localization of Path Integrals. XII, 315 pages. 2000.

Vol. m 64: I. G. Avramidi, Heat Kernel and Quantum Gravity. X, 143 pages. 2000.

Vol. m 65: M. Hénon, Generating Families in the Restricted Three-Body Problem. Quantitative Study of Bifurcations. XII, 301 pages. 2001.

Vol. m 66: F. Calogero, Classical Many-Body Problems Amenable to Exact Treatments. XIX, 749 pages. 2001.

Vol. m 67: A. S. Holevo, Statistical Structure of Quantum Theory. IX, 159 pages. 2001.

Vol. m 68: N. Polonsky, Supersymmetry: Structure and Phenomena. Extensions of the Standard Model. XV, 169 pages. 2001.

Vol. m 69: W. Staude, Laser-Strophometry. High-Resolution Techniques for Velocity Gradient Measurements in Fluid Flows. XV, 178 pages. 2001.

Vol. m 70: P. T. Chruściel, J. Jezierski, J. Kijowski, Hamiltonian Field Theory in the Radiating Regime. VI, 172 pages. 2002.

Vol. m 71: S. Odenbach, Magnetoviscous Effects in Ferrofluids. X, 151 pages. 2002.

Vol. m 72: J. G. Muga, R. Sala Mayato, I. L. Egusquiza (Eds.), Time in Quantum Mechanics. XII, 419 pages. 2002.

Lightning Source UK Ltd.
Milton Keynes UK

171906UK00003B/33/P

9 783642 077432